Digital Image Processing

Principles and Applications

Gregory A. Baxes

John Wiley & Sons, Inc.
New York • Chichester • Brisbane • Toronto • Singapore

About the cover:

The front cover images are digitally processed microscope views of a thin-film technology recording head. These miniaturized coils are part of the read/write head subsystem used in IBM's computer hard disk drives. Each image was processed using a different brightness slicing operation and then false-colored to highlight the internal spiral structure. Subsequent machine vision operations can follow these operations to automatically measure the important pattern dimensions, freeing the human from the tedious task while improving the inspection accuracies two- or three-fold. These images appear courtesy of International Business Machines Corporation.

This text is printed on acid-free paper.

Library of Congress Cataloging-in-Publication Data

Baxes, Gregory A.
Digital image processing : prinicples and applications / Gregory A. Baxes.
p. cm.
Includes index.
ISBN 0-471-00949-0 (paper)
1. Image processing—Digital techniques I. Title.
TA1637.B39 1994
621.36'7—dc20 94-9744
CIP

Printed in the United States of America

10 9 8 7 6 5 4 3 2 1

To my friend and colleague
Bert Kadzielawa
your memory lives forever

Contents

Appendix A—Further Reading/References 417

Appendix B—Digital Image Processing: Hands-On User's Guide 421

Foreword

by Kenneth R. Castleman

Digital image processing, as we now know it, started in the 1950s, when computers first became powerful enough to deal with pictorial data in a meaningful way. Prior to that, images had been transmitted by digital means, but any significant processing had to wait for the emergence of modern digital computers. Those early days required that images be digitized very crudely to avoid overloading the available machines. A few farsighted individuals, however, spotted the potential of this fledgling discipline and began laying groundwork for decades of growth.

In the early days, only large, government-funded laboratories could afford to develop the expensive hardware required to digitize, process, and display an image accurately enough to produce visually satisfying results. At the Jet Propulsion Laboratory (JPL), for example, we had an entire floor of a building devoted to the equipment and personnel of the Image Processing Lab. Many hours of expensive computer time were required to process each of the thousands of images that were returned from spacecraft. But in terms of increased scientific value, the additional cost of employing digital image enhancement and analysis techniques was a bargain.

Although digital image processing had its beginnings in the space program and the military, many developments also came out of medical and character recognition applications. As a result of some widely publicized successes, more people began to see the potential of the technology, despite its high cost.

Digital image processing next began to spring up on campuses, as large computing systems became more affordable and more common. University researchers began to contribute more rigorous application solutions to the growing list of the technology's capabilities. As a result, image processing began to find its way into some of the more expensive commercial and medical products.

Finally, digital image processing techniques went commercial, entering the market as a viable technology. Many companies began adapting image processing hardware and software into their products. With the evolution of minicomputers, it became practical for an engineer to have a complete image processing workstation for personal use. The technology also went public, as motion pictures and television embraced it for broad public distribution. Revolving logos and satellite-produced color weather maps invaded the living room, and gargantuan space ships, born of nothing more than bits, bytes, and imagination, cruised through theaters—all due to this burgeoning technology.

This migration has now reached the desktop. Many home offices now house a microprocessor-based image processing system comparable in power to those massive machines we used at JPL to enhance the first images from space. Computer power is so inexpensive that almost any medium-priced commercial product can employ image processing methods. Public awareness of image processing technology has also reached new highs, along with an increasing demand for practical solutions to previously unsolved problems.

The computer industry has not yet reached the end of the current "smaller-cheaper-faster" technological rocket ride, so the 40-year trend of image processing progress will certainly continue. There are improved circuits to design and better algorithms to develop, to be sure, but the area of greatest impact will undoubtedly be the application of digital imaging to real-world problems. This is where the activity is greatest and the reward most handsome. The pacing function now is not hardware or algorithms, but innovative solutions developed by people with a working knowledge of digital image processing technology and an understanding of how to apply it to real-world problems.

In 1984, Gregory Baxes' book, *Digital Image Processing: A Practical Primer*, placed this technology in the hands of those whose expertise and educational backgrounds fell outside the electrical engineering and advanced mathematics required by college-level textbooks. The book presented the functional basics of the field in a manner that could be understood and applied by users from a wide variety of backgrounds. These users, expert in their fields of application rather than the rigors of electrical engineering or computer science, became armed with the technology of digital image processing.

Now, ten years later, the need for such a treatment of the subject is even more acute. In the coming years, we will depend heavily upon workers from widely diverse fields for the solutions to emerging problems and applications. This second-generation book, *Digital Image Processing: Principles and Applications*, introduces digital image processing as a problem-solving tool while giving the guidance for the skillful application of that tool. This book is greatly expanded and refined from Baxes' earlier work, and I believe that it, like the earlier book, will prove quite valuable to the continued evolution of digital image processing applications.

Kenneth R. Castleman
Vice-President for Research and Development
Perceptive Scientific Instruments, Inc.

Kenneth R. Castleman is renowned for his early and continuing contributions to the digital image processing field. He began his work in digital image processing research in the late

1960s while at NASA's Jet Propulsion Laboratory in Pasadena, California. He has taught digital image processing courses at the California Institute of Technology and has served as a Research Fellow at both the University of Southern California and the University of California at Los Angeles. Dr. Castleman authored one of the preeminent academic texts on digital image processing and has been involved in projects ranging from biomedical and space science imagery to numerous forensic image processing applications.

About the Author

GREGORY A. BAXES has worked in the design, development, and engineering management of digital image processing systems and software for over 15 years. His work has concentrated on real-time machine vision, photo-interpretation image analysis, video special-effects generation, and, most recently, synthetic scene rendering for visualization and reality-simulation applications. Based in the Denver, Colorado area, Mr. Baxes has worked with several notable companies, including Martin Marietta Aerospace, Time-Warner Cable, Ampex Corporation, U S WEST Advanced Technologies, and Intergraph Corporation. He is currently involved with projects in the fields of medical imaging and computer scene modeling and rendering.

Preface

Ten years ago, when I wrote the preface to my preceding book, *Digital Image Processing: A Practical Primer,* I began with the sentence, "Digital image processing is a fascinating field that is just now beginning to reach into our everyday lives." Today, this dynamic field truly touches us all, playing an important role in our world. From medical imaging to machine vision and more, digital image processing continues to evolve with broad and innovative applications. In the 1960s, digital image processing was used almost exclusively to enhance and correct NASA spacecraft imagery. Now, digital image processing techniques are used to do everything from reading your checks at the bank to automatically inspecting the fill-level of your pop bottle.

Like all new scientific disciplines, digital image processing has its theoretical roots. The field began as a mathematically complex subject taught primarily to graduate-level electrical engineering students. The individual lacking significant higher mathematics and signal processing backgrounds often found the complexities of the subject difficult to grasp and virtually impossible to apply.

Digital image processing systems once cost hundreds of thousands of dollars. Now, the hardware and software are affordable and widely available for costs starting at under a few thousand dollars. Anyone with a basic personal computer can painlessly implement real problem-solving applications. As a result, the interest in digital image processing and its applications is growing at an incredible rate. This trend is fueling its expansion into many new and diverse fields.

The driver of an automobile doesn't need to master the workings of the internal combustion engine to drive the vehicle efficiently. The operator of a handheld calculator doesn't have to understand the solid-state physics behind the chips in the device. Similarly, the digital image processing user doesn't need to be burdened with the subject's theoretical underpinnings to make perfectly good use of the technology.

Like my earlier book, this book provides an elementary, practical introduction to digital image processing. The mathematical proofs typically found in rigorous digital image processing texts have been reduced to an intuitive level. Functional basics are presented and illustrated through many pictorial graphics and photographic images, with a minimum of theory. This book will give you the foundation to understand and apply digital image processing techniques. Also, you will be fully prepared to pursue additional, more rigorous development of the topics if you so desire.

This book has four primary sections. Part I, "Introduction to Image Processing," introduces the field. The digital image processing reference frame—its definition, history, and applications—is presented. Further, this section introduces the fundamental classes of digital image processing operations, along with an overview of the generic digital image processing system. Part II, "Processing Concepts," examines in some depth the concepts behind the processing of digital images. This section discusses the concept of a digital image and the myriad operations that can be applied to it. This part of the book methodically brings to light, in a straightforward manner, the basic principles and techniques for processing digital images.

Part III, "Processing Systems," covers digital image processing system components and architectures. This section discusses methods and equipment for acquiring and displaying digital images as well as the hardware and software used to process them. The final section of the book, Part IV, "Processing in Action," presents a reference guide to commonly used digital image processing operations. Each image operation study contains a complete review of a single operation, along with comprehensive "before" and "after" images. A concise description, summary of uses, and method of implementation are also included for each operation.

An important part of learning the techniques of a new field is gaining a good understanding of its applications. Throughout the book, references are made to real-world uses for digital image processing techniques. In most cases, real images, taken from industry, are used to illustrate processing concepts.

Additionally, this book contains a unique learning tool. The enclosed disk contains *Digital Image Processing: Hands-On*, a working digital image processing software application. The software is built on the *Global Lab/Image Function Library* provided by Data Translation, Inc., and was specifically developed as a supplement to this book. It runs on any Intel processor-based workstation with an 80386-class (or better) processor running the Microsoft Windows operating system. *Digital Image Processing: Hands-On* will allow you to display and experimentally process the digital images contained on the disk, giving you hands-on experience with digital image processing techniques. Appendix B contains specific information on the use of the software.

This book introduces the field of digital image processing. It is intended to provide a complete overview of the fundamental topics. But don't stop here! Research and practical applications of digital image processing techniques continue to expand and are changing daily. I encourage you to seek out additional information sources, such as the many textbooks on the subject and technical and trade journals (see Appendix A). By applying the information contained in this book and continuing with further study, you can become a member of the digital image processing community and contribute significantly to the digital image processing field.

I wish to thank and express my appreciation to several organizations and individuals who helped make this book possible. First, the people of the following organizations gave their support by graciously providing equipment, software, images, and/or information: Aldus Corp.; Autodesk, Inc.; Coreco, Inc.; Datacube, Inc.; Data Translation, Inc.; DIPIX Technologies, Inc.; EDS/Unigraphics, Inc.; Earth Observation Satellite Co.; Electronic Imagery, Inc.; EPIX, Inc.; G W Hannaway & Associates; Image West; Imaging Technology, Inc.; Intergraph Corp.; Jandel Corp.; Lucas Digital, Ltd./Industrial Light and Magic; Matrox Electronic Systems, Ltd.; Media Cybernetics, Inc.; National Aeronautics and Space Administration; National Oceanic and Atmospheric Administration; Noesis Vision, Inc.; Optimas Corp.; S & M Microscopes, Inc.; Signal Analytics, Inc.; Silicon Graphics, Inc.; Visual Numerics, Inc.; and Vital Images, Inc.

Second, I am indebted to the following individuals for their generous contributions of time and expertise. Kevin Rizza, John Geis, Deborah Strange, and Bill Wolfe read the early manuscript for technical accuracy and treatment; their feedback was essential to this book's final quality and is much appreciated. Gregg Crockett, Don Lund, and Rich Siegel each read and offered important critiques of this book's predecessor, an important contribution to both books that I failed to acknowledge in the earlier book. Further, the comments of Fred Shippey and Ken Castleman helped to refine this book's ultimate content. Julie Baxes copy edited the final manuscript, helping my sometimes feeble writing to look good. Duayna Wing created the line-art figures, artistically transforming my sketches into finished art. Craig Abramson wrote the *Hands-On* software application, and Dick Nowels contributed to the creation of several of the visualization images. I am grateful for the interest and support of Curt Lipkie, at the ANA Tech division of Intergraph Corporation, where I was employed while writing this manuscript. I must also express my thanks and appreciation to this book's editor, Paul Farrell of John Wiley & Sons. Paul's interest in the title and his gentle, albeit persuasive, time management helped get this book started and completed on time (almost).

Finally, I would like to again thank my wife, Julie, and my family for their support, encouragement, and acceptance of the long days and nights throughout this project.

GREGORY A. BAXES
Littleton, Colorado
September 1994

Gregory A. Baxes can be reached with your comments or questions via the following links:

Internet: gbax@netcom.com or 76506.1755@compuserve.com
Voice: 303/979-5255
FAX: 303/979-0585
Mail: P.O. Box 621291
Littleton, CO 80162-1291 USA

1 What Is Image Processing?

Image Processing Defined

Image processing has become a familiar, almost routine expression recognized by a large percentage of the general public. Some people have seen it in their work in industrial settings, others in the laboratory, and some just through consumer use of personal computer paint and document-processing programs.

Television network and newspaper news have covered image processing on numerous occasions. Contemporary processed space imagery, the Hubble Space Telescope's incredible images, and military smart-bomb and missile-guidance technologies have been repeatedly presented during the past several years. Image processing is becoming a widely acknowledged technology. People are becoming aware that many everyday processes use automated vision systems, all of which rely upon image processing techniques.

Image processing, in general terms, refers to the manipulation and analysis of pictorial information. In this case, pictorial information means a two-dimensional visual image. Any operation that acts to improve, correct, analyze, or in some way change an image is called image processing.

We can find simple instances of image processing around us all the time. Perhaps the most common form of man-made image processing is the prescription eyeglass. Corrective eyeglasses geometrically predistort a viewed image to counteract the distortion created by the wearer's eye. Eyeglasses perform three primary geometric corrections, or predistortions, upon images. By measuring the distortions in the wearer's eyes, optometrists can create eyeglasses to provide virtually perfect eyesight.

As an example, let's say that a person is plagued, in one eye, with a common visual defect known as astigmatism. In that eye, the person sees images that are slightly magnified in one direction. Let's further say that the direction, or axis, of the astigmatism is perfectly vertical. This causes the eye to see images that are vertically distorted—everything looks skinnier and taller than it is in real life. If the person's other eye is perfect, the brain receives confusing visual information because each eye is viewing the scene slightly differently. This discrepancy leads to impaired overall vision.

Through the use of a prescription eyeglass that, in this case, shrinks the image in the vertical direction before the image enters the eye, the perceived image can be corrected. The image is predistorted to match the inverse of the distortion caused by the eye so that the eyeglass–eye combination forms a perfect image. This corrective process is illustrated in Figure 1.1.

Television receiver controls provide another everyday form of image processing. The viewer can adjust the "brightness," "contrast," "tint," and "color" of the image appearing on the display. These controls affect the video black-level offset, gain, color hue, and color saturation of the electrical signals being received and processed for display by the television receiver. We don't always think much about which control to use when we perceive a deficiency in the television image, but by varying the controls, maybe even randomly, we can usually converge on an image that looks good to us.

Other forms of everyday image processing are the distortions we see in the reflection of an image in a pond of water. The image is geometrically reversed and often distorted by the water's motion. Haze in the sky degrades the perceived image of a mountainside by reducing the scene's contrast and, hence, the detail we see within it. Perspective distortion created by basic viewing geometry makes a person standing 10 feet away look twice as large as a person standing 20 feet away.

We usually don't perceive all of these distortions as degradations, because our brains add surmised information to make up for the physical circumstances. For instance, we "know" that all people are roughly the same height and that, regardless of their distance from us, they maintain that height. Likewise, we have learned what moving water does to a reflected image, and thus tend not to perceive its induced distortions.

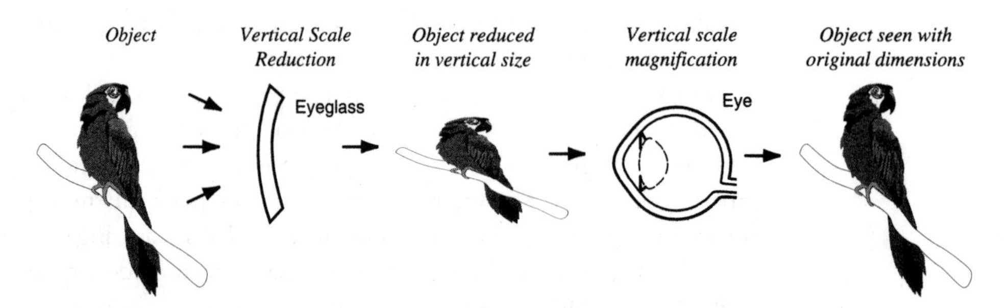

Figure 1.1 *The corrective eyeglass process. The optical image is predistorted to match the inverse of the distortion created by the eye. The net result is that the eye sees a perfect image.*

Certainly the most powerful image processing system we see and use every day is the one composed of the human eye and brain. This biological image processing system focuses, acquires, enhances, restores, analyzes, compresses, and stores images at astounding rates. Amazingly, we don't even consciously control more than a fraction of this system's processes.

The above examples illustrate image processing functions that occur around us every day. Simply, some form of image processing occurs whenever any process operates upon pictorial information.

Methods of Image Processing

In the preceding image processing examples, an image was changed, either for the better or for the worse. Generally, the objective of image processing is to transform or analyze an image so that new information about the image is made evident. There are different ways that we can apply an image processing operation.

Most images originate in an *optical form*. An optical image may be converted to an electrical signal with a video camera or similar device. This conversion changes the representation of the image from an optical light form to a continuously varying electrical signal. This electrical signal is called the *analog signal form*. Further, the analog image can be digitized and turned into a digital data form. Image processing operations can be applied to an image in the optical, analog, or digital form, as shown in Figure 1.2.

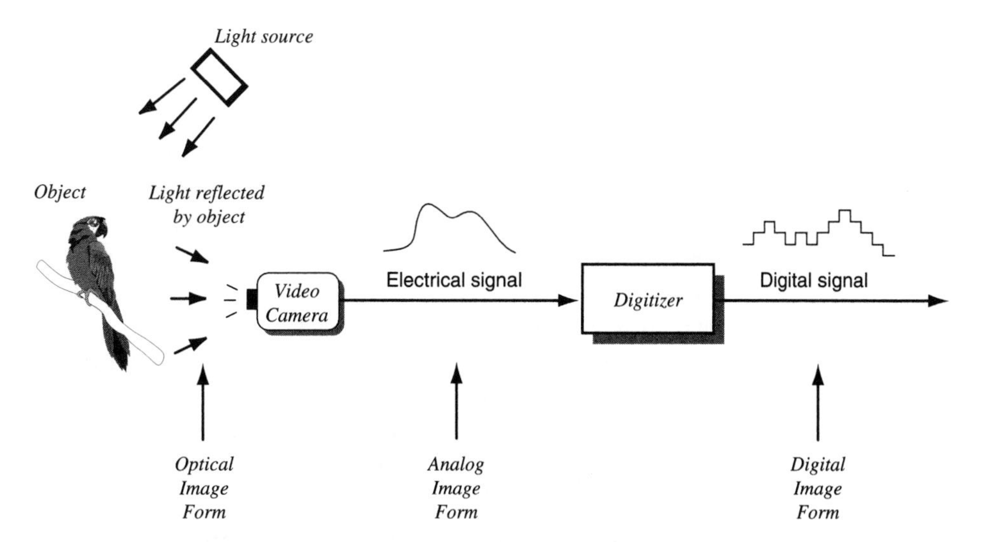

Figure 1.2 *Images generally begin in an optical form. The image can be converted to an analog signal form, and finally to a digital data form.*

Optical image processing uses an arrangement of optical elements to carry out an operation. Eyeglasses are a form of optical image processing. When a process is applied to an image that is in the form of transmitted or reflected light, we refer to it as an *optical process*.

An important form of optical image processing has evolved from the use of the photographic camera and darkroom. For years, the photographer has enhanced, manipulated, and abstracted images from one form to another to create a better or more favorable printed image. Image color and contrast filtering, dodging and burning, sharpening, and vignetting are all photographic optical image processing techniques. This classical form of image processing has been refined through years of improving technology, and through numerous hours of trial-and-error exploration. Today's photographer is armed with a broad base of rules and techniques to produce quick and predictable results. Camera and darkroom pioneers were the first to use defined image processing techniques regularly in their everyday professional pursuits.

Analog image processing uses analog electrical circuits to carry out an operation. When a process is applied to an image that is in the form of an analog signal, we refer to it as an *analog process*.

The earlier example of television receiver controls illustrated analog image processing techniques. The black-and-white portion of the standard commercial television signal is a voltage level that varies to represent brightness throughout the transmitted images. When we change the electrical characteristics of this signal, a corresponding change in the final displayed images occurs. The "brightness" and "contrast" controls of the television receiver modify the overall voltage amplitude and black-level offset of the video signal. This action directly results in changes in the brightness and contrast range of the displayed images.

Digital image processing uses digital circuits, computer processors, and software to carry out an operation. It is a method of image processing enabled by the advent of the digital computer. Digital techniques provide precise implementation of image processing functions, along with great flexibility and power for general-purpose image processing applications.

Within the digital domain, an image is represented by discrete points of numerically defined brightness. By manipulating these brightness values, the digital computer can carry out immensely complex operations with relative ease. Furthermore, the flexibility of the computer programming process allows us to adapt and modify digital image processing operations quickly.

Digital image processing has become a significant form of image processing because of continuing improvements in sophisticated semiconductor technologies. This has led to computer hardware performance increases and declining costs. Digital image processing algorithm research also continues at a fast pace, spurred on by increasing commercial demands. When coupled with an expanding application base, digital image processing techniques are becoming key tools in diverse new industries.

Digital Image Processing—An Historical Evolution

The roots of significant achievement in digital image processing can be traced back to the early 1960s. The United States, through the National Aeronautics and Space Administration (NASA), was energetically pursuing its lunar science program. NASA's interest was in characterizing the lunar surface to support the upcoming Apollo manned lunar exploration program.

The Ranger program was established, in part, to image the lunar surface, relaying the pictures to earthbound scientists for evaluation and composition into lunar maps. After four previous Ranger missions, during which the video equipment failed to function, Ranger 7 returned the first video image from the moon. Ranger 7 went on to return thousands of images in what became a very successful mission.

The original Ranger images were created and transmitted to Earth in an analog signal form. They were recorded and then converted to a digital form. Subsequent digital processing of the images by NASA's Jet Propulsion Laboratory (JPL) in Pasadena, California, was used to correct various camera geometric and response distortions successfully. This digital image processing was carried out at significant expense using large mainframe computers. This processing of Ranger 7 imagery ushered the digital computer into the image processing world.

With its early success in digital image processing, NASA continued to fund research and development to support its other space programs. A series of planetary exploration probes, where digital image processing techniques were used to support the imaging systems, followed the Ranger program. The Mariner project returned imagery from the inner planets—Mars, Venus, and Mercury. Mariner spacecraft were the first to use an on-board digital imaging system and returned images in a digital form. Project Surveyor soft-landed cameras on the lunar surface. Pioneers 10 and 11 spacecraft returned fly-by images of Jupiter and Saturn. Two Viking spacecraft, equipped with color cameras, orbited and soft-landed on the surface of Mars. Voyagers 1 and 2 encountered the outer planets Jupiter, Saturn, Uranus, and Neptune, returning a wide range of imagery that enhanced the scientific studies of those planets.

Digital imagery and image processing have become an integral part of NASA's planetary exploration programs. The technology has been used in many subsequent programs, including the Hubble Space Telescope program. We can expect to see NASA's use of digital image processing well into the future.

Although NASA space programs provided the initial impetus and funding for research and development of digital image processing systems, other applications for the technology began to emerge. In the late 1960s, the medical diagnostic imaging field began to apply digital image processing techniques to X-ray imagery. Medical uses for digital image processing techniques have expanded throughout the 1980s and 1990s to include Computed Tomography (CT), Magnetic Resonance Imagery

(MRI), Positron Emission Tomography (PET), and diagnostic ultrasound imaging applications. In fact, these medical imaging applications would not even be possible without the use of extensive and complex digital image processing techniques.

The U.S. Landsat earth-orbiting satellites began acquiring and returning digital multispectral earth imagery in the early 1970s, and the program continues today. Landsat imagery is used extensively in agricultural land-use analysis. Landsat image data uses digital transmission techniques and digital image processing techniques to restore, analyze, and archive the data. Other earth-orbiting imaging satellites launched by the United States since the 1960s include TIROS, NIMBUS, and GOES systems. Each of these satellites has returned various forms of daily meteorological imagery. The GOES satellites provide the U.S. weather images that we see regularly on our local television newscasts. A GOES image is illustrated in Figure 1.3. Likewise, a variety of military surveillance satellite systems rely heavily on digital image processing methods to transmit and analyze returned imagery.

Digital biological image analysis techniques were first applied in the 1970s and, with the advent of the microprocessor-based personal computer, increased in popularity in the 1980s. Methods of automated cell detection, counting, and classification helped provide faster and more accurate analysis and detection of studied cell formations.

In the 1980s, digital image processing techniques found uses in broadcast television special-effects processors. These processors could geometrically rotate and warp television images in real-time, about 33 milliseconds per color image. This technology quickly migrated into other special-purpose digital hardware systems to support the use of digital image processing techniques in motion picture production.

While the entertainment industry was using digital image processing techniques to improve artistic expression, the military began using it to improve guidance systems for the accurate delivery of missiles and bombs. At this time, the hardware had become small enough and required little enough power to allow the digital imaging equipment to fly directly aboard the weapon systems.

During the mid-1980s, uses for digital image processing techniques expanded to include various machine vision applications. In particular, factories employed digital image processing techniques to automate manufacturing processes. Instead of a

Figure 1.3 *A GOES weather image over the western United States. These images are transmitted to Earth hourly. (Image courtesy NOAA/NWS.)*

human operator inspecting parts, a computerized digital image processing system could autonomously look and make measurements to accept or reject an inspected item. Many other automated inspection tasks have evolved into the 1990s, and all rely on digital image processing methods. Now, these machine vision applications almost exclusively use relatively inexpensive microprocessor-based personal computers.

From the 1960s through today, the evolution of the digital computer has certainly been largely responsible for enabling the proliferation of digital image processing applications. Costly mainframe computers are no longer a requirement of the digital image processing equation like they were in the 1960s. The advent of microprocessors, leading to the personal computer, has allowed stand-alone digital image processing applications to become viable.

Applications of Digital Image Processing

With an appreciation for the historical roots of digital image processing, let's look at its contemporary uses.

Digital image processing techniques can be used to analyze a digital image or to process it to a new, improved image. Any situation requiring the enhancement, restoration, analysis, or creation of a digital image is a candidate for these techniques. Here are some of the major applications of digital image processing technology in use today.

Biological Research Bioresearch and biomedical laboratories use digital techniques to visually analyze components of a biological sample. In some cases, digital image processing techniques provide totally automated systems for specimen analysis.

- Image enhancement—various techniques for improving the visibility of features that are not evident or clear in the original image, such as contrast balancing and edge sharpening.
- Bone, tissue, and cell analysis—automatic counting and classification of cell structures and other objects meeting prescribed characteristics.
- DNA typing—analysis, classification, and matching of DNA material.

Defense/Intelligence The military has made widespread use of digital image processing techniques for various applications. Defense and intelligence agencies of the United States, and of other governments, have poured vast resources into the research and implementation of the technology.

- Image enhancement—various techniques for improving the visibility of features that are not evident or clear in the original image, such as contrast balancing and edge sharpening.

- Reconnaissance photo-interpretation—automated interpretation of earth satellite imagery to look for sensitive targets or military threats such as airports, sea vessels, missile launches, or military installations.
- Target acquisition and guidance—recognizing and tracking targets in real-time for smart-bomb and missile-guidance systems.

Document Processing Acquisition and processing of documents and drawings have helped to automate many industries that were classically paper-driven, such as banking (check processing) and insurance-claim processing.

- Scanning, archiving, and transmission—converting paper documents to a digital image form, compressing the image, and storing it on magnetic or other media for archiving.
- Document reading—automatically detecting and recognizing printed characters so that documents like bank checks, tax forms, and so forth can be intelligently processed by computer.

Factory Automation Vision systems in the manufacturing environment provide automated quality inspection and process monitoring. These systems free the human operator and inspector, while improving overall process accuracies and reliability.

- Visual inspection—automatically analyzing predetermined features of manufactured parts on an assembly line to look for defects and process variations.

Law Enforcement Forensics Law enforcement agencies process large volumes of images for mug shots, evidence, and fingerprints. Various forms of digital enhancement, archiving, and classification processing are part of the modern operation.

- Image enhancement—various techniques for improving the visibility of features that are not evident or clear in the original image, such as contrast balancing and edge sharpening.
- Fingerprint feature analysis—automated fingerprint-classification and -identification operations.
- DNA matching—biological material analysis and matching between multiple samples.

Materials Research Multidisciplinary laboratories involved in materials research use digital image processing techniques to visually analyze components of a material sample.

- Image enhancement—various techniques for improving the visibility of features that are not evident or clear in the original image, such as contrast balancing and edge sharpening.
- Material feature analysis—automatic counting and classification of objects, such as impurities and grain sizes meeting prescribed characteristics.
- Surface and structural rendering—creating three-dimensional surface and internal structure renderings for visualization of features.

Medical Diagnostic Imaging Medical radiological imaging looks at the internal components of the human body. X-ray imaging and computed tomography techniques make intensive use of digital image processing.

- Image enhancement—various techniques for improving the visibility of features that are not evident or clear in the original image, such as contrast balancing and edge sharpening.
- Digital subtraction angiography—enhancing blood vessel imagery by subtracting a baseline X-ray image from a second image with an X-ray-opaque liquid in the bloodstream.
- Computed tomography—creating images using multiple image projections. This method is used in CT, MRI, and PET scanners.

Photography Digital image processing techniques have augmented and, in some cases, replaced methods used by the photographer for image composition and darkroom processing.

- Image enhancement—various techniques for improving the visibility or artistic rendering of features that are not acceptable or clear in the original image, such as contrast balancing, edge sharpening, color balancing, or retouching of defects.
- Multiple-object scene compositing—adding and subtracting objects to and from a scene to create illusions that did not originally exist.
- Special effects—warping, blending, and other visual effects to convert existing imagery into new visual forms.

Publishing/Prepress The desktop publishing and prepress industries use digital image processing techniques to enhance and lay out digital images for publication. Most publications use digital image and typography techniques.

- Image enhancement—various techniques for improving the visibility or artistic rendering of features that are not acceptable or clear in the original image, such as contrast balancing, edge sharpening, color balancing, or retouching of defects.

- Layout compositing—mixing of image, text, and graphical elements into final film suitable for printing.
- Color separation—creating cyan, magenta, yellow, and black film separates for the four-color printing process.

Remote Sensing/Earth Resources Orbiting satellites image every square mile of the Earth's surface on a regular basis for resource management purposes. This image data is used to analyze crop yields and damage due to disease, early frost, and other factors.

- Landcover analysis—measuring various vegetation features such as water content, temperature, chlorophyll absorption characteristics, and geometric features.
- Terrain rendering—creating three-dimensional Earth terrain renderings for analysis, based on elevation data returned by remote sensing satellites.

Space Exploration/Astronomy Digital imaging systems are used almost exclusively on board exploratory spacecraft and earthbound telescopes. This equipment makes extensive use of digital image processing techniques to enhance, restore, and analyze extraterrestrial imagery.

- Image enhancement—various techniques for improving the visibility of features that are not evident or clear in the original image, such as contrast balancing and edge sharpening.
- Imaging system deficiency correction—techniques for correcting known and unknown defects in the original image, such as sensor response nonlinearities, geometric distortions, image noise, and motion blur.
- Automatic event detection—detecting features that are changing over time, such as solar activity and other cosmic events.
- Terrain rendering—creating three-dimensional planetary terrain renderings, based on elevation data returned by exploratory satellites and space vehicles.

Video/Film Special Effects The video and film production industries use various digital image processing techniques for creating and hiding artifacts of special visual effects.

- Multiple-object scene compositing—adding and subtracting objects to and from a scene to create illusions that did not originally exist.
- Scene creation—fabricating synthetic imagery when physical creation would be costly or impossible.

- Special effects—warping, blending, and other visual effects to convert existing imagery into new visual forms.

Video Programming Distribution Methods for reducing the data size of images improve both video archiving and transmission processes. The television program distribution industries such as cable television, direct-broadcast satellite, and electronic video movie distribution can all benefit greatly from image compression techniques.

- Video archiving and transmission—removing redundant information from a digital image or sequence of images; this reduces data size, making transmission faster, archiving more efficient, and both processes less expensive.

These applications represent many of the uses for digital image processing at the time of this writing, and the opportunities remain endless. With declining costs of general computing equipment and of specific devices for digital image acquisition, processing, and display, new applications emerge daily. Digital image processing is, and will be, the key technology for many innovations yet to come.

In the following pages, we will explore and evaluate the techniques and methods of digital image processing. This practical introduction will discuss how applications, such as those discussed above, can be turned into reality through digital image processing.

2 *Fundamentals of Digital Image Processing*

This chapter explores several fundamental topics of digital image processing. We start out with a description of the human visual system; an understanding of how humans perceive images is essential to many of our digital image processing tasks. Next, the five fundamental classes of digital image processing operations are presented. We'll then look at the structure of the processing solution and the differences between a digital image processing application, class, operation, and process. Finally, we'll review the basic components of a digital image processing system. The chapter introduces these subjects as a foundation upon which future chapters will build.

The Human Visual System

In general, many digital image processing applications are designed to produce an improved image for use by a human observer. Because, in these cases, a human is the end-user, it is important to understand the *human visual system*. Comprehending the various characteristics and limitations of the eye–brain system can help us to maximize the effectiveness of digital image processing operations.

Physical Structure of the Eye

The human visual system is made up of the eye and a portion of the brain that processes neural signals from the eye. Together, the eye and brain convert optical information to a perception of a visual scene. The eye is the camera portion of the human visual system. It converts visual information into nerve impulses used by the brain. The major functional components of the eye are shown in Figure 2.1.

Light rays generated or reflected by a scene first strike the eye at the *cornea*. The cornea acts as a convex lens, refracting the rays. This refraction forms the initial

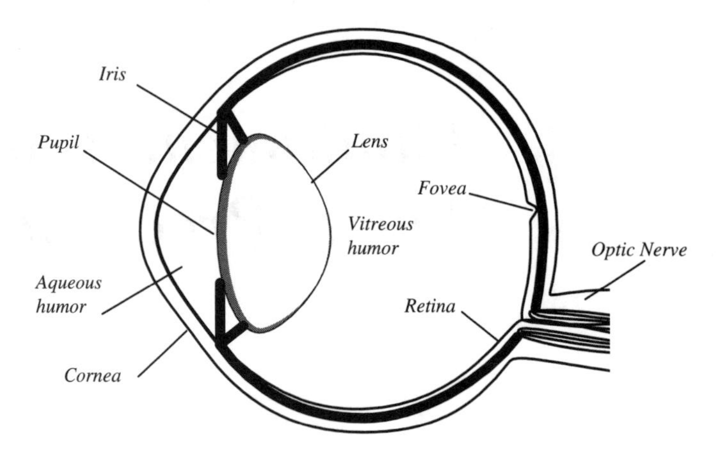

Figure 2.1 *A cross-sectional view of the human eye.*

focusing of light entering the eye. The cornea also forms a clear protective covering to the eye. Following the cornea, the rays pass through a clear liquid called the *aqueous humor*, and then through the *iris* and *lens*. The iris acts as a variable aperture to control the amount of light allowed to pass through to the lens. The iris is controlled by muscles that open and close based on the average intensity of the scene being viewed. At night, the iris is generally wide open. In daylight, it is usually closed quite small.

The lens performs the second focusing of light, projecting it upon the *retina*. The lens is also controlled by muscles, which allow the focal length of the overall optical system to vary depending upon the observer's distance from the viewed scene. Just like a camera, the eye must be focused based on how far it is from the scene. Light rays leaving the lens pass through another clear liquid, called the *vitreous humor*, and are ultimately focused upon the retina. The vitreous humor maintains the structure of the eye while optically matching the lens to the retina.

The retina is the eye's image detector. The retina is composed of *photoreceptors* that convert the intensity and color of light to neural signals. The eye has on the order of 100 million of these photosensitive elements.

Two types of photoreceptors exist, *rods* and *cones*. The rods are the most proliferous (about 100 million) and are also the most responsive to light. Rods respond to broad-spectrum color light, and therefore cannot discriminate color, as shown in Figure 2.2. The rods are used mainly for low-light vision, such as at night.

The cones are much less abundant than rods (about 6 to 7 million) and are somewhat less responsive to light. The cones are used for bright-light vision, such as during daylight. Three different types of cones exist; each responds to a distinct spectral band of color light. This allows the eye and brain to discriminate color through a process called *trichromacy*. Basically, each cone responds differently to an arbitrary color, thus generating a unique set of responses for every unique color of light. With these signals from the three cone types, the brain has the information

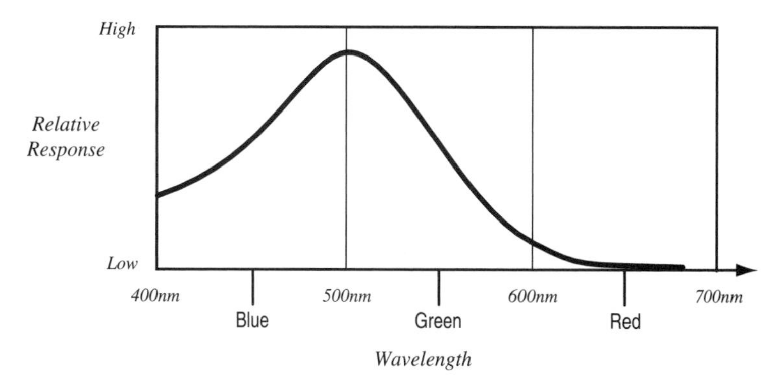

Figure 2.2 *Relative response curve for rods. Rods are highly responsive to light, but see only a single spectral band of light, and therefore cannot discriminate color.*

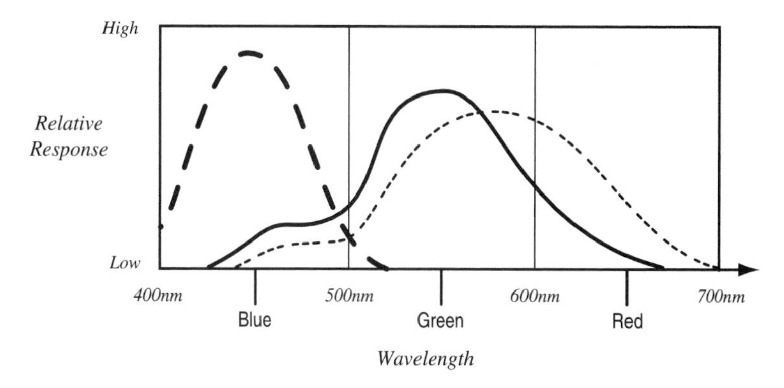

Figure 2.3 *Relative response curves for cones. Cones are less responsive, but see three distinct color spectral bands of light, enabling color vision.*

with which to form a distinct perception of a large number of different colors. Figure 2.3 shows the three response curves of the three different cones.

Rods and cones are distributed across the retina in inverse proportions. Cones are most proliferous around an area called the *fovea*. The fovea is the part of the retina where light is focused when we look straight forward. Virtually no rods exist at the fovea. Instead, the rods are distributed everywhere else across the retina. As shown in Figure 2.4, the fovea has most of the cones and few rods, whereas the rest of the retina has most of the rods and few cones.

The differences between rods and cones, and their distributions across the retina, are responsible for several aspects of vision. Because the color-sensing cones are concentrated at the fovea, our color perception is best for the object that we are viewing directly forward. Conversely, we have minimal perception of color for objects in our peripheral vision. Because the highly sensitive rods are abundant everywhere but the fovea, our low-light-level perception is best in our peripheral

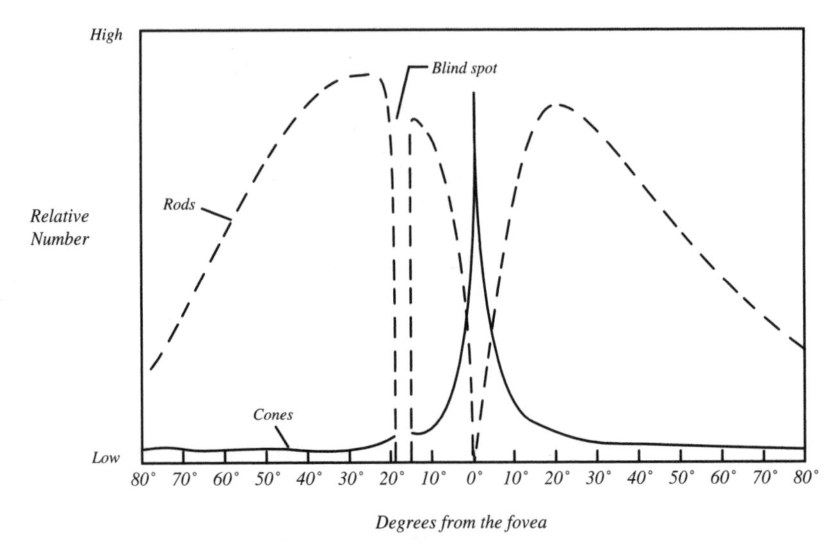

Figure 2.4 *Distribution of rods and cones across the retina. Most of the cones are at the fovea. Rods are spread just about everywhere except the fovea.*

vision. Hence, at night we see the best detail slightly off from our direct forward view. The relative insensitivity of cones also accounts for our inability to perceive color under low-light conditions, such as at night.

As light strikes the rods and cones, it causes an electrochemical reaction that generates neural impulses. These impulses are passed to the brain via the *optic nerve*. The optic nerve is an extension of the retina that connects it to the brain. A small blind spot is created on the retina where the optic nerve attaches. The neural impulses are received by the brain and processed by the visual cortex. The perception of vision is created within the processes of the visual cortex.

Image Processing in the Eye

Before the visual impulses of the eye are processed by the visual cortex, some interesting preprocessing of intensity information occurs in the eye itself. Knowledge of these phenomena can affect the way we use digital image processing techniques to improve an image for the human observer. The primary characteristics of interest are the way in which the eye's photoreceptors respond to differences in light intensity, the way in which they interact with one another, and their response characteristics to fine details.

The relationship between the intensity of light entering the eye and its perceived brightness is not a linear function. This means that as the intensity of a viewed feature is changed, the viewer will not perceive an equal change in brightness. The eye's actual intensity response is more logarithmic, appearing as a curve

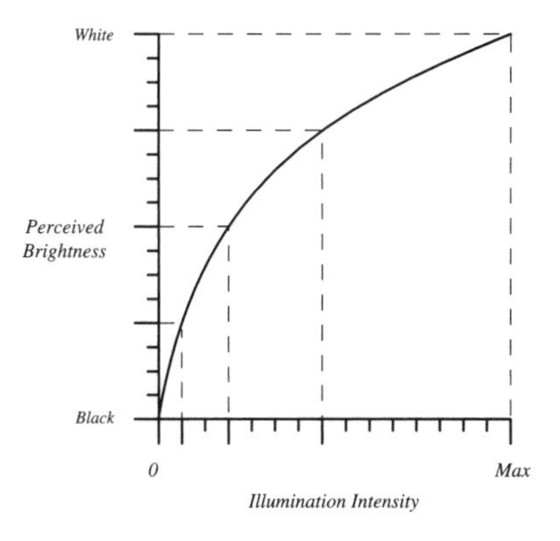

Figure 2.5 *Logarithmic intensity response of the eye. The eye perceives the next noticeable difference in perceived brightness each time the intensity of a viewed object doubles.*

similar to that in Figure 2.5. In fact, it has been shown that the intensity of a feature must nearly double before the eye can detect that the intensity has changed. Therefore, slight changes in intensity in dark regions of a scene tend to be more perceptible than identically slight changes in bright regions. This relationship of intensity to perceived brightness is known as *Weber's law.*

Figure 2.6 illustrates the logarithmic intensity response of the eye. Two images are shown along with their actual intensity curves. In Figure 2.6a and b, the intensity of the bars ascends from left to right in equal steps. The intensity steps span the full range of grays from black to white. As we would expect from the curve in Figure 2.5, the steps in the dark region of the image are easily detectable, whereas the steps in the bright region of the image tend to be indistinguishable. Two phenomena are evident: (1) the difference in perceived brightness of the steps does not appear equal, and (2) the eye cannot see the same intensity increments in the bright regions that it sees in the dark regions.

In Figure 2.6c and d, the intensity of the bars ascends from left to right in steps that match the eye's logarithmic response. The intensity steps still span the full

Figure 2.6a *Weber's law—gray-scale steps with equal intensity steps. The steps appear compacted to the dark region of the scale.*

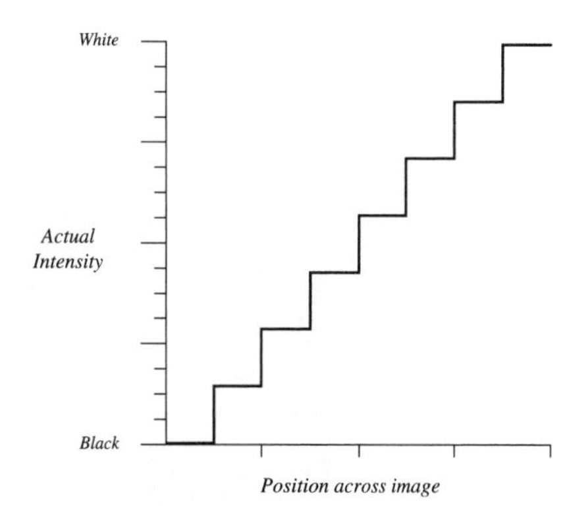

Figure 2.6b *The actual intensity of the gray-scale steps in (a).*

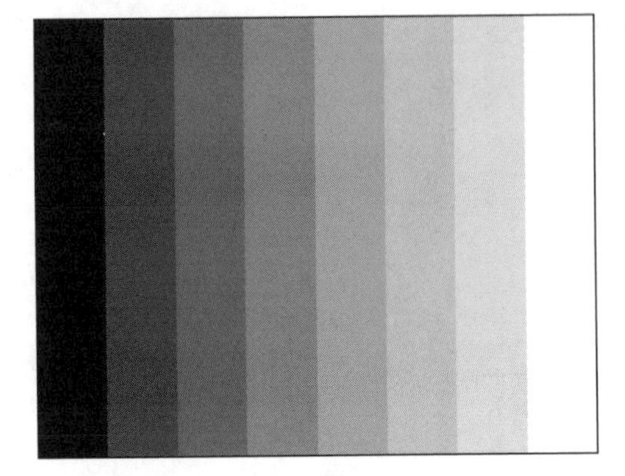

Figure 2.6c Gray-scale steps that match the eye's logarithmic response. The steps appear to have uniform increments in intensity.

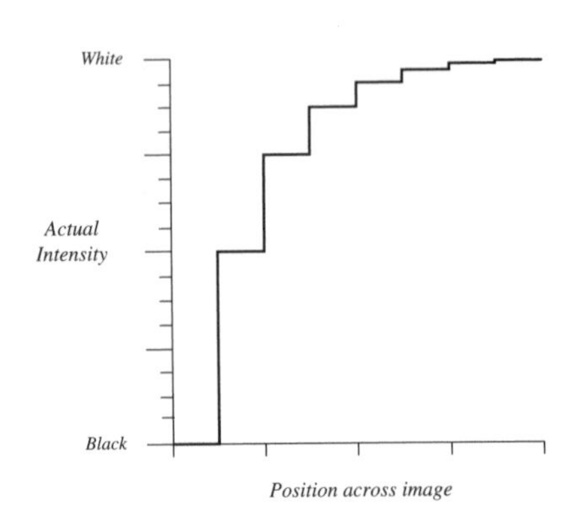

Figure 2.6d The actual intensity of the gray-scale steps in (c).

range of grays from black to white. However, in this figure, the perceived brightnesses of the steps tend to appear equally spaced and equally well defined in the bright regions of the image as in the dark regions.

The important point is that the logarithmic intensity response of the eye makes it more sensitive to intensity change in the dark regions than in the bright regions of an image. In the digital processing of an image, simple darkening of bright regions can make undetectably minute intensity changes perceptible.

Interactions between photoreceptors, known as *lateral inhibition*, also cause important visual phenomena to occur. Two of these effects illustrate lateral inhibition's role in the perception of brightness. One effect, referred to as *simultaneous contrast*, is an illusion whereby the perceived brightness of a region depends on the intensity of the surrounding area. This effect is demonstrated in Figure 2.7. The two squares have identical intensities, yet the one on the left appears brighter than the one on the right. This is because the area surrounding the square on the left is darker than the area surrounding the square on the right. The visual system adjusts its intensity response based on the average intensity surrounding the viewed feature. Since the left side of the

Figure 2.7 Simultaneous contrast—the two small squares have the same intensity, but appear as different brightnesses because the eye adapts to the surrounding areas of different intensities.

image has a darker average intensity (because of the darker background area), its square appears brighter. The right side's brighter average intensity makes its square appear darker. Hence, there is an difference in the apparent brightness of the two squares.

A second phenomenon caused by lateral inhibition is the *Mach band effect*. With this effect, the visual system accentuates sharp intensity changes. Figure 2.8 illustrates the effect. As the eye views a sharp change in intensity, it adds apparent brightness undershoots and overshoots to its response. In other words, the brightness appears to dip a little prior to the transition and peak a little following the transition. This makes the transition appear to have greater amplitude than it really does. It turns out that this is the eye's way of adding edge enhancement to intensity transitions. The visual system actually sharpens everything we view, giving us improved visual acuity.

The visual system has fundamental *frequency response* limitations. As in any optical system, the eye has a limit to how fine a detail, or intensity transition, it can resolve. The limiting factors are the number and organization of photoreceptors in the retina, the quality of the eye's optics (cornea, aqueous humor, lens, and vitre-

Figure 2.8a *Mach band effect—when you view this gray-scale step image from left to right, the apparent brightness dips just before each step, and appears to increase after each step.*

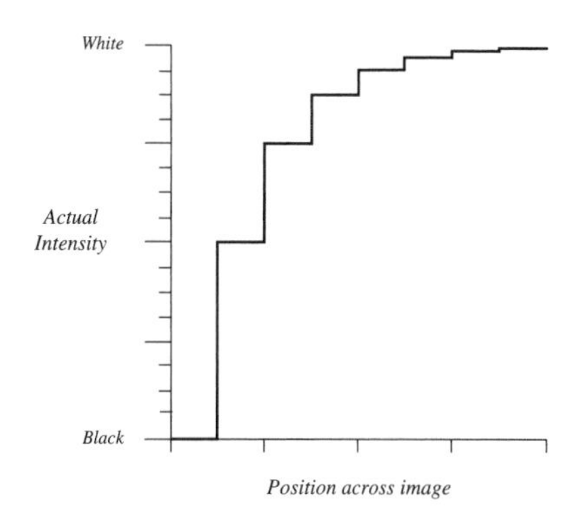

Figure 2.8b *The actual intensity of the gray-scale steps.*

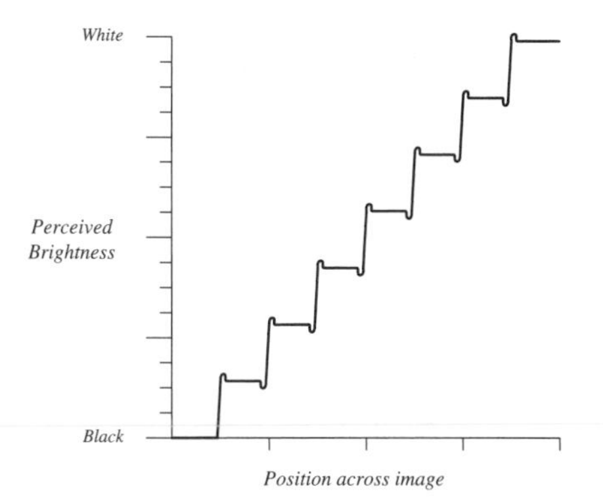

Figure 2.8c *The perceived brightness of the gray-scale steps.*

Figure 2.9 Ascending sinewave pattern—this pattern increases in frequency from left to right. It also decreases in contrast from top to bottom.

ous humor), and the transmission and processing of visual information to the brain. Generally, the eye's frequency response falls off as viewed intensity transitions get finer and finer in size, as shown in Figure 2.9. The contrast, or difference between gray levels, of the intensity transition is also a factor. The higher the contrast, the finer the detail that the eye can resolve. Ultimately, when the transitions are too fine, or the contrast is too low, the eye can no longer resolve them. At this point, the eye can perceive only an average gray-level of the detail area.

The phenomena discussed illustrate the complex processes that occur within the human visual system. By combining the concepts of nonlinear intensity response, photoreceptor interaction, and frequency response of the eye, we can make a few observations:

1. The intensity of a viewed object is related to the average intensity surrounding the object. The object appears darker if the surrounding area is bright, or brighter if the surrounding area is dark.
2. Subtle intensity changes are more apparent in dark regions than in bright regions of an image.
3. Sharp intensity transitions within an image are accentuated.
4. The response to image detail falls off as the details get too fine to resolve. Details with high contrast are more resolvable than those with low contrast.

With these observations in mind, we can tune the action of digital image processing operations to best exploit the human visual system. The result is images that are better optimized for viewing by a human observer.

Fundamental Classes of Digital Image Processing

Numerous and widely varying types of digital image processing operations have been developed during the past 30 years. While one operation may improve the quality of

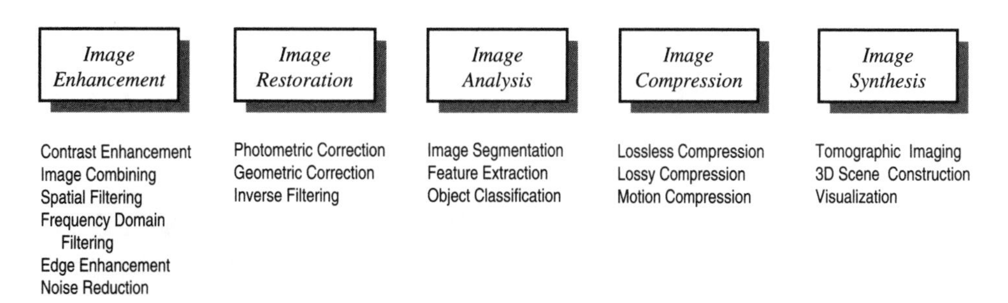

Figure 2.10 *The five fundamental classes of digital image processing and some of their representative operations.*

an image, another may automatically extract information from it. Whatever the operation, the same steps are followed: A digital technique is applied to a digital image to form a digital result, such as a new image or a list of extracted data.

Digital image processing operations can be broadly grouped into five *fundamental classes*: image enhancement, restoration, analysis, compression, and synthesis. Each class contains specific operations, as shown in Figure 2.10. A certain application may call upon the operations of at least one, but perhaps several, of these classes. In this section, we introduce these classes and describe some of their uses and needs. Part II of this book will go on to examine many of the individual operations in depth.

Image Enhancement

By far the most ubiquitous digital image processing class of operations is image enhancement. Virtually all digital image processing tasks involve some sort of enhancement, either as a desired end result or as a preprocess to some other operation.

Image enhancement operations improve the qualities of an image. They can be used to improve an image's contrast and brightness characteristics, reduce its noise content, or sharpen its details. What is considered an improvement in the image is often subjective and is generally dependent upon the application, as well as on the judgement of the observer.

For instance, as shown in Figure 2.11, one application may require balancing the contrast of an image to make its washed-out appearance more pleasing to a human observer. Another application, however, may require that the contrast of a well-balanced image be dramatically increased so that specific features are enhanced for an automated image analysis operation. Because each of these applications has an entirely different goal, the enhancement operations appear contrary to the other's interest. In short, one application's enhancement is another's degradation.

As in the second case, image enhancement techniques are often used to improve an image for a subsequent image analysis operation. In a case like this, the enhanced

Figure 2.11a *An image with washed out contrast has details that are hard to see.*

Figure 2.11b *An image enhancement operation balances the image contrast, improving the visibility of its details.*

Figure 2.11c *An image with good contrast balance requires yet higher contrast to aid a following analysis operation in measuring its features.*

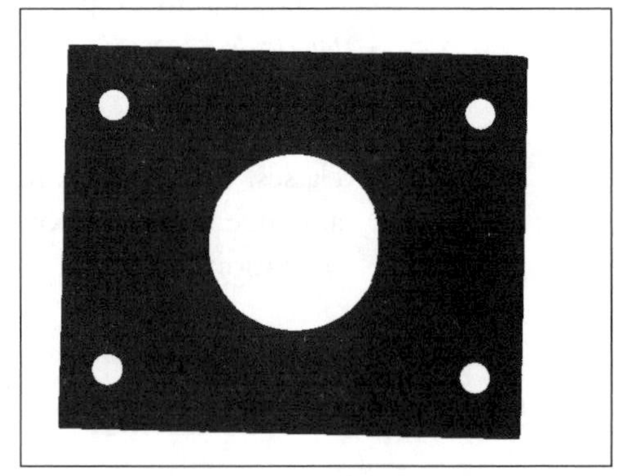

Figure 2.11d *An image enhancement operation stretches the image contrast to its limit, improving the discrimination of the holes in the part from the background.*

image may not look at all better to the human observer, but may be more suited for the digital image processing operation that follows.

Image enhancement techniques may be grouped as either *subjective enhancements* or *objective enhancements*. Subjective enhancements are used to make an image more visually appealing or more appropriate for subsequent processing. They may be applied in various forms until this goal is achieved. In this way, subjective enhancement techniques are usually ad hoc and applied with user discretion. For instance, if an image shows poor detail, a subjective enhancement technique might be used to apply a sharpening operation. This operation may be repeatedly applied until the observer feels that the image yields the detail necessary for the particular application.

Objective image enhancements, on the other hand, correct an image for known degradations. These enhancements do not necessarily attempt to make the image more appealing. Objective enhancements are applied based on known distortions to the original image and cannot be used arbitrarily based on subjective evaluations of the image's quality. For instance, if an image is acquired from a system that has a known blur that makes the image appear fuzzy, an objective enhancement operation may be applied to correct for the particular blur based on measurements from the original system. The enhancement would not be repeatedly applied until the image looks good. Rather, it would be carefully computed from measurements taken from the system and then applied once. In this way, objective image enhancements are used to correct known degradations back to their mathematically correct points, regardless of appearance. Objective image enhancements cross over into the image restoration class of digital image processing.

Contrast Degradations/Enhancements

Contrast degradations in an image are associated with poor brightness distribution characteristics. This means that the image may appear either washed-out or harshly saturated. A washed-out image exhibits undesirably low contrast. Gray tones in the image do not span the overall range from black to white. Scene details appear subdued, making the viewing of image details tedious. A harshly saturated image is a result of high contrast. Tones in the image tend to be either black or white without many in-between tones. A high-contrast image lacks the naturally occurring, smooth distribution of gray tones. Generally, neither low-contrast nor high-contrast qualities are visually pleasing in an image.

A well-balanced image of good contrast is composed of gray tones stretching from the dark blacks, through the middle gray tones, to the bright whites. Operations to enhance an image of poor contrast are relatively simple. Contrast-enhanced images exhibit a more natural distribution of gray levels.

As mentioned earlier, achieving good contrast balance within an image may not be an appropriate goal for some applications. Image enhancement techniques may be used to make some arbitrary feature within an image more visible, perhaps a feature that was previously hidden. In an application of this sort, a resulting image of high contrast, where a desired feature is highlighted, may be the goal. The results of this sort of image enhancement application do not always produce visually pleasing images; rather, they amplify a feature of interest.

Spatial Degradations/Enhancements

Spatial degradations in an image are generally associated with the presentation of scene details. The term *spatial* relates to the two-dimensional nature of an image.

An image may be said to exhibit poor spatial characteristics if detailed areas are either blurred or over-accentuated. Often edge details, such as black-to-white transitions, may be blurred, lacking the sharpness generally associated with good detail. Generally, an image exhibiting either characteristic is difficult to view, and is therefore undesirable.

Additional spatial degradations include image noise, such as the black-and-white speckle noise we often see in poor television images. Geometric deficiencies, such as improper sizing, can also be corrected by geometric image enhancements.

As with contrast enhancements, improving an image so that it's visually pleasing may not be the objective. Some applications require that spatial details be enhanced to an extreme, making the underlying structural features of a scene more visible. Edge enhancement, where only edge details are highlighted, is a common spatial enhancement task for many machine vision applications.

Image Restoration

Image restoration operations, like enhancement operations, also improve the qualities of an image. All restoration operations, however, are strictly objective. That is, they are based on known, measured, or accurately surmised degradations of the original image.

Usually, images that are candidates for image restoration have suffered some form of degradation in the imaging system. Image restoration techniques may be used to restore images with such problems as geometric distortion, improper focus, repetitive noise, and camera motion.

Sometimes images will exhibit degradations created by the geometry of the imaging system. As an extreme example, cameras on earth-orbiting spacecraft will rarely look perfectly straight down on the Earth's surface. Therefore, images from these systems often have a perspective distortion in the image; the image appears tilted either forward or backward, as illustrated in Figure 2.12. As long as the spacecraft–Earth geometry is known at the exact time the image is acquired, a corrected image may be computed to restore a downward-looking view. This type of restoration is essential when satellite images are used for mapmaking or other applications requiring accurate ground referencing.

Image restoration operations can be used to correct images for known degradations, as in the example above. Sometimes, however, the degradation induced into an image may not be known. In these cases, it may be possible to arrive at an estimate of how the image was degraded. Often, clues within an image can yield important information about the factors that led to the image's degradation. Sometimes, with these clues, an estimate of the degradation may be determined

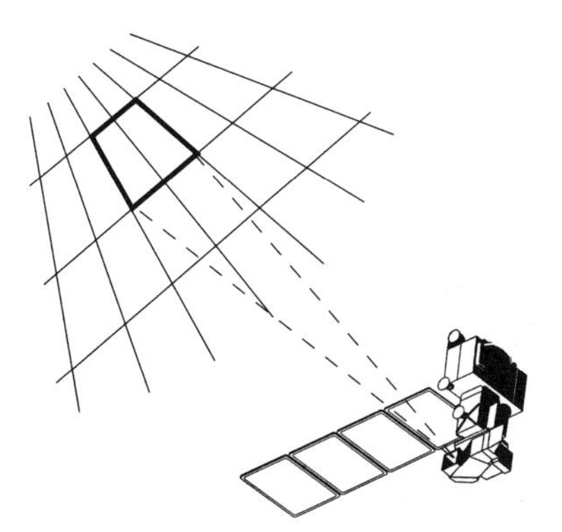

Figure 2.12 *A spacecraft-bound camera is rarely pointed straight down toward a target such as a planetary surface. Using the known attitude of the spacecraft, the geometric distortion can be removed, restoring the image to a perfect downward view.*

and used to inversely filter the image and recreate a close approximation of the original.

System Deficiency Correction

All imaging systems create various forms of distortion. Camera systems usually involve a lighting system, lens, photodetector, and digitization function. Each of these components can contribute degradations to the final image. If the distortions can be measured, either prior to or following the acquisition of an image, the distortion can usually be removed from the image.

Photometric restoration corrects an image for intensity response deficiencies. Often, lighting or photodetector response characteristics will cause parts of an image to suffer brightness anomalies. These may be in the form of light rolloff due to the lens or brightness striping due to inconsistent photodetector response.

Some geometric distortions can be caused by the physical geometry of the imaging system, such as the earlier spacecraft example. In that case, the known orbital and attitude parameters of the spacecraft can be used to restore the image. Others can be caused by distortions that can be premeasured for later restoration. For instance, camera lenses often have geometric distortions, one of which is known as "pincushion" distortion. This distortion makes the image appear bloated in the center. By acquiring an image of a known square grid target, we can take measurements and create a warping equation to correct for the distortion. By applying the warping operation to the grid image, we can restore the grid to appear perfectly square. Likewise, other images acquired through the same camera can be restored by applying the same warping operation.

Another significant restorable system deficiency centers around unwanted noise in an image. Images acquired from real-world sources are occasionally exposed to certain situations or malfunctions that can introduce distracting forms of noise. If the noise is repetitive, we can usually characterize and successfully remove it. If it is random, we can use techniques like multiple-image averaging to reduce its appearance.

Inverse Filtering

Images with misfocus or motion blur appear fuzzy, with a lack of detail. By making assumptions about the blurring function that caused the degradation, such as an out-of-focus lens, it is possible to perform an inverse filtering operation to remove the degradation. Sometimes, even if the degradation function is not known, the image itself may hold clues. By evaluating the appearance of known image features, such as points of light and edges, it is often possible to derive an inverse filtering function that will reverse the process of the original degradation.

Further, techniques exist whereby "blind" estimates can be used to characterize an image's degradation. We can apply a corresponding inverse filter and use the results to converge on the best estimate of the offending distortion.

Image Analysis

Image analysis operations generally do not produce pictorial results. Instead, they produce numerical or graphical information based on characteristics of the original image. Image analysis operations break an image into discrete objects and then classify those objects using some measurement process. Additionally, an analysis operation can produce image statistics. Common image analysis operations include extraction and description of scene and overall image features, automated measurements, and object classification.

Image analysis methods are primarily used in machine vision applications. These are applications where the imaging system must make immediate decisions regarding, for instance, the dimensions of a manufactured part traveling on a conveyer. Additionally, image analysis operations can determine the overall or regional qualities of an image's appearance. Information of this sort can be used to apply enhancement or restoration operations.

An example of image analysis is automated measurement. Manufactured parts moving down a conveyer can be imaged by a camera and analyzed to determine whether the parts are good or bad. This type of automated inspection can free human operators from the tedious, and sometimes inaccurate, manual inspection process. Let's say that it is important that a certain part have a specific width, within some predefined tolerance. Each time a part arrives under the camera, an image is acquired. The image is processed first by enhancement operations to, say, improve its contrast and edge definition. Then, the distances between the sides of the part are measured by analysis operations, resulting in information that characterizes the important physical parameter that the part must meet. This process is shown in Figure 2.13. If the part is within tolerance, it is allowed to progress down the conveyer. If not, it is rejected and diverted to a reject or rework area.

Figure 2.13a *An image of a machined part to be analyzed.*

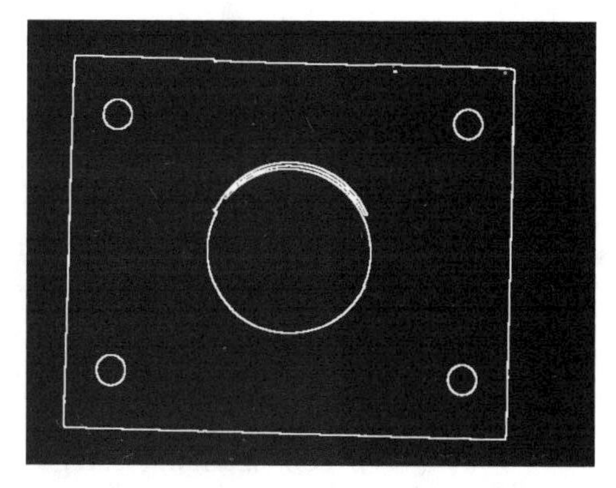

Figure 2.13b *Contrast enhancement operations, followed by edge enhancement operations, yield the outline of the part.*

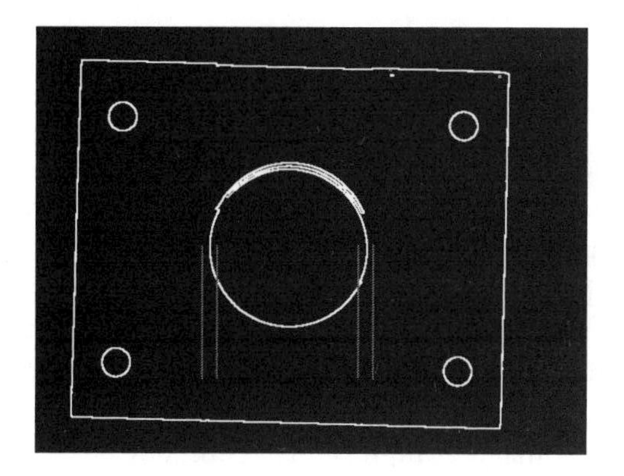

Figure 2.13c *An analysis operation verifies that the part's center hole diameter is within minimum and maximum tolerances—it's a good part.*

Measurements and Statistics

Measuring the parameters of objects is an important image analysis operation. These measurements include object shapes, sizes, relative locations, textures, gray tones, colors, and other parameters. Shapes, for instance, can take on certain qualities, such as round or square, long or short, big or small, and so on. The characteristics used to classify an object can vary, depending upon the requirements of the application. These can be as simple as rough shape characteristics or as complex as precisely measured dimensions and geometries.

Image analysis operations also yield various image statistics. One important statistic, the *brightness histogram*, is the distribution of gray levels in an image. This distribution is displayed in a graphical or tabular form. The histogram describes the overall or regional contrast attributes of an image, and can therefore be used to determine contrast enhancement parameters. Other image statistics, such as the brightness mean and mode and the frequency content, can also be useful information in carrying out subsequent image enhancement and restoration operations.

Image Segmentation, Feature Extraction, and Object Classification

Generally, the process of analyzing objects in an image begins with image segmentation operations, such as image enhancement or restoration operations. These operations are used to isolate and highlight the objects of interest. Then, the features of the objects are extracted, resulting in object outlines and other object measures. These measures describe and characterize the objects in the image. Finally, the object measures are used to classify the objects into specific categories.

As an example, a biological research application may be required to analyze a slide sample of cells. Let's assume that the results need to be reported as a list of cell-size range categories and the number of cells that fall into each group. Here, the analysis operation highlights and outlines each cell, isolating the objects of interest. Then, the area of each outlined cell is computed. The sizes are classified into cell-size ranges and presented as a tabular result.

Image Compression

Image compression and decompression operations reduce the data content necessary to describe an image. Image compression is possible because most images inherently contain large amounts of redundant information. Elimination of these redundancies is the goal of all image compression operations.

Generally, images are compressed to a smaller form so they can be efficiently stored or electronically transported. Later, the compressed image can be decompressed with an inverse operation that restores the image to its original form.

It is desirable to pursue image compression techniques because of the large amounts of data necessary to represent an image. A standard black-and-white television image occupies about 300,000 bytes of data. A color television image is up to three times as large. Finite limits to the sizes and costs of storage media and transport links create excellent reasons to use image compression. Image compression becomes the most compelling when large sequences of images must be handled, such as in video processing or transmission services.

Two general forms of image compression exist. *Lossless image compression* techniques preserve the exact data found in the original image. *Lossy image compression* techniques do not exactly represent the data of the original images but strive only to maintain a particular level of subjective image quality. Figure 2.14 shows one type of lossy decompression of a compressed image.

Lossless Compression/Decompression

Lossless compression and decompression techniques are used when the data in an image must be exactly preserved. Such images often include ones from medical

Figure 2.14a *An image to be compressed using a lossy compression technique.*

Figure 2.14b *Decompressed image. Some distortions are evident in parts of the image.*

and other scientific sources. Whenever subtle image features may contain valuable information, it is preferable to use a lossless compression technique. Lossless techniques are usually used when the image may later be processed with objective enhancement or restoration techniques.

Lossless compression techniques provide roughly a two-to-one compression factor. This factor varies depending upon the content of the image and the exact operation used.

Lossy Compression/Decompression

Lossy compression and decompression techniques are used when the quality of the reconstructed image must be maintained at some level, but need not be precisely identical to that of the original image. This means that only the subjective quality, or how the observer sees the quality of the image, is important. Lossy techniques are used on images where exact measurements or objective processing will not be used. Images appropriate for lossy compression include those in video programming, videoconferencing, and some printing applications.

Lossy compression techniques can provide compression factors from 10-to-1 to 100-to-one and above. The compression factor depends upon image content, operation used, and acceptable quality of the decompressed image.

Image Synthesis

Image synthesis operations create images from other images or non-image data. These operations are used when a desired image is either physically impossible or impractical to acquire, or does not exist in a physical form at all.

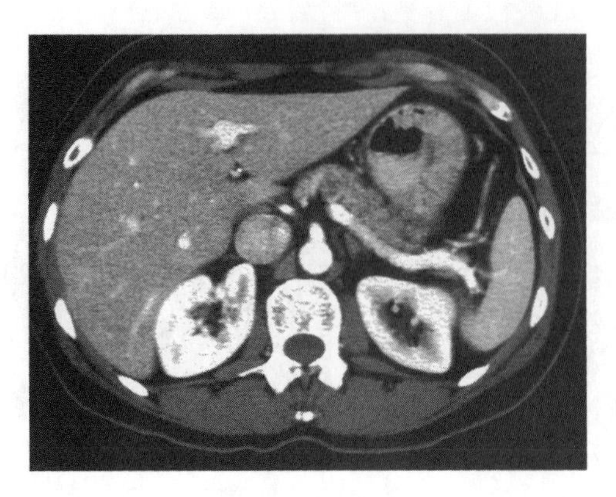

Figure 2.15 *A human abdominal cross-sectional CT image created using computed tomography techniques.*

There are two primary forms of image synthesis operations. The first is the reconstruction of an image using multiple projection images. This form has been exploited primarily by the medical diagnostic imaging field through the technique known as computed tomography. Figure 2.15 shows a typical medical image created using computed tomography techniques.

Visualization is the second form of image synthesis applications. These applications create images for presentation purposes that may or may not be based on physical objects at all. The intent of these operations is to create images that convey important features found in the data that might otherwise be very difficult or impossible to detect. This area of image synthesis is a part of the computer graphics domain.

Reconstruction from Projections

We find the most prolific image synthesis operation in use in various computed tomography medical scanners. Such devices include Computed Tomography (CT), Magnetic Resonance Imaging (MRI), and Positron Emission Tomography (PET) scanners. These devices synthesize cross-sectional images, generally of humans, from multiple transaxial image projections taken around the object of interest. The exciting aspect of this synthesis operation is that the cross-sectional images are physically unobtainable (without slicing the person into numerous sections). The resulting synthetic image is digitally derived from the physically obtainable image data.

CT scanners use X-ray transmission methods to gather projection images, whereas PET scanners use the emission properties of implanted isotopes. Other imaging methods, such as radar-reflection techniques, can also use reconstruction operations to create images.

Creation from Non-image Data

Often, the presentation of data in the form of an image can be desirable. Visualization applications take physical or nonphysical data and use it to create images. The multidimensional characteristics of an image can provide unique ways of interpreting complex data. These applications are being exploited using financial, atmospheric, and numerous other complex data sets.

For instance, a colorful image of airflow across a proposed new airfoil design, visually showing several points of turbulence and thermal properties, may be desired. We could use a computer-aided design (CAD) system to create a computer model of the airfoil and to compute airflow and thermal data under modelled atmospheric conditions. Using these results, we could graphically render a three-dimensional image of the airfoil, showing the simulated lines of airflow. The local temperatures on the airfoil could be shown as differing colors. The entire simulation could be done and the results presented in the form of an image, without physically building the airfoil, much less taking it to a wind tunnel.

Other forms of data, such as three-dimensional elevation data, can be used to create synthetic images of landforms or material samples. This type of image synthesis offers computer visual simulation of physical, three-dimensional objects.

Hierarchy of a Processing Solution

Now that we have a feeling for the classes and applications of a wide set of operations, let's look at how we can systematically apply them to a digital image processing problem.

A digital image processing problem may be solved in a methodical, structured manner. A hierarchy can be defined that provides this structure and traces the problem-solving flow from the statement of an application's problem down to the specific techniques or processes used to solve it. The primary elements in the digital image processing hierarchy are:

- Applications
  - Fundamental Classes
    - Operations
      - Processes

Let's use an example to go through this problem-solving process. Don't worry at this point about what these operations and processes are; they are the essence of following chapters.

At the top, there is an *application*. An application is the statement of a problem that is to be solved. An example application might go as follows:

1. Acquire an image of an automobile license plate from a video camera.
2. Process the image, and extract from it the numbers and letters on the plate as a text string.
3. Transmit the characters to a central database to determine whether the vehicle is wanted for any outstanding driving violations.

This application could be used to detect illegal vehicles automatically as they pass by a video camera along the roadside. The digital image processing portion of the application is step 2, above: "Process the image, and extract from it the numbers and letters on the plate as a text string." We may state the digital image processing application as:

Application: Read license plate characters

Next, we need to analyze the application's problem and break it into rudimentary problems that can be addressed by the fundamental classes of digital image processing techniques. For this application, we might determine that the image must first be processed by an enhancement operation to improve the image quality. Following that, an analysis operation might be used to determine the characters imprinted upon the plate. We may state the fundamental classes of operations to be used as follows:

Application: Read license plate characters
    Image Enhancement: Improve image quality
    Image Analysis: Determine characters on plate

We then select appropriate *operations* from the classes. Here, we determine precisely which operations to use from the fundamental classes chosen. For the enhancement operation, we could chose two operations—a contrast enhancement operation to make the characters stand out more from the background, followed by an edge enhancement operation to derive the outlines of the characters. For the analysis operation, we could use a feature extraction operation to evaluate the outlines and an object classification operation to determine their best matches to stored character templates. We may state the operations to be used as follows:

Application: Read license plate characters
    Image Enhancement: Improve image quality
        Contrast Enhancement: Improve contrast
        Edge Detection: Create character edge outlines
    Image Analysis: Determine characters on plate
        Feature Extraction: Trace character outlines
        Object Classification: Match outlines to character templates

At this point, we have determined the operations necessary to solve the application's problem, but have not yet defined the precise methods by which we will do it. The final step is to chose digital image processing *processes* to carry out the chosen operations. A process is the final mathematically specific technique used to accomplish a particular operation. For instance, we could decide that the contrast enhancement operation should use a contrast stretch pixel point process and the edge enhancement operation should use a Sobel edge enhancement pixel group process. The feature extraction operation could use a contour following process to characterize the character outlines and a line segment comparison process to classify and compare the outlines against stored templates for final character determination. We would ultimately state the final hierarchy, the path from application to processes, as follows:

Application: Read license plate characters
    Image Enhancement: Improve image quality
        Contrast Enhancement: Improve contrast
            Contrast stretch pixel point process
        Edge Detection: Create character edge outlines
            Sobel edge enhancement pixel group process
    Image Analysis: Determine characters on plate
        Feature Extraction: Trace character outlines
            Contour following process
        Object Classification: Match outlines to character templates
            Line segment comparison process

This final flow shows how we will solve this digital image processing application's problem. By applying a structured approach to the solution, we have created a clear statement of task. The developer of the application can immediately see the processes that must be implemented as well as the thought process behind their selection.

The Digital Image Processing System

A digital image processing system is a collection of hardware and software components that can acquire, store, display, and process digital images, as shown in Figure 2.16.

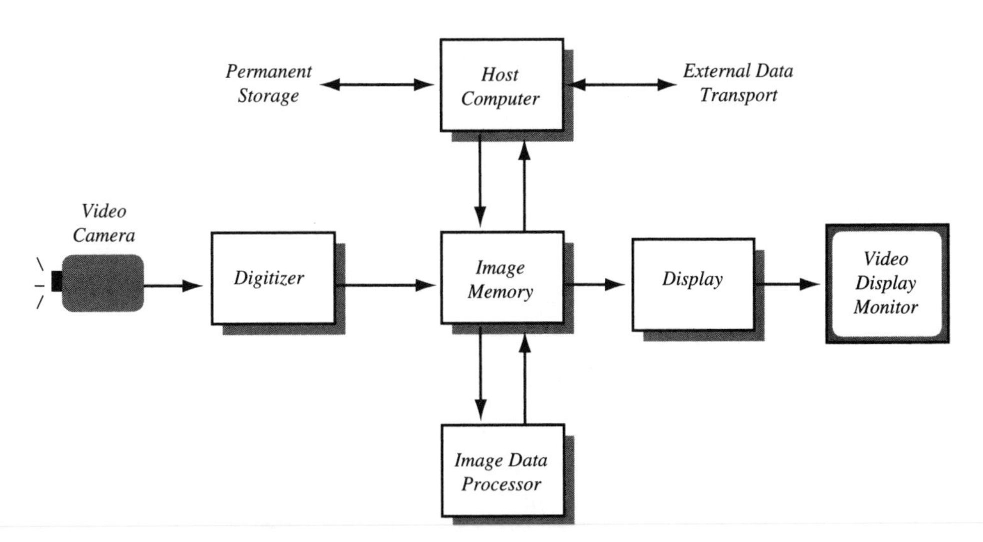

Figure 2.16 *The fundamental components of a digital image processing system.*

Although these components may be physically separated in space and time, each is fundamentally necessary to complete the digital image processing cycle.

In Part III of this book, we will come back to this topic in depth. For now, let's look briefly at the generalized system basics.

The first stage in any digital image processing system is to acquire an image. This is a two-step process requiring some sensor device and digitization function. Many different types of sensors exist. Essentially, though, the sensor must create an electrical signal representing the two-dimensional array of brightnesses of the image being viewed. A video camera is a commonly used sensor. For this discussion, we will assume that a video camera is our image acquisition sensor.

A video camera converts the optical image to an analog signal. It uses a lens to focus light rays to a two-dimensional photodetector. The photodetector converts the light energy into a proportional electrical signal. The electrical signal is the analog signal form of the image.

The digitization function follows the camera. Here, the analog signal is converted to a digital data form through an analog-to-digital conversion function. The digital image is then stored in digital memory, generally fast semiconductor devices. Once the image is in digital memory, it is physically accessible for subsequent digital image processing operations.

To display an image stored in memory, the image data is repetitively read from memory. The data is reconstructed back to an analog signal form, through a digital-to-analog conversion function, and delivered to a video monitor for display.

A host computer oversees the entire system. It provides the interface to the user along with the sequencing of acquisition, storage, display, and processing actions. The digital image stored in memory is freely accessible for processing by the host computer. Further, the host may transfer the image from memory to other computers, various networked devices, or permanent memory media such as high-capacity disk or tape. For small stand-alone processing systems, the host computer is often a microprocessor-based personal computer. For more complex applications, sophisticated workstations take on the task.

Although the host computer has the full ability to carry out any conceivable operation upon a stored image, its execution speed can be limited. After all, the host is usually a general-purpose computer, and as such, is not optimized for digital image processing operations. To augment the host, specialized high-speed digital image processing processors are usually a part of the processing system. These processors, like the host computer, have free access to the digital image stored in memory.

This additional processing hardware can take the form of high-speed hardware circuits or secondary microprocessors, optimized to handle common digital image processing operations. For applications that must run fast enough to keep up with real-time events, like a moving conveyer line of parts, the high-speed hardware approach is often essential.

Many off-the-shelf digital image processing software packages and function libraries are available. These programs contain various digital image processing operations and generally run on the host computer. Many packages support the use of specialized digital image processing processors, as well. The processing software is often the part of the system that "glues" all the other components together into a complete system.

Digital image processing systems can take on many different architectures. The actual configuration may mix various combinations of hardware and software components. The choice, of course, affects the cost of the hardware and is usually driven specifically by the application requirements. All systems must, however, contain the primary components to acquire, store, display, and process images. Part III of this book develops, in depth, the varieties and trade-offs of different processing system configurations.

3 *The Digital Image*

Digital image processing tasks, by definition, operate upon pictorial information of a digital form. While some images may originate in the digital form, such as those initially created by computer, the great majority don't. Before we can process an image using digital techniques, it must exist in the digital form.

The most common image originates in the form of optical light energy. This is the image form that we deal with every day. Optical light energy images are the type we perceive with our eyes, capture with a video camera, and photograph onto film with a conventional camera. Optical images also include images originating in other energy forms such as X-ray, infrared, radar, and acoustic images.

No matter what the origin of an image, it must ultimately exist in the form the digital image processor understands—the digital form. In this chapter, we explore the essence of a digital image, examining its creation, conventions, terms, and limitations.

Creating a Digital Image

A natural image begins as a continuously varying array of shades and colors. In the case of a photograph, shades vary from light to dark and colors vary from reds through yellows to blues. An image of this sort is known as a *continuous-tone* image. This means that the various shades and colors blend, with no disruptions, to form a faithful reproduction of the original scene.

For the time being, we will base this discussion on the concept of a gray-scale continuous-tone image. That is, we will assume that all images are made up of levels of gray spanning from black to white. Later, we will add the concept of color as a continuation of this initial discussion.

A *digital image* is composed of discrete points of gray tone, or *brightness*, rather than continuously varying tones. To make a digital image from a continuous-tone

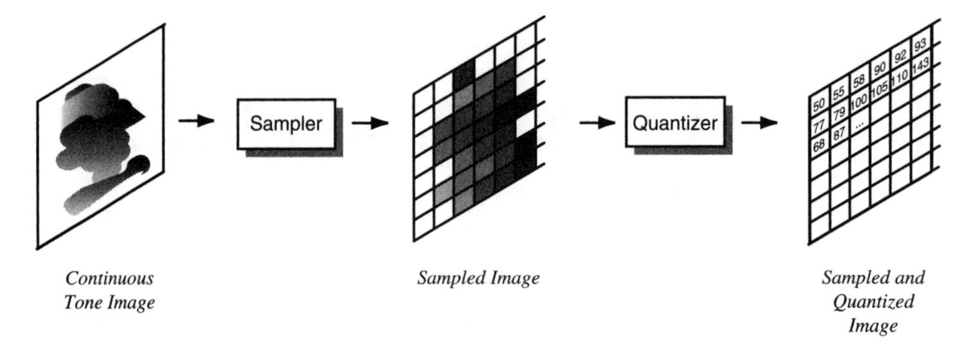

Figure 3.1 *An image is converted from a continuous-tone form to a digital form through the processes of sampling and quantization.*

image, it must be divided up into individual points of brightness. Additionally, each point of brightness must be described by a digital data value. The processes of breaking up a continuous-tone image and determining digital brightness values are referred to as *sampling* and *quantization*, and are illustrated in Figure 3.1. The sampling process samples the intensity of the continuous-tone image at specific locations. The quantization process determines the digital brightness value of each sample, ranging from black, through the grays, to white. A quantized sample is referred to as a picture element, or *pixel*, because it represents a discrete digital element of the digital image. The combination of sampling and quantization processes is referred to as *image digitization*.

An image is generally sampled into a rectangular array of pixels. Each pixel has an (x,y) coordinate that corresponds to its location within the image. The x coordinate is the pixel's horizontal location; the y coordinate is its vertical location. We can think of the x location as the column in which the pixel is located and the y location as the row in which the pixel is located. In most cases and for the purposes of this book, the pixel at location (0,0) is in the upper left corner of the image. As an example, the pixel with coordinates (200,150) would be located 200 pixels to the right of the left-hand side of the image and 150 lines down from the top of the image. This is shown in Figure 3.2. Often, the x location is referred to as the *pixel number* and the y location as the *line number*.

We now have a digital image. It is composed of a rectangular array of pixels, each representing a particular (x,y) location, and each with a digital brightness data value. The digital image exists merely as a large array of numbers (data values) that, when arranged properly and displayed as brightnesses, forms an image.

One question that arises is, how good is the digital image's representation of the original scene? The quality of the digital image is directly related to the number of pixels and lines, along with the range of brightness values, in the image. These aspects are known as *image resolution*. Image resolution is the capability of the digital image to resolve the elements of the original scene. For digital images, the res-

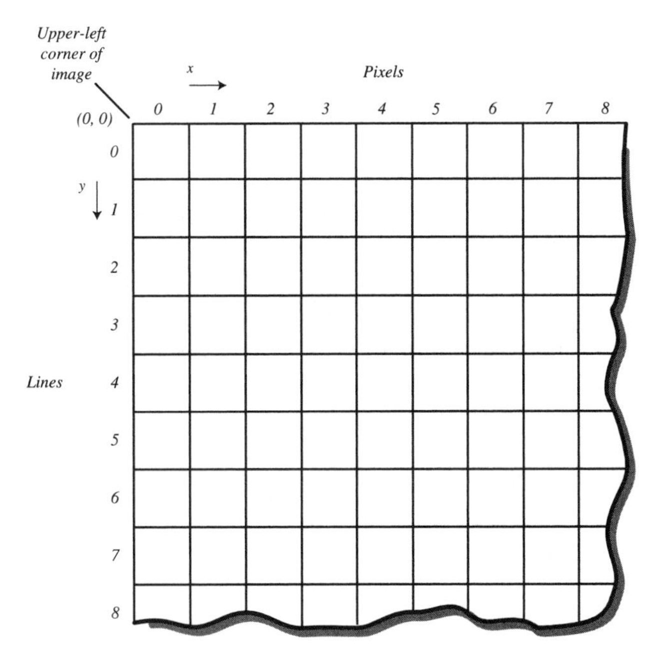

Figure 3.2a *The discrete pixel numbering convention.*

Figure 3.2b *A 640 pixel × 480 line digital image showing the location of pixel (200,150).*

olution characteristics can be broken into two primary parts—the spatial resolution and the brightness resolution (or color resolution, if the image is in color). In the case of a motion sequence of several images, frame rate also plays a role in the quality of the represented image sequence.

Spatial Resolution

The term *spatial* refers to the concept of space—in our case, two-dimensional image space. We use the term *spatial resolution* to describe how many pixels comprise a digital image. The more pixels in the image, the greater its spatial resolution. The number of pixels in a digital image depends on how finely we sample, or divide, the image into discrete pixels.

Before continuing, let's clarify our usage of the term "spatial resolution." Strictly speaking, the measurement of a digital image's spatial resolution is related to two distinct measurements—its *spatial density* and its *optical resolution*. Spatial density is a measure of the number of pixels in a digital image. Optical resolution, on the other hand, is a measure of the capability of how well the entire physical imaging system can resolve the spatial details of an original scene. It therefore relates to the quality of the imaging system's optics, photosensor, and electronics *as well as the spatial density*. Whichever is less—the spatial density or optical resolution—dictates an image's spatial resolution. If the optical resolution of an imaging system is superior to the spatial density, then the image's spatial resolution is limited only by the spatial density. For the purposes of this book, we will assume that the optical resolution is always better than the spatial density, and that the spatial resolution of a digital image is therefore limited by the spatial density alone.

Optimally, we'd like to take a continuous-tone image, like a photograph or video-camera image, and break it into enough discrete pixels so that the digitized image contains all the information of the original. This means that, to an observer, the displayed digital image would look identical to the original continuous-tone image. This criterion holds true when the digital image is intended for use by a human observer. Sometimes, though, if the digital image is to be used for computer analysis purposes, it may be necessary to have more pixels, or sufficient to have fewer pixels.

The concept of *spatial frequency* can explain how finely we should sample an image. All images contain details, some fine details and some coarse details. These details are made up of brightness transitions that cycle from dark to light and back to dark. The rate at which brightnesses cycle is the spatial frequency. The higher the cycle rate, the higher the spatial frequency. Figure 3.3 illustrates several distinct spatial frequency patterns.

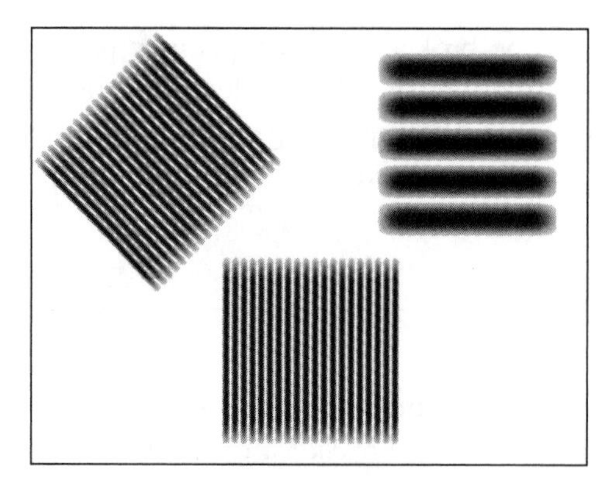

Figure 3.3 *Several distinct spatial frequency patterns. The wider ones are low-frequency patterns, and the tighter ones are high-frequency patterns.*

Figure 3.4a *A scene of varying spatial frequency detail.*

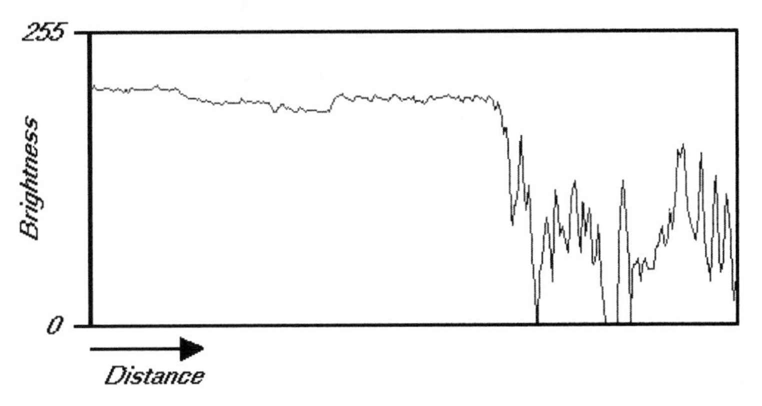

Figure 3.4b *The brightness plot taken along the highlighted line.*

Figure 3.4 is an image of varying detail with a plot showing pixel brightnesses across the image. The details in this image vary depending on which region of the image we look at. In the cloudy sky area, the image details vary smoothly. The area with the trees has minute details where the brightnesses vary rapidly. The minute tree details constitute a high rate of brightness change, or high spatial frequency. The smooth sky details have low rates of brightness change, or low spatial frequency.

We must determine the necessary sampling rate so that our digital image adequately resolves all the spatial details of the original continuous-tone image. This is done by using the classical *sampling theorem*. This theorem tells us mathematically that in order to represent fully the spatial details of an original continuous-tone image, we must sample the image at a rate *at least twice as fast* as the highest spatial frequency contained in it. This means that to capture an image's finest dark-to-light-to-dark detail, sampling must occur at a rate fast enough so that at least two samples fall upon the detail. This guarantees that both the dark and light portions of the detail are sampled, and hence preserved, in the resulting digital image. Whatever sampling rate we use, the highest spatial frequencies that can be contained in the resulting digital image will not exceed one-half the sampling rate. This frequency is referred to as the *Nyquist rate*.

If sampling occurs at a slower rate than required by the sampling theorem, high spatial frequency details will be missed in the digital image. Hence, the digital image will appear to have less spatial resolution than the original image. This is because there are not enough pixels in the digital image to represent adequately all of the spatial details of the original. On the other hand, if sampling occurs at a rate higher than that required by the sampling theorem, extra pixels will be created. According to the sampling theorem, these superfluous pixels do not theoretically contribute anything to the spatial resolution of the resulting digital image. In practice, however, oversampling can help improve the accuracy of feature measurements taken from a digital image.

Some applications do not require all of the details in an original image to be captured in the digital image. Other applications may have very strict requirements for maintaining details. In either case, once a continuous-tone image is digitized, any spatial resolution limits caused by an inadequate sampling rate become part of the digital image forever.

In real-life systems, the sampling rate of a particular image acquisition system is generally fixed; it is not adjusted image by image depending on the current image's spatial frequency content. A camera is combined with a digitizer to acquire digital images with a fixed number of pixels. The camera and digitizer are selected to meet the minimum spatial-resolution requirements of the application. If the sampling rate happens to be at a higher rate than necessary for a particular image, then that digital image may contain more data than needed, but at least no spatial information will be lost.

Let's look at the effect of an image sampled at different spatial resolutions. Figure 3.5 shows the same image sampled at six different spatial resolutions. The image with the least spatial resolution is the one with a resolution of 20 pixels ×

Figure 3.5a *An image with spatial resolution of 640 pixels × 480 lines.*

Figure 3.5b *Resolution reduced to ½ in both the* x *and* y *dimensions: 320 pixels × 240 lines.*

Figure 3.5c *Resolution reduced to ¼: 160 pixels × 120 lines.*

Figure 3.5d *Resolution reduced to ⅛: 80 pixels × 60 lines.*

Figure 3.5e *Resolution reduced to 1/16: 40 pixels × 30 lines.*

Figure 3.5f *Resolution reduced to 1/32: 20 pixels × 15 lines.*

15 lines. The image is virtually unintelligible. The *pixel blocking* effect, sometimes referred to as *pixelation*, occludes most of the features of the image. As the resolution is increased, the structure of the image becomes more apparent. The highest-resolution version of the image is 640 pixels × 480 lines.

The spatial-resolution requirements of an image are based on its intended application. Looking again at the images of Figure 3.5, the 320 pixel × 240 line image appears to be virtually identical to the 640 pixel × 480 line version. This means that for the halftone printing process of an image the size of those in this book, a spatial resolution of 320 pixels × 240 lines is sufficient to represent the original continuous-tone image adequately.

Digital images for use in television production applications require a spatial resolution of about 640 pixels × 480 lines. Motion picture applications require about ten times this spatial resolution in both the x and y directions. These differences are due to the fact that the relative quality of the images displayed is dramatically

Table 3.1 Image Size Versus Spatial Resolution

NUMBER OF PIXELS	NUMBER OF LINES	TOTAL PIXELS IN IMAGE
20	15	300
40	30	1,200
80	60	4,800
160	120	19,200
320	240	76,800
640	480	307,200

different, even though we tend to perceive the images as about the same. The television image is displayed on a small screen at a considerable distance from the observer. A movie image is displayed on a very large screen that is relatively close to the observer. As a result, the observer is exposed to considerably more spatial detail in a movie presentation than in a television presentation. To accommodate this fact, the digital film production process requires a much higher spatial resolution than the digital television production process.

Table 3.1 shows the size (in number of pixels) of a digital image sampled at the spatial resolutions shown in Figure 3.5.

As the spatial resolution increases, the number of pixels climbs exponentially. It is therefore important to consider the spatial resolution required by a given application. Digital image storage and processing times can be significantly reduced by selecting the minimum resolution necessary. In practice, common spatial resolutions range from 320 pixels × 240 lines to 1024 pixels × 768 lines. All images in this book were sampled with a spatial resolution of either 320 pixels × 240 lines or 640 pixels × 480 lines.

Viewing Geometry

The physical dimensions of an image and the distance that it appears from an observer determine how much scene detail will be visible. This is because the eye has a fixed resolving ability that depends directly on image size. For instance, a photograph held 12 inches away shows much more detail than if it were held 5 feet away. Therefore, as the display–observer distance increases, or display size decreases, required spatial resolution in an image may decrease. Let's return to Figure 3.5. Place this book 10 feet away from you, and view these images of varying spatial resolution. The blocking effect of the lower-resolution images diminishes. By moving the images farther and farther away, you will finally get to a point where no pixel blocking will be evident for any of the images—even the one with the lowest resolution. This is because you have finally reached a point where the eye cannot resolve any more detail than the lowest-resolution image contains, based on the image's size and distance from you.

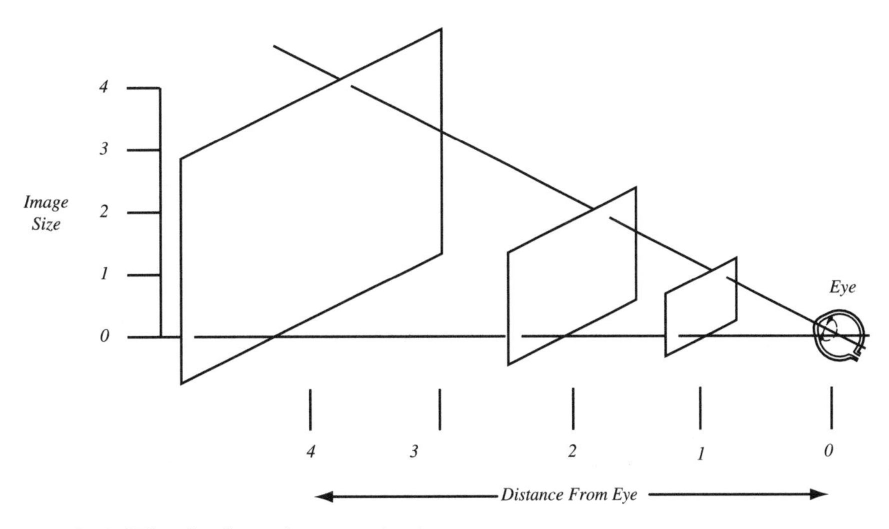

Figure 3.6 *The display–observer viewing geometry.*

The display–observer geometry is depicted in Figure 3.6. As the distance between an image and observer doubles, or as image size is reduced by one-half, the image's spatial resolution can be decreased by one-half without causing any spatial detail loss for the observer.

Aspect Ratio

The *aspect ratio* is a measure of an image's rectangular form. It is calculated by dividing the image's horizontal width by its vertical height. In the case of commercial broadcast television and common video equipment, images have an aspect ratio of 1.333. Commonly, this aspect ratio is denoted as 4:3 (read "four to three"). This means that the horizontal dimension of the image is 1.33 times wider than the vertical dimension. An image with a 1:1 aspect ratio, on the other hand, appears to be perfectly square.

When a continuous-tone image is sampled, pixel dimensions in the digital image also take on an aspect ratio quality. It is important, for most digital image processing operations, that pixels have a 1:1 aspect ratio. Pixels with a 1:1 aspect ratio represent an area of the original image that is made up of identical width and height dimensions. These are referred to as *square pixels*. Figure 3.7 illustrates the dimensions of image and pixel aspect ratios. When sampling an image with a 4:3 aspect ratio, more samples must be taken in the horizontal direction of the image than in the vertical direction. In fact, for every three samples taken in the vertical direction, precisely four must be taken in the horizontal direction. This was illustrated back in Figure 3.5, where each image has 1.333 times more pixels than lines. The images presented in this book have aspect ratios of 4:3 and pixel aspect ratios of 1:1.

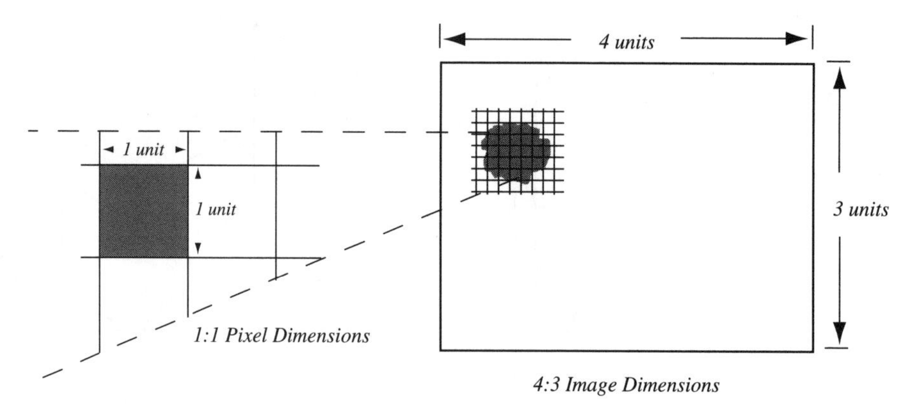

Figure 3.7 *Image and pixel aspect ratio dimensions.*

Spatial Aliasing

One final topic related to spatial resolution is important to discuss—the phenomenon of *spatial aliasing*. Spatial aliasing appears in a digital image when the details in an image are sampled at a rate less than twice their spatial frequency. Not only are these high-frequency details missed, as we discussed earlier, but they are also corrupted to appear as new frequencies, as shown in Figure 3.8.

When a detail within an image has a frequency greater than twice the sampling rate, the detail is said to be *undersampled*. The high-frequency detail ends up being translated to a lower frequency because some of its brightness transitions are missed in the sampling process. When significant spatial aliasing occurs, the resulting digital image can look terrible, as shown in the image of Figure 3.9.

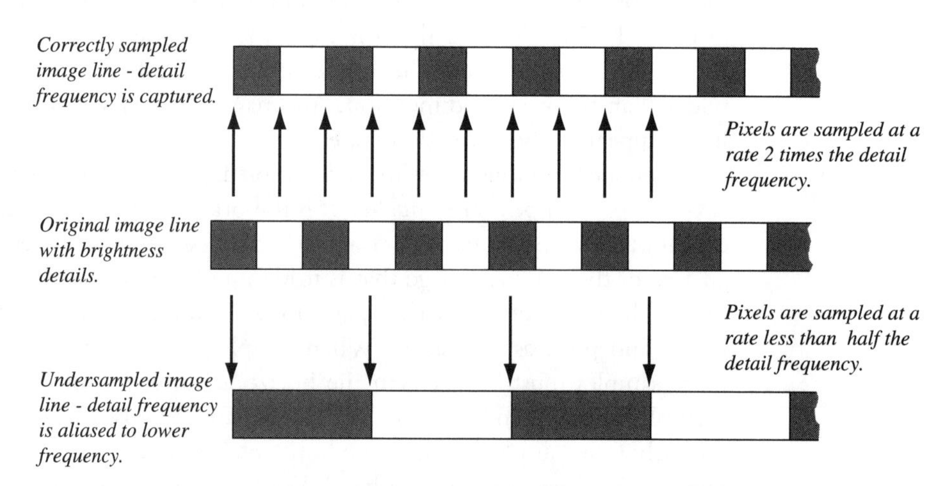

Figure 3.8a *The spatial aliasing effect caused by undersampling.*

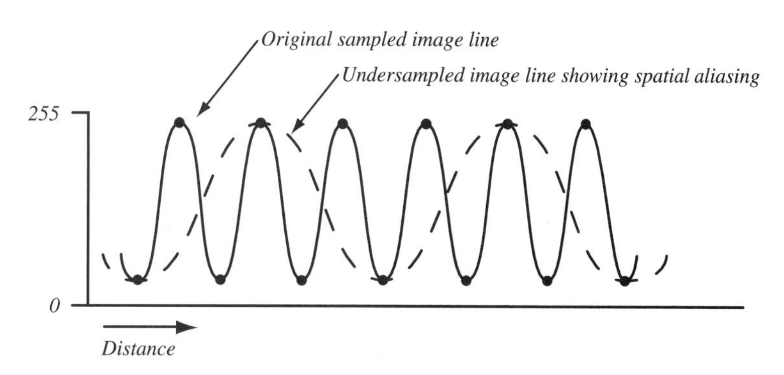

Figure 3.8b *Spatial aliasing shown on an image-line brightness profile.*

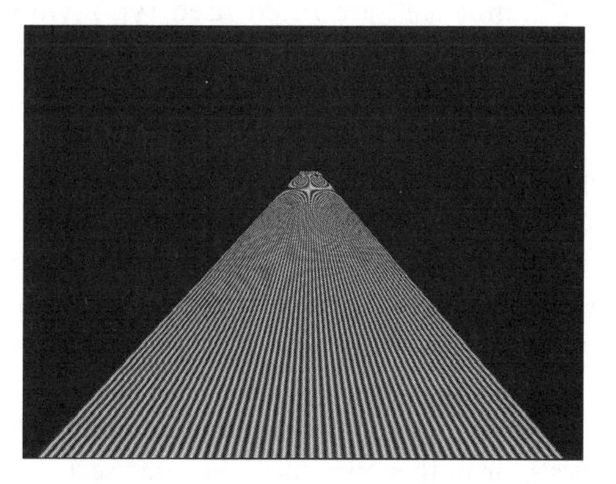

Figure 3.9 *An image showing significant spatial aliasing due to undersampling. The line pattern increases in frequency as it extends into the background. As the spatial frequency crosses one-half the sampling rate, Moiré patterns caused by spatial aliasing become evident.*

You can see an everyday example of spatial aliasing by looking through a window screen at a scene with repetitive high-frequency details, such as a picket fence in the distance. As you move slowly from side to side, the pickets tend to disappear and reappear. Because the pickets are a finer detail than the screen holes, you see an aliasing artifact.

The worst aliasing artifacts tend to occur when repetitive patterns are undersampled. The spatial frequency of the pattern ends up appearing different, and may sometimes contain distracting Moiré patterns in the digitized image. However, unless an image contains very repetitive high-frequency details, the visual effects of spatial aliasing are generally minimal.

Brightness Resolution

Every pixel in a digital image represents the intensity of the original image at the spatial location where it was sampled. The concept of *brightness resolution* addresses how accurately the digital pixel's brightness can represent the intensity of the original image. When the numeric range of a pixel's brightness is increased, so is the pixel's brightness resolution.

Digressing momentarily, let's clarify our definitions of the terms *intensity* and *brightness*. "Intensity" (or, more appropriately, *radiant intensity*) refers to the magnitude, or amount, of light energy actually reflected or transmitted from a physical scene. The term "brightness" (or, more appropriately, *luminous brightness*) refers to the measured intensity after it is acquired (say, using a video camera), sampled, quantized, displayed, and observed (with our eyes). The brightness of a pixel accounts for all the effects induced by the entire imaging system. This may seem like a somewhat trivial distinction, but it accounts for the fact that the measured brightnesses in our digital image are only representations of the actual energy radiated from the original physical scene. Additionally, the terms "intensity," "brightness," "radiance," and "luminance" are often used synonymously to mean the same thing. They are, however, distinctly different terms relating to measures of the quantity of light. As we discuss digital image processing operations, we will refer to digital pixels as having a brightness property.

Following the sampling process, each sample is quantized. This quantization process converts the continuous-tone intensity, at the sample point, to a digital brightness value. The accuracy of the digital value is directly dependent upon how many bits are used in the quantizer. If three bits are used, the brightness can be converted to one of eight gray levels. In this case, gray level "0" represents black, gray level "7" represents white, and gray levels "1" through "6" represent the ascending gray tones between black and white. The eight gray levels comprise what is called the *gray scale*, or in this case, the 3-bit gray scale.

With a 4-bit brightness value, every pixel's brightness is represented by one of 16 gray levels. A 5-bit brightness value yields a 32-level gray-scale range. An 8-bit brightness value yields a 256-level gray-scale range. Every additional bit used to represent the brightness doubles the range of the gray scale. The range of the gray scale is also referred to as *dynamic range*. An image with 8-bit brightness values is said to have an available dynamic range of 256 to 1. Several gray scales are shown in Figure 3.10. Notice how the smoothness of the gray scale improves as more bits are used to represent brightnesses.

Figure 3.11 shows an image quantized to various brightness resolutions. The image quantized to eight bits of brightness resolution appears very natural and continuous. As the brightness resolution decreases, the image appears coarser and more mechanical. This effect is known as *brightness contouring*, or *posterization*. Contouring occurs when there are not enough gray levels to represent the actual brightness in the original image adequately. Gradual brightness changes end up quantized a little brighter or dimmer than their original intensities. Brightness contouring is the effect of insufficient brightness resolution.

Figure 3.12 shows the same image broken into individual *bit-planes*. Each bit-plane represents the on or off level of the particular bit's contribution to the overall pixel brightness. Clearly, much of the structure of the image is conveyed

Figure 3.10a *The 3-bit gray scale.*

Figure 3.10b *The 4-bit gray scale.*

Figure 3.10c *The 5-bit gray scale.*

Figure 3.10d *The 6-bit gray scale.*

Figure 3.10e *The 7-bit gray scale.*

Figure 3.10f *The 8-bit gray scale.*

through the higher-order, most significant bits. The lower-order bits tend to carry the important subtle shading aspects of the digital image.

Figure 3.11a *An image with brightness resolution of 8 bits, or 256 gray levels.*

Figure 3.11b *A 7-bit, or 128 gray-level, image.*

Figure 3.11c *A 6-bit, or 64 gray-level, image.*

Figure 3.11d *A 5-bit, or 32 gray-level, image.*

Figure 3.11e *A 4-bit, or 16 gray-level, image.*

Figure 3.11f *A 3-bit, or 8 gray-level, image.*

Figure 3.11g *A 2-bit, or 4 gray-level, image.*

Figure 3.11h *A 1-bit, or 2 gray-level, image.*

Figure 3.12a *Bitplane 7 (most significant bit) of an 8-bit image.*

Figure 3.12b *Bitplane 6.*

Figure 3.12c *Bitplane 5.*

Figure 3.12d *Bitplane 4.*

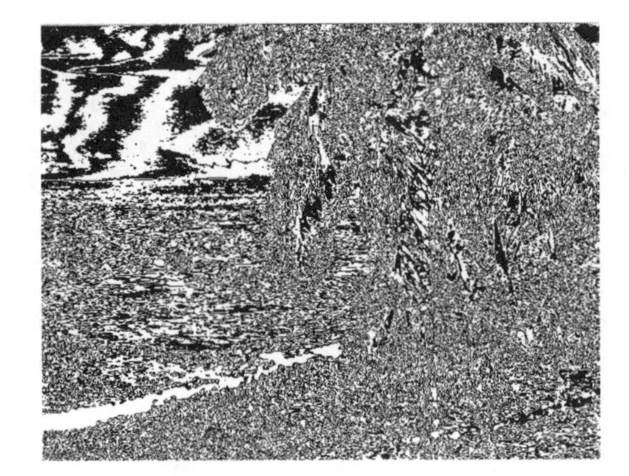

Figure 3.12e *Bitplane 3.*

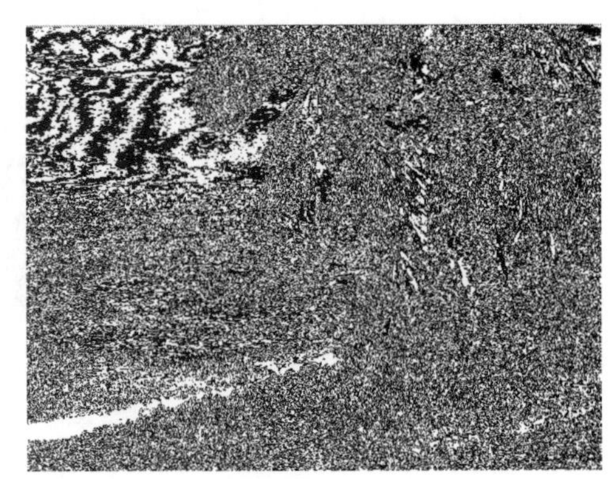

Figure 3.12f *Bitplane 2.*

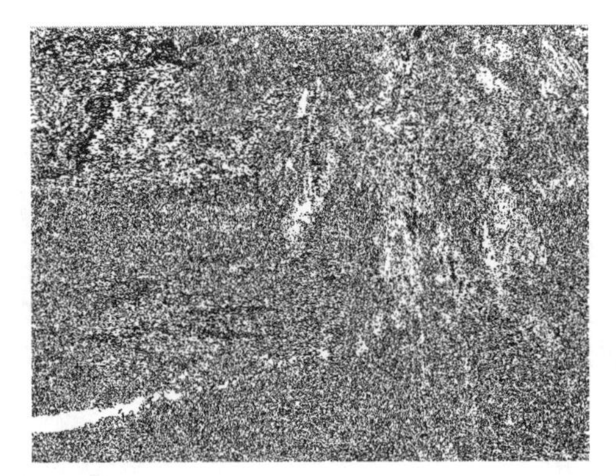

Figure 3.12g *Bitplane 1.*

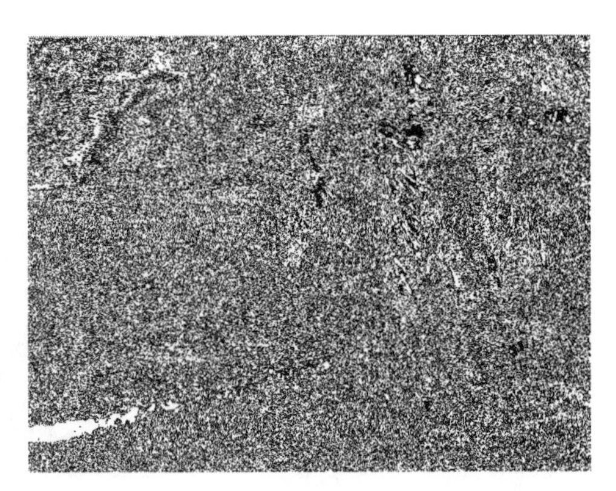

Figure 3.12h *Bitplane 0 (least significant bit).*

In Chapter 2, we discussed the logarithmic intensity response of the eye. When an image is digitized for use by a human observer, sometimes it is preferable to quantize the image on a logarithmic scale rather than a linear scale. This results in higher brightness resolution for darker pixels, where the eye is more responsive to subtle brightness changes. Conversely, brighter pixels are quantized with a lower brightness resolution. General-purpose image digitization equipment usually quantizes samples on a linear scale, however. This is because many digital image processing operations are designed specifically for images with equal-step linear quantization ranges.

The number of bits used in quantizing an image depends on how the image will be used. 8-bit quantization is very common and generally sufficient for most applications. However, some applications may require fewer bits and others may require more. For instance, medical X-ray images usually require 12-bit quantization. This is because very subtle brightness changes exist in the original X-ray images. Like spatial resolution, brightness resolution is generally fixed for a particular imaging system. All images in this book were linearly quantized with a brightness resolution of 8 bits.

Like spatial resolution, the brightness resolution can be reduced or increased depending upon the image–observer viewing geometry. The smaller the image and farther it is from the observer, the smaller its brightness resolution needs to be. When viewing the images in Figure 3.11, the brightness contouring effect diminishes as the images are moved farther and farther away. Eventually the eye cannot resolve any more brightness shading than is present in the lower-resolution images, based on their size and distance from the observer.

Color Resolution

When we work with color images, the same concepts of sampling, quantization, and spatial and brightness resolution hold true. But, instead of a single brightness value, color digital images have pixels that are generally quantized using three brightness components. The concept of trichromacy introduced in Chapter 2 can also be applied to acquiring and displaying color images. In displaying color, three independent color emitters are used, each emitting a unique spectral band of light to generate all colors in the spectrum. Likewise, a sensor can acquire color images by using three independent sensors, each responding to a unique spectral band of light. Color video cameras work very much the way our eyes do to discriminate color through trichromacy.

Following this discussion of color images, we will restrict our treatment to focus primarily on the processing of gray-scale images. Generally, all operations presented can be extended to process color images simply by applying them to each color component of the image.

Additive and Subtractive Color Spaces

If you look very closely at a color video display screen, whether it's a cathode ray tube (CRT), liquid crystal display (LCD), or another type, you will notice individual dots of solid colors. These dots emit light in the colors of red, green, and blue. As you move away from the display, the dots tend to fuse together to the point where the individual dots are no longer visible. Instead, their summation of color is perceived as a single color.

All the colors in the spectrum can be created with the *primary colors* of *red, green,* and *blue* (*RGB*). This is called the *additive color* property, and it works for the mixing of primary colors that are emitting light. If only the red component is on, the perceived color will be red. As the red is varied from dark to light, the different brightnesses of red will be seen. Likewise, if only the green or blue component is used, the same effect will be seen. If the red and green components are both turned on, yellow will be perceived. If they are varied up and down together, the colors of dark brown through bright yellow will be seen. If the red and green components are varied independently, a wide array of colors can be created including red, green, yellow, brown, and everything in between.

The same kind of color ranges can be created by mixing red with blue and blue with green. When red, green, and blue are all mixed together, an entire spectrum of colors can be created. The red, green, and blue colors make up a *color space* (or *color gamut*) of colors that can be represented by a cube. The RGB color space cube is shown in Figure 3.13. There are three axes, each representing one of the primary colors of red, green, or blue. The lower left corner of the cube is the absence of all three colors, or black. The upper right corner is the full brightness mix of all three colors, or white. Each face of the cube contains the colors created by mixing the two respective primaries. The volume inside the cube contains all the colors of the spectrum created by mixing differing proportions of all three primaries.

The primary colors chosen to represent a color space do not have to be red, green, and blue. Variations of red, green, and blue—or totally different primary colors—can be used. The resulting color gamut, however, is constrained by the choice of the primary colors. For instance, if red, yellow, and green are used, the color gamut will not be able to represent colors with blue components such as blue, magenta, and violet. Similarly, if washed-out red, green, and blue primary colors are used, vivid, highly saturated colors cannot be represented. The red, green, and blue primary colors are typically used to represent color because they best approximate the response of the human eye's color receptors.

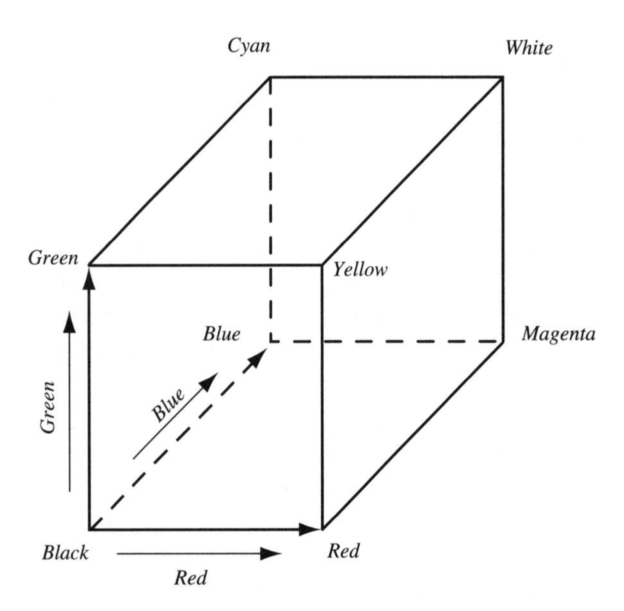

Figure 3.13 *The RGB color space cube.*

Similarly, the property of *subtractive color* exists. Subtractive color mixing is based on reflective colors rather than emissive colors. Instead of emitting light like a video display, subtractive colors reflect the light shined upon them. The subtractive colors, called *secondary colors*, are *cyan, magenta,* and *yellow* (*CMY*). Subtractive color is used primarily in the printing industry. By printing the three subtractive colors in differing proportions on white paper, the printing process can create all the colors in the spectrum. The process is simply the subtractive version of what a video display monitor does to create color.

Since the color cyan is the mix of green and blue light, as shown in Figure 3.14, it reflects green and blue light only, while absorbing red light. When cyan is printed on white paper and illuminated by a white light source, the red portion of light that the paper would otherwise reflect is subtracted out by the ink. Likewise, magenta ink subtracts out green light and yellow ink subtracts out blue light. For instance, if cyan and magenta are overlapped, the cyan first subtracts out red and then the magenta subtracts out green, leaving only blue. Yellow and magenta mix to reflect only red; yellow and cyan mix to reflect only green. By mixing the three subtractive colors, the entire spectrum of color can be created.

When mixing additive colors, it is always important that the display system be calibrated so that the correct proportions of red, green, and blue light mix to reproduce the expected colors faithfully. Because black is the absence of all three colors, it is always accurate—simply the black face of the display screen. White, on the other hand, is a mix of all three colors and is usually adjusted so that it appears pure without any color casts. This process is called *white balancing*. For subtractive colors, the white paper and light source control the white balance. The concern for the printer is the black balance. In practice, it is very difficult to accurately mix and maintain the correct proportions of cyan, magenta, and yellow inks to create

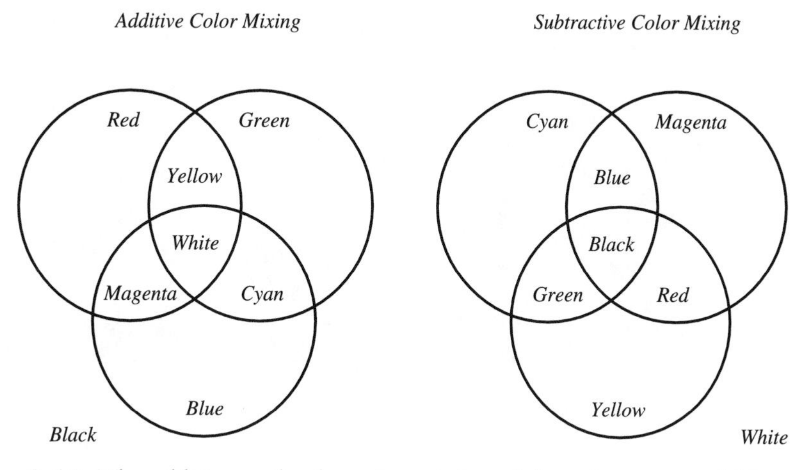

Figure 3.14 *The additive and subtractive mixing of the red, green, and blue primary colors.*

black. Instead, a fourth color component, black, is used to create the black, or total absence of color, in the printing process.

In digital image processing applications, the RGB color space is used to acquire and display color images. Processing of color images is often done in RGB space or an alternate color space that is more appropriate for the application (see the next section). When an application requires an image in another color space, it is simply converted mathematically from the RGB space to the appropriate color space. For instance, if an image is to be printed, it is acquired as an RGB image, processed, and then converted to a CMY image.

A digital image processing system must handle the red, green, and blue components of each pixel in an image. Figure 3.15 shows the three component images of a color image. Generally, each component must be quantized at a resolution equivalent to the brightness resolution used in a gray-scale image. This means that for most applications, each component would be quantized to 8 bits, yielding pixels of 24 bits. Some applications may require fewer bits of color resolution, while others will require more.

Other Color Spaces

Although RGB color space is the fundamental color space used to physically detect and generate color light, other derivative color spaces can be created to aid color image processing. These alternate color spaces are merely different methods of representing color. They can be arbitrarily defined to more appropriately fit the needs of a particular application.

Figure 3.15a *The red component image of a color image—the left bird's yellow front appears bright, along with the right bird's red body.*

Figure 3.15b *The green component image—the left bird's yellow front and cyan wing appear bright, as does the green plant in the background.*

Figure 3.15c *The blue component image—the left bird's cyan wing appears bright.*

Hue, Saturation, and Brightness Space

The most important derivative color space is the *hue, saturation, and brightness* (*HSB*) space. This color space represents color as we tend perceive it. Instead of describing the red, green, and blue primary colors, HSB space describes the intuitive components of color. Whenever an application requires a human to interpret or control the colors of an image, HSB space is well suited. The *hue* component controls the color spectrum from red through the yellows, greens, blues, and violets. The *saturation* component controls the purity of the color, or how washed out the color is with white light. For instance, a hue of red can have numerous saturation levels ranging from deep red (fully saturated) to pink and finally white (no saturation of red at all). The *brightness* component controls how bright the color appears.

HSB space appears as shown in Figure 3.16. In this space, color is still represented by three components, like RGB space. But, unlike RGB space, the three components are not neatly lined up on the sides of a cube. Instead, they make up the axes of a double inverted cone. The circumference of the cone is the hue axis, around which the pure colors of the spectrum are represented. The distance from the center of the cone is the saturation axis. The distance from the point of the bottom cone to the point of the top cone is the brightness axis.

The conversions to and from RGB and HSB space are straightforward mathematical operations. In processing color images, it is often useful to convert an RGB image to HSB space, process it, and then convert it back to RGB space for display.

Color spaces that are similar to HSB include *hue, saturation, lightness* (*HSL*) space, *hue, saturation, value* (*HSV*) space, and *hue, saturation, intensity* (*HSI*) space. The terms *brightness, lightness, value,* and *intensity* are often used synonymously, but actually represent distinctively different representations of how bright a color appears.

Numerous other color spaces exist as well. Each one, like HSB color space, provides a representation of color that is tailored for its particular application.

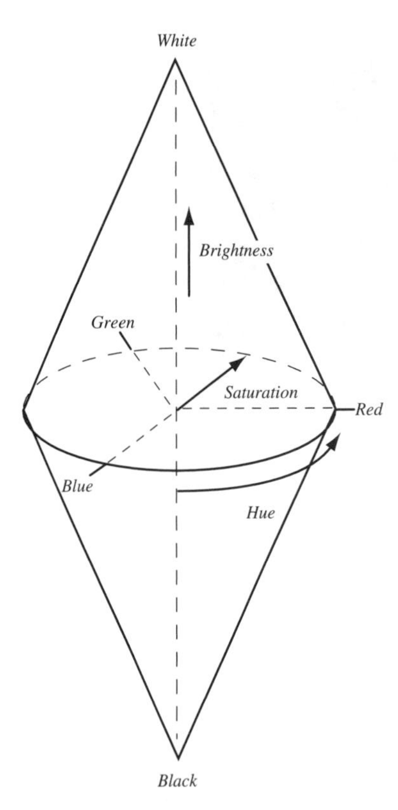

Figure 3.16 *The HSB color space double inverted cone.*

Multispectral Spaces

In some digital image processing applications, color images will be acquired through sensors that do not use the RGB components of light. For instance, infrared sensors detect heat energy. Other sensors detect other energy sources. These sensor systems are called *multispectral* sensors. The images that they produce are not color in the way we visually think of color. Instead, they are composed of multiple detectors, each responding to a discrete spectrum of energy. In this way, they represent their own kind of color space. Display of these images can be done by mapping their color components to the RGB color space through an appropriate mathematical transformation. Processing of these images, however, is generally done on the raw, untransformed multispectral images.

Frame Rate

A more subtle form of image resolution is manifest as the *frame rate*. This is the rate at which images are acquired and displayed.

Image Acquisition

For discussing the repetitive acquisition of image sequences, the frame rate is an important resolution characteristic. The higher the frame rate, the more accurately motion is depicted in the acquired image sequence. In this respect, frame rate refers to the *temporal resolution*, or time-related resolution, of a sequence of images.

Frame rate is a digital sampling term that is applied to digital and other non-digital forms of sequential image acquisition and display. Motion pictures and broadcast television image sequences are not intrinsically digital images, but because their sequences are made up of discrete images, they have an associated frame sampling rate. These images represent instances in time when the image was sampled.

For example, if the frame rate of a system is 24 frames per second (as used in motion pictures), an image frame is acquired and displayed every 1/24th of a second. If an object being imaged moves across the image frame at a faster rate, it may never be caught in an individual image. This is because the frame sampling rate is not fast enough to capture motion that is faster than the frame rate itself.

Broadcast television uses a 30 frame-per-second rate. Both 24 and 30 frame-per-second rates are fast enough to acquire and display general live-action scenes for the human observer. Other systems can provide frame rates in excess of 1000 frames per second. Systems of this sort are used for high-speed image analysis of physical events, such as high-speed machinery and ballistics.

Temporal Aliasing

Just like spatial resolution, a form of aliasing exists with the frame sampling rate. *Temporal aliasing* refers to the time-related representation of motion within an image sequence. The classical example of temporal aliasing is that of the wagon wheel in old "Western" movies. At times, the spokes of the wheel appear to be going backwards, and at other times, the spokes even appear to stop. This is because the frame rate is slower than the spokes' rate of motion. For instance, if the wheel contains eight spokes and turns so that the wheel makes exactly one-eighth rotation each time a frame is acquired, the wheel will not appear to move at all. This is because our motion cue for a rotating wheel is based on our perception of spoke motion, and in this case, the spokes look as though they have not moved from one frame to another. In fact, if the wheel turns one-quarter, one-half or any other multiple of one-eighth of a turn, it will appear motionless.

The sampling theorem that we discussed earlier holds true for temporal sampling as well. The sample rate must be at least twice as fast as the highest rate of motion in the sequence. If it is not, temporal aliasing will appear in the image sequence. In the case of the eight-spoke wagon wheel, the frame rate must be fast enough to capture the wheel rotating no more than 1/16th revolution between

No Aliasing - wheel turns less than 1/16th rotation

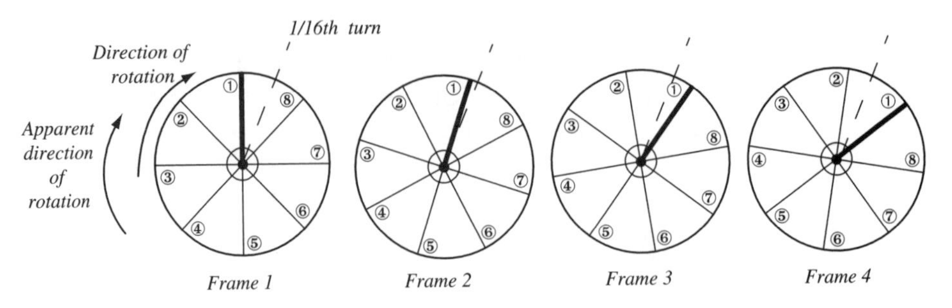

Temporal Aliasing - wheel turns more than 1/16th rotation

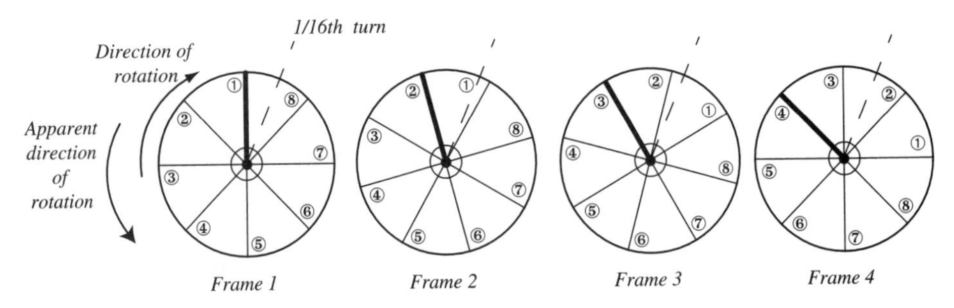

Figure 3.17 *Temporal aliasing—as the wagon wheel turns more than 1/16th rotation, the spokes appear to move backward.*

each succeeding frame. If the wheel turns a little more than 1/16th revolution, it will appear to turn backward. Varying wheel speeds above 1/16th revolution will lead to different aliasing effects including reduced forward motion, backward motion, and no motion at all. This effect is shown in Figure 3.17.

Image Display

In a discussion of the display of still images, the concept of frame rate relates to how often an image is updated on the viewing display. Because the normal display mechanism is a video display monitor, images must be repeatedly refreshed. The rate at which images are refreshed can cause display flicker, and therefore human eye fatigue.

Progressive Versus Interlaced Scan

Display flicker also depends on how the image is scanned on the display monitor. Common commercial broadcast television equipment uses a technique known as

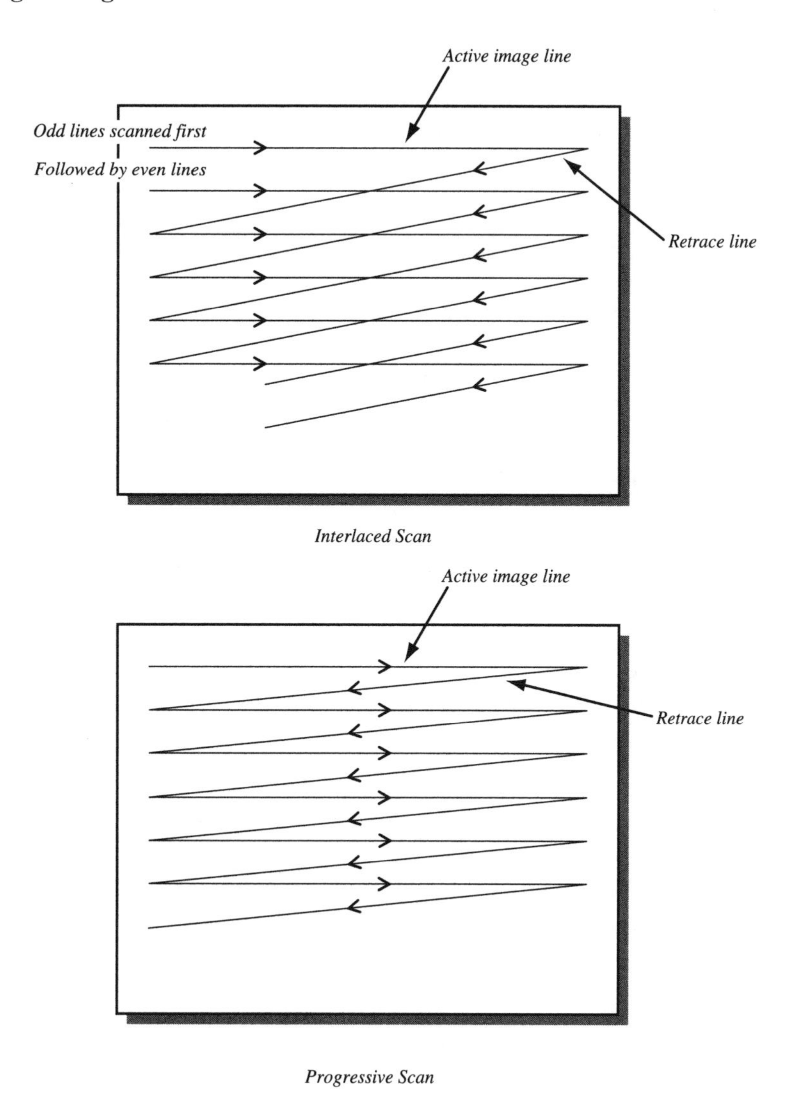

Figure 3.18 *Interlaced and progressive scan image display.*

interlaced scan display, as shown in Figure 3.18. This means that the odd-numbered lines of the image are displayed first, followed by the even-numbered lines. The effect is to interleave, in time, the two interlaced halves of the image, one after another. Interlacing gives the impression to the observer that a new frame is present twice as often as it really is. This technique was used originally for television broadcast signals because the display could be refreshed less regularly without noticeable image flickering, although some minor line-to-line flicker occurred. Systems using a standard commercial broadcast television display monitor for image display typically have a 30 frame-per-second frame rate and interlaced scan.

In motion image sequences, interlaced scan displays can show noticeable motion defects because the odd and even halves of each image are separated in time by one-half the frame rate. The result is a tearing effect that appears on objects with a fast rate of motion through the image frame.

The *non-interlaced* method of image display is known as *progressive scan* display. Progressive scan means that the entire image is displayed in one pass. In this case, the frame rate must be twice that of an equivalent interlaced display, or image flickering will be noticeable. Progressive scan eliminates line-to-line flicker and motion artifacts in displayed images. Systems using a progressive scan display monitor for image display typically have a 72 frame-per-second frame rate.

Qualities of the Digital Image

To determine appropriate processing steps, it is often desirable to assess the overall qualities of an image. Of particular interest are the qualities that describe the brightness and spatial frequency characteristics. Two important tools, the brightness histogram and spatial frequency transforms, can uncover telltale signs of an image's overall deficiencies and strengths.

Brightness Histograms

The brightness characteristics of an image can be concisely displayed with a tool known as the *brightness histogram*. In general terms, a histogram is a distribution graph of a set of numbers. The brightness histogram is a distribution graph of the gray levels of pixels within a digital image. It provides a graphical representation of how many pixels within an image fall into the various gray-level steps.

A histogram appears as a graph with "brightness" on the horizontal axis from 0 to 255 (for an 8-bit gray scale) and "number of pixels" on the vertical axis. To find the number of pixels having a particular brightness within an image, we simply look up the brightness on the horizontal axis, follow the bar graph up, and read off the number of pixels on the vertical axis. Because all pixels must have some brightness defining them, the number of pixels in each brightness column adds up to the total number of pixels in the image.

The histogram gives us a convenient, easy-to-read representation of the concentration of pixels versus brightness in an image. Using this graph we are able to see immediately whether an image is basically dark or light and high or low contrast. Furthermore, it gives us our first clues about what contrast enhancements would be appropriately applied to make the image more subjectively pleasing to an observer, or easier to interpret by succeeding image analysis operations.

Contrast and Dynamic Range Indications

Contrast is a term that is often used to describe the brightness attributes of an image. We intuitively understand contrast to mean how vivid or washed-out an image appears with respect to gray tones. Contrast in an image is clearly illustrated in the brightness histogram. Low contrast appears as a tightly grouped mound of pixel brightnesses in the gray scale, leaving other gray levels minimally or completely unoccupied. High contrast appears as a bimodal histogram—two peaks exist at the outer brightness regions, leaving the gray levels in between unoccupied.

The histogram also tells us how much of the available dynamic range is used by an image. The actual dynamic range of an image is represented by how many gray levels in the gray scale are occupied. For instance, a mound of pixels falling between gray values 50 and 100 (within a range of 0 to 255), with none in the other regions, indicates a small dynamic range of brightness, whereas a wide gray-scale distribution shows large dynamic range. An image with small dynamic range does not occupy all the available spread of gray levels; the image has really only been quantized to a gray scale composed of the occupied range. This indicates low brightness resolution, along with low contrast. Large dynamic range generally implies a well-balanced image, except if it is a bimodal distribution, in which case the image is high contrast. Three common histograms, along with their original images, are illustrated in Figure 3.19.

Color Histograms

Histograms for color images are a threefold version of the regular, gray-level brightness histogram. Three histograms are computed and displayed, one for each color component, as shown in Figure 3.20. The color histograms can represent

Figure 3.19a *Image with low contrast and low dynamic range.*

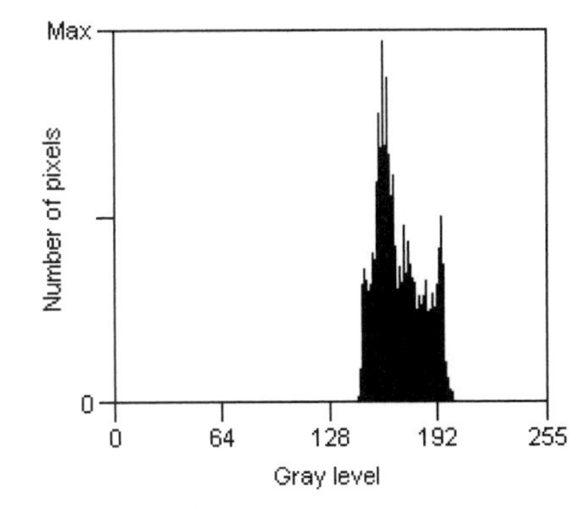

Figure 3.19b *Low contrast and low dynamic range histogram.*

Figure 3.19c *Image with high contrast and high dynamic range.*

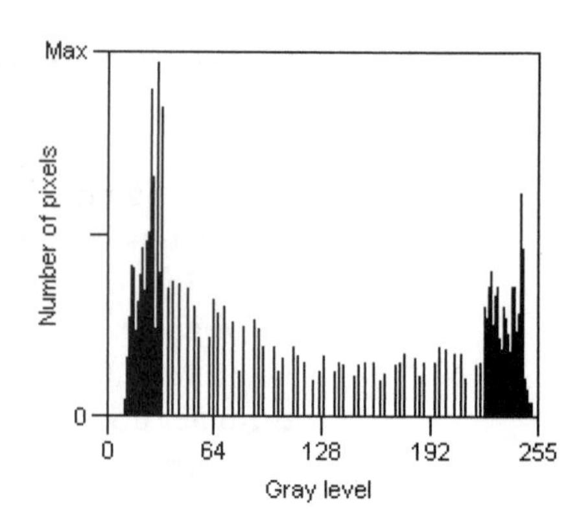

Figure 3.19d *High contrast and high dynamic range histogram.*

Figure 3.19e *Image with well-balanced contrast and high dynamic range.*

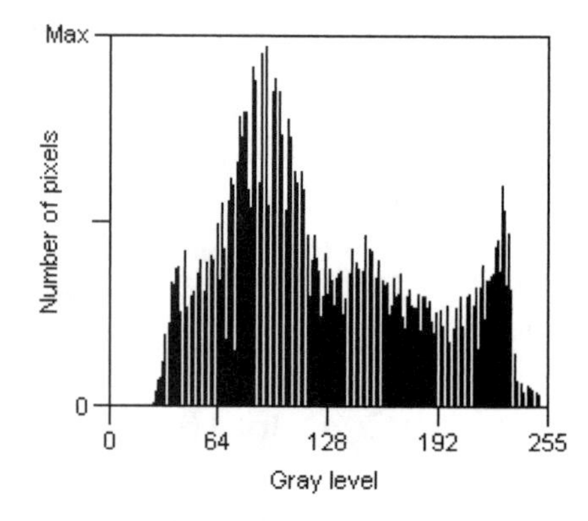

Figure 3.19f *Well-balanced contrast and high dynamic range histogram.*

RGB space, HSB space, or any other color space. Each histogram can help us determine the brightness distributions, contrast, and dynamic ranges of the individual color components.

Spatial Frequency Transforms

Earlier we discussed the concept of image spatial frequency content. We noted that details within an image comprise low or high spatial frequency components, depending on the rates at which the details transitioned from dark to light and back to dark. *Frequency transforms* provide a pictorial view of these spatial frequency components. The most common frequency transform is the Fourier transform; however, many others also exist.

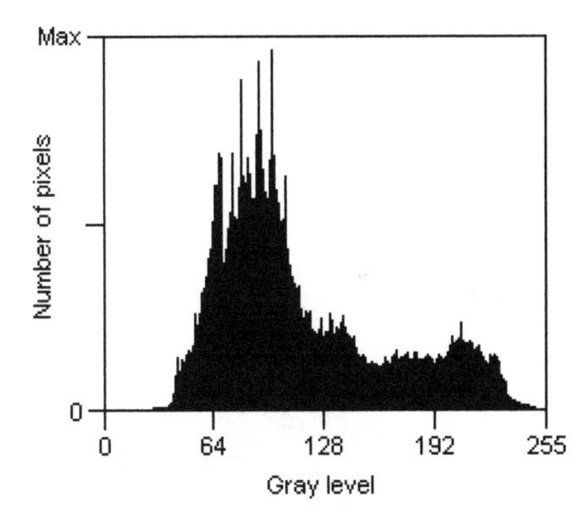

Figure 3.20a *The red component histogram of Figure 3.15a image.*

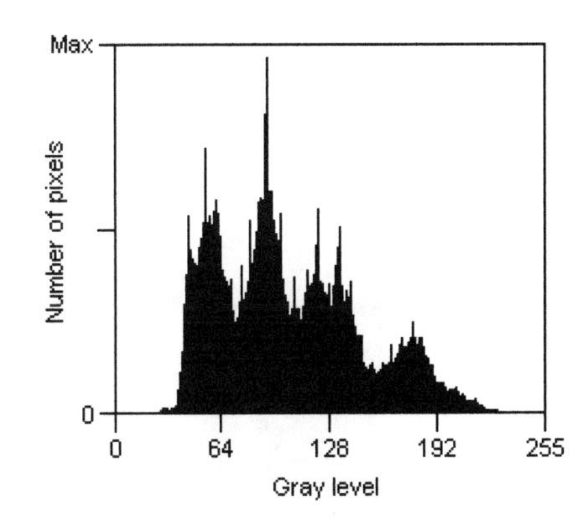

Figure 3.20b *The green component histogram of Figure 3.15b image.*

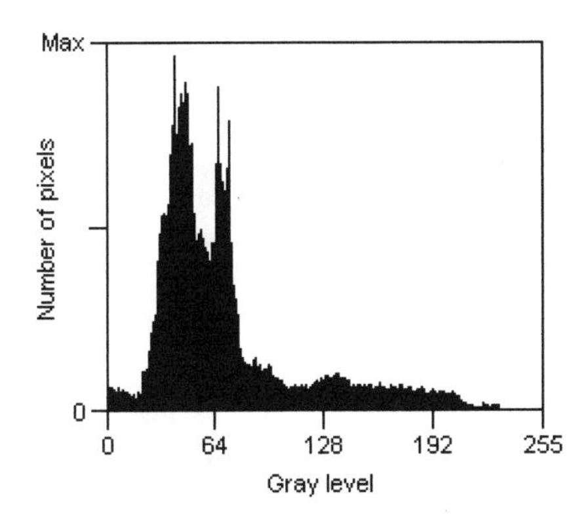

Figure 3.20c *The blue component histogram of Figure 3.15c image.*

Essentially, a digital image is made up of fundamental spatial frequency components that, when combined, make up the form of the image. These two-dimensional spatial frequency components have varying orientations—horizontal, vertical, diagonal, and so on. A frequency transform converts an image from the spatial domain of brightnesses to the frequency domain of frequency components. This frequency decomposition results in a new image that displays the array of spatial frequency components present in the original image, as shown in Figure 3.21.

The frequency image shows the original image's spatial frequency components by the brightness of points at respective locations. "Horizontal frequency" is defined along an imaginary x-axis that passes horizontally through the center of the image. Likewise, "vertical frequency" is defined along an imaginary y-axis that passes vertically through the center of the image. Diagonal frequencies are all the points in between. The brightness of a point in the frequency image corresponds to the magnitude of the frequency component represented by the point's coordinates. The negative frequencies—those to the left of the y-axis and below the x-axis—are merely a mirroring of the positive frequencies, and are therefore not relevant in our quality assessment.

Figure 3.21a Original lighthouse image.

Figure 3.21b Fourier-transform image showing spatial frequency components. The magnitude of each component is displayed by its relative brightness in the frequency image.

Repetitive Noise and Spatial Frequency Characteristics

Frequency transforms allow us to quickly assess the spatial frequency content of an image. The frequency image will clearly show if repetitive noise patterns exist in the original image. For instance, a diagonal noise pattern will show up as a distinct point in the frequency-transform image, as shown in Figure 3.22. This information can be used to filter out the noise and restore the image to a more accurate rendition of its original form. Likewise, an absence or abundance of high, low, or mid-frequency detail can also be seen in a frequency-transform image. Although we will discover important additional uses for frequency transforms in Chapter 4,

Figure 3.22a Original jet fighter image corrupted with a diagonal periodic noise pattern.

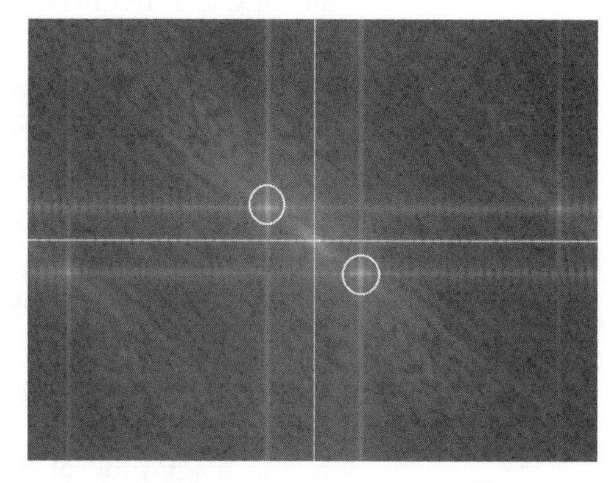

Figure 3.22b Fourier-transform image showing diagonal noise pattern as bright spots circled in the image.

they are a fundamental tool for the understanding of the spatial frequency characteristics of any image.

Color Frequency Transforms

Frequency transforms for color images, like color histograms, are a threefold version of a gray-level transform. Three transforms are computed and displayed, one for each color component. Generally, the spatial frequency of the RGB components of an image will be nearly identical. Therefore, it is usually more appropriate to work with HSB color spaces when creating color frequency transforms. Each transform can be examined to determine the spatial frequency characteristics and potential repetitive noise content of each color component of the image.

4 Image Enhancement and Restoration

Throughout the next four chapters, we will visit the five classes of digital image processing operations—enhancement, restoration, analysis, compression, and synthesis. Many references to related "image operation studies" will be made, as appropriate. These studies comprise a reference compendium, and are compiled in Part IV, "Processing in Action."

The processing techniques presented in the next four chapters represent the fundamental techniques of digital image processing. These techniques are exciting because they change the form or appearance of an image through computerized, numerical techniques. Unfortunately, we cannot study exhaustively every known technique and variation thereof; virtually an endless number exist. Rather, the techniques discussed here will provide you with the essential underpinnings for designing and applying operations of all sorts. You are encouraged to seek additional depth, where necessary, through continued study or practical application. Many references to excellent texts, journals, and periodicals are provided in Appendix A.

By far the most commonly used digital image processing operations involve image enhancement and restoration, the subject of this chapter. These operations improve the visual quality of an image using known or unknown attributes of the image degradation. Operations of this class can be used to improve image contrast and spatial characteristics, reduce annoying image noise, remove some forms of misfocus and motion blur, modify or correct image geometry, and combine multiple images.

Before we jump into image enhancement and restoration techniques, let's add to the conventions of the digital image, discussed in Chapter 3. As you recall, a digital image is a rectangular array of pixels. Each pixel is physically located in the image by an x (pixel) and a y (line) coordinate. These Cartesian pixel coordinates are represented as (x,y). Pixel (0,0) is located in the upper right of the image. Pixel coordinates ascend from the left side of the image to the right. Line coordinates ascend from the top of the image to the bottom.

We now take this convention a little bit further: We must discriminate between input and output images. An *input image* is defined as an image that is used as data to be processed. Any resulting image is referred to as an *output image*. In referring to the coordinates of an image pixel, a prefix of either I or O is used to denote an input or output image, respectively. When multiple input images are used in an operation, a subscript may be appended to the I prefix.

The general-case flow diagram of a digital image processing operation is depicted in Figure 4.1. This basic diagram illustrates the fundamental representation of all image processing operations. Input images are denoted by $I_1(x,y)$, $I_2(x,y)$, and so on. In an operation requiring a single input image, no subscript is used. The output image, if present, is denoted by $O(x,y)$. In some cases, no output image is produced, such as in most image analysis operations. Instead, quantitative numerical data is often the result.

A typical image processing system with reasonable image quality representation is based on an image resolution of 640 × 480, linearly quantized to 8 bits. When necessary, we will assume these resolutions throughout the proceeding development of topics.

Pixel Point Processing—Single Image

Pixel point processes make up the most primitive, yet essential, image processing operations. Used primarily in contrast enhancement operations, the pixel point process is a very simplistic tool. Point processes alter the gray levels of image pix-

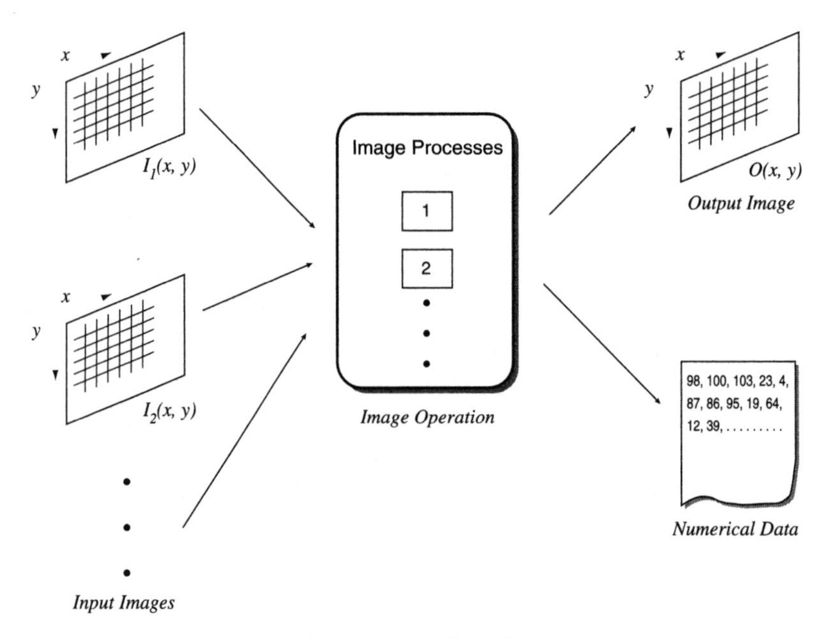

Figure 4.1 *The flow diagram for a general digital image processing operation.*

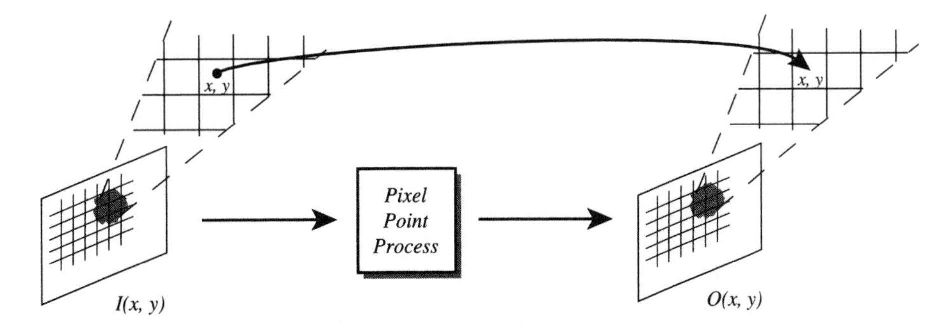

Figure 4.2 *Each pixel in the input image is modified by the pixel point process and placed at the same (x,y) spatial location in the resulting output image.*

els. On a one-by-one basis, the gray level of each pixel in the input image is modified to a new value, often by a mathematical or logical relationship, and placed in the output image at the same spatial location. All pixels are handled individually. For instance, the pixel at coordinates $I(x,y)$ in the input image is modified and returned to the output image at coordinate $O(x,y)$, as shown in Figure 4.2. With this in mind, we note that point operations process pixel brightness attributes with no action on spatial attributes. Spatial processing, as we will see later, is handled by pixel group processing.

The general equation for a point process is given by the equation

$$O(x,y) = M[I(x,y)]$$

where M is the *mapping function*. The mapping function gets its name because it maps (or converts) input brightnesses to output brightnesses. This means that the brightness of an output pixel residing at coordinates (x,y) is equal to the brightness of the respective input pixel, at coordinates (x,y), after being converted by the function M. It is implied that all pixels in the input image are mapped, through the mapping function, to the output image.

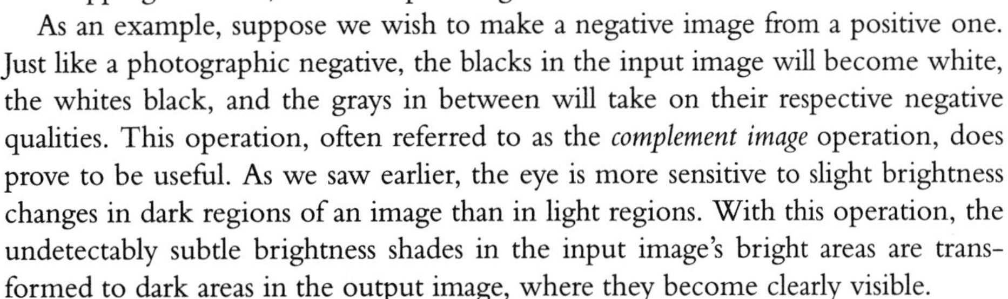

As an example, suppose we wish to make a negative image from a positive one. Just like a photographic negative, the blacks in the input image will become white, the whites black, and the grays in between will take on their respective negative qualities. This operation, often referred to as the *complement image* operation, does prove to be useful. As we saw earlier, the eye is more sensitive to slight brightness changes in dark regions of an image than in light regions. With this operation, the undetectably subtle brightness shades in the input image's bright areas are transformed to dark areas in the output image, where they become clearly visible.

Figure 4.3 illustrates this operation, along with its mapping function. By locating the input pixel gray level on the map's horizontal axis, moving up to the map point and across to the vertical axis, we acquire the respective output gray level. As expected, black (0) maps to white (255) and vice versa. All of the intermediate gray levels are correspondingly mapped, yielding the final complemented image.

Figure 4.3a Original lighthouse image.

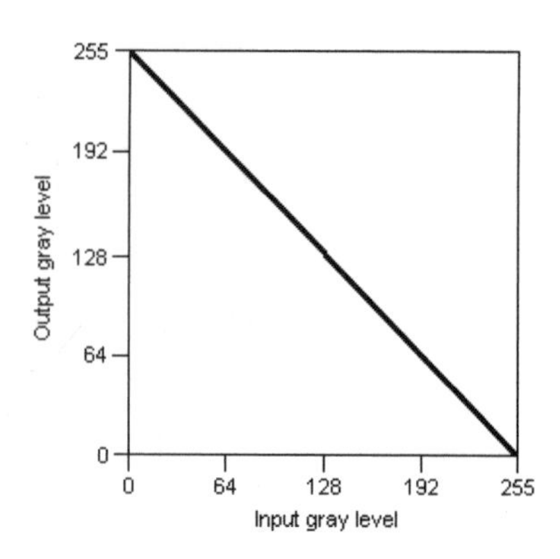

Figure 4.3b The complement operation mapping function. The input brightness value is followed up to the mapping function curve and then over to the corresponding output brightness value.

Figure 4.3c Resulting image with complemented brightnesses.

As mentioned, the most prevalent use of point processes is for contrast enhancement. This set of point processes changes gray levels in an image, with the effect of adding or subtracting image brightness, or stretching or shrinking image contrast. Another operation, known as *photometric correction*, corrects brightness response distortions caused by photosensor incongruities. One such application restores spacecraft and other solid-state image-sensor response nonlinearities caused by size, weight, and cost constraints. Numerous graphic and photographic art applications add artistic effects to images through point processes. These methods include color balancing and general brightness- and color-related special effects.

Point processing is a simple but truly fundamental element of digital image processing. No matter what the application is, some form of point processing is probably involved, even if it is only to clean up undesired artifacts left behind by another process.

Contrast Enhancement

Histogram Sliding and Stretching

Contrast enhancement is a point process involving the addition, subtraction, multiplication, or division of a constant value to every pixel of an image. The image histogram, introduced in Chapter 3, is useful in determining the operations to be employed and in measuring the results. Two operations are commonly implemented—*histogram sliding* and *histogram stretching*. These operations redistribute the brightnesses in an image, enhancing its contrast characteristics.

By looking at an image's histogram, we can generally determine the contrast deficiencies. As we saw in Chapter 3, the types of histograms most frequently encountered are those showing characteristics such as low contrast/low dynamic range, high contrast/high dynamic range, and well-balanced contrast/high dynamic range (see Figure 3.19). By sliding and stretching these histograms, we can make an image of poor contrast quality look considerably better.

Looking back at Figure 3.19a and b, we can see that the pixel gray-scale distribution is clumped in one area of the graph. By sliding the clump to the left and then stretching it out to the right, we can produce a higher-contrast image that will appear to be more well-balanced and natural.

The histogram sliding operation is simply an addition or subtraction of a constant brightness to all pixels in the image. The sliding operation is sometimes referred to as adding an *offset* to the image brightness. The histogram stretching operation is the multiplication or division of all pixels by a constant value. The stretching operation is sometimes referred to as adding *gain* to the image brightness.

Let's look at the entire process of the histogram sliding and stretching operations, as illustrated in Figure 4.4. By looking at the original image's histogram, we see that we can slide the histogram to the left by 140 gray levels, placing the darkest pixel brightness at a brightness of 5. This is done by subtracting the brightness value of 140 from all pixels by using a point process with the appropriate map. The resulting image has no more contrast than the original; we have simply relocated the pixel brightness range so that the darkest pixel brightness of the original image truly becomes a dark value. Now comes the stretching. To make the clump of occupied gray levels stretch a wider range of grays, we can multiply every brightness by the value 3.8. Black pixels (0) remain black, because $0 \times 3.8 = 0$. A pixel of the darkest brightness of 5 becomes 19, and so forth. The pixels in the input image with the maximum brightness of 65 are accordingly stretched to $65 \times 3.8 = 247$, which is very close to the limit of the 256-level gray scale. Again, a pixel point process and appropriate map are used to do the stretch. Our image now appears considerably better with good contrast characteristics; brightnesses span the entire range, from black to white. The final histogram shows this.

Figure 4.4a Original low-contrast image.

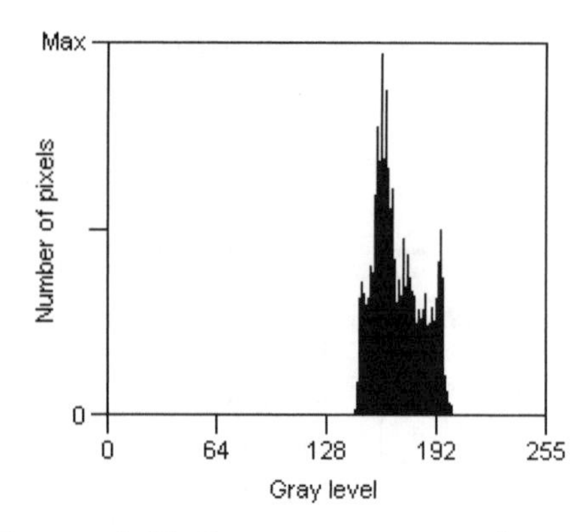

Figure 4.4b Low-contrast image histogram.

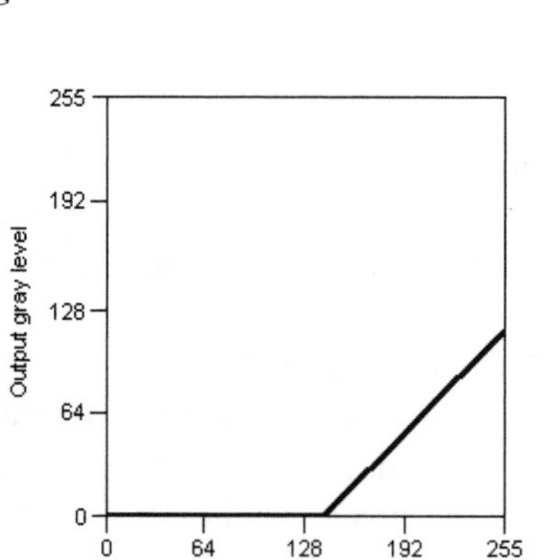

Figure 4.4c Histogram slide-mapping function.

Figure 4.4d Image after histogram slide.

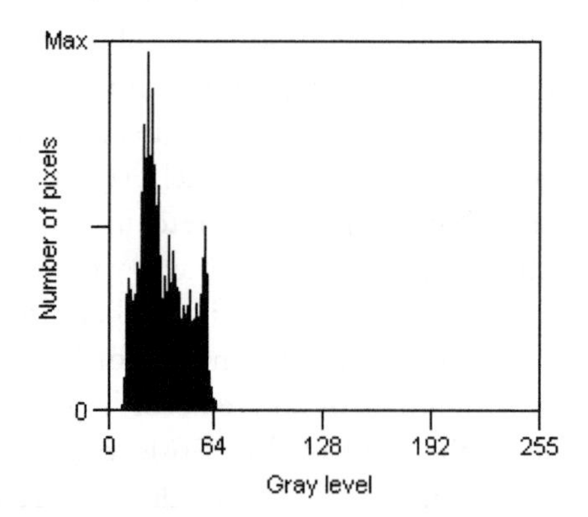

Figure 4.4e Histogram after slide.

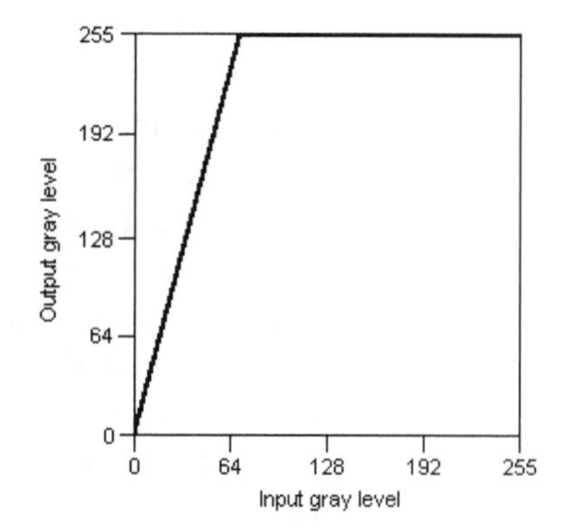

Figure 4.4f Histogram stretch-mapping function.

Figure 4.4g *Image after histogram stretch.*

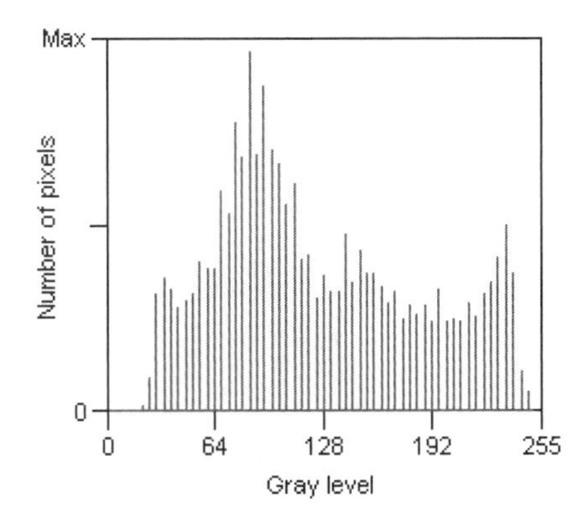

Figure 4.4h *Histogram after stretch.*

Any image can be adjusted in the above manner to achieve a histogram that appears well balanced. But what if the desired result is a high-contrast image? Or, what if the original scene had low contrast, and the goal was to recreate the exact characteristics of that scene? Often it is desirable to produce an image that does not have well-balanced contrast attributes, in order to make some other attribute clearer. For instance, a high-contrast image may show some feature that is not clearly evident in the original. Or, recreating the low-contrast characteristics of an original scene may be important to analyze some aspect of it. In short, it is important to realize that contrast enhancement, like many digital image processing operations, does not have an absolute goodness quality for which we can always aim. The subjective criteria for judging an image's contrast as good or bad are based entirely on the image's intended application.

Binary Contrast Enhancement

Some images are very low in contrast as a function of their origin, such as a paper document that is severely faded. In cases like this, perhaps only a few gray values separate the background from the object of interest. Figure 4.5 shows an image of a map document exhibiting these conditions. To see the lettering and map lines more clearly, we can implement a *binary contrast enhancement*, or *thresholding*, operation. The mapping function illustrates that all pixels of brightness less than that of a selected threshold brightness will be set to black (0), and those equal to or above will be set to white (255). By choosing an appropriate threshold value, we can map the lettering and lines to become black and the background to become white. This is possible because in the original, the letters and lines all appear slightly darker than the background. The processed image is characterized by very high contrast with sharply highlighted lettering and map lines appearing on a white background. In this case, we have taken a low-contrast original and created a very high-contrast image.

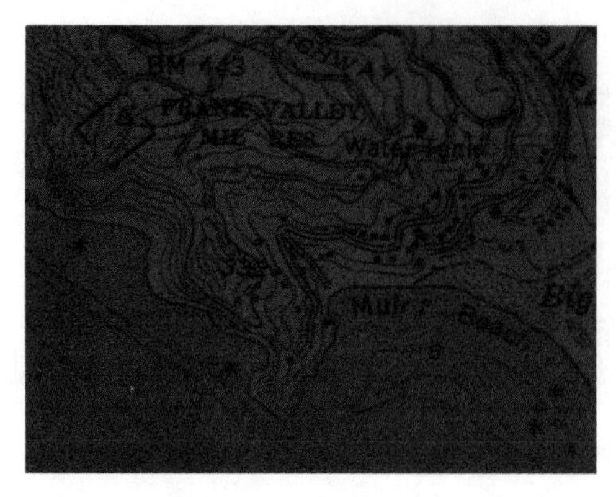

Figure 4.5a Original map document image exhibiting low contrast.

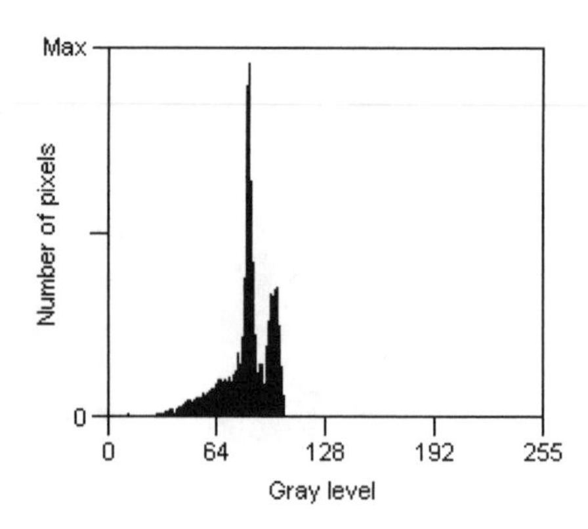

Figure 4.5b Image histogram.

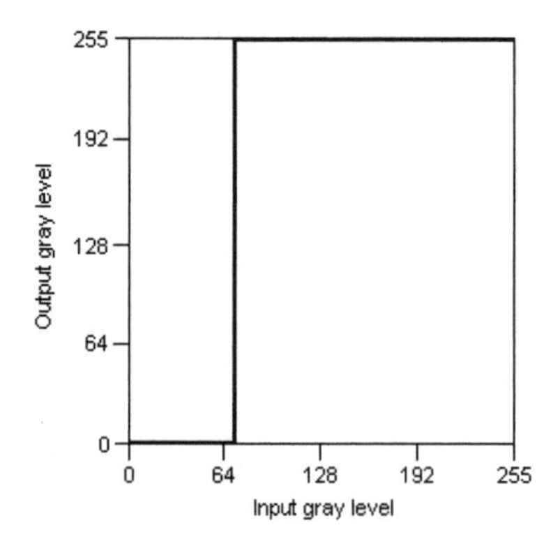

Figure 4.5c Binary contrast enhancement mapping function.

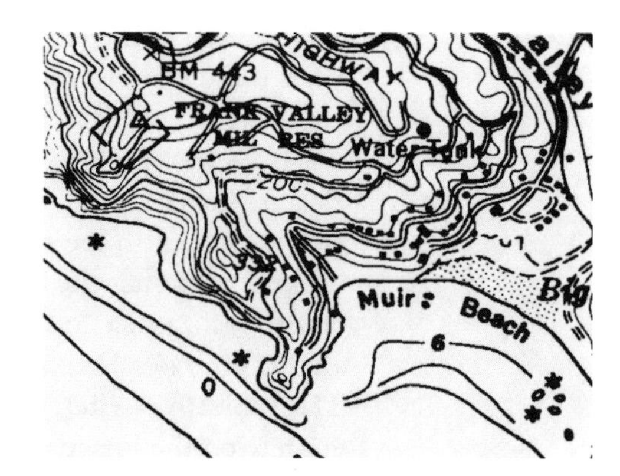

Figure 4.5d Image after binary contrast enhancement, showing text characters and map lines as black (0) and the background as white (255).

Adaptive Thresholding

Sometimes the appropriate threshold value may need to change throughout the image being processed. This might occur with an image of a large document that has faded nonuniformly. As we defined the point process, the mapping function is used for every pixel in the image. Therefore, if the right threshold value must vary, a strict point process will not do the job. In this case, an *adaptive thresholding* technique can be used. The term "adaptive" simply means that the values controlling the point process can adapt to the circumstances of the image; the point process map will change depending on the qualities of the image in the area being processed. In areas where the image is darker, the threshold value decreases. Likewise, in areas where the image is lighter, the threshold value increases. As long as a few gray levels separate the darker text from the lighter background, there is always a threshold value that can distinguish between them.

The adaptive threshold value is usually determined by computing the average brightness in the area being processed and using it to modify the original point-process mapping function. To create this adaptive information, however, we must use techniques discussed in the upcoming pixel group processing section.

Brightness Slicing

Often independent elements of an image can be highlighted using a point process. The technique of *brightness slicing* effectively does a double binary contrast enhancement. Both a lower and an upper brightness threshold are selected, so that brightnesses in between are mapped to white, and brightnesses below the lower value and above the upper value are mapped to black. This method works well when the features to be highlighted occupy brightnesses that are in between the brightnesses of other features, as shown in Figure 4.6.

Additionally, multiple features can be highlighted to different gray levels, or to different colors, providing clear distinction. This works well as long as each feature tends to be made up of gray levels that are different from one another. Instead of using a single brightness slicing map that maps brightnesses to either black or white, a map is created where ranges of brightnesses are mapped to perhaps three or four distinct gray levels. Each gray level in the resulting image then represents one of the multiple brightness features.

As in any digital image processing operation, undesirable artifacts can result if the conditions of the original image are not appropriate for the operation. For brightness slicing, if objects share gray levels with other objects, then they cannot be classified from each other. Brightness slicing works best on simple scenes with clear brightness delineations.

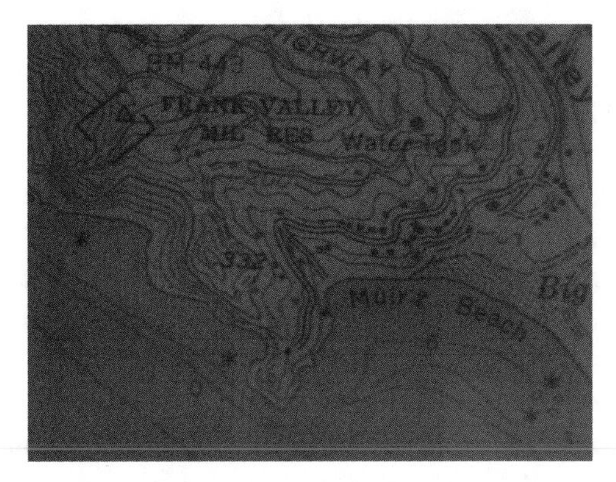

Figure 4.6a *Original map document image.*

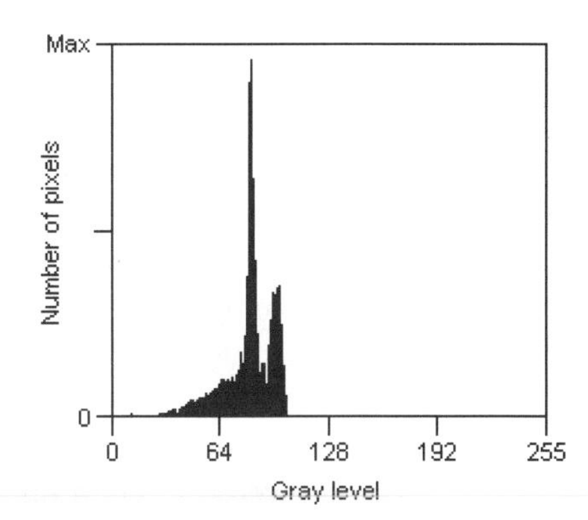

Figure 4.6b *Image histogram.*

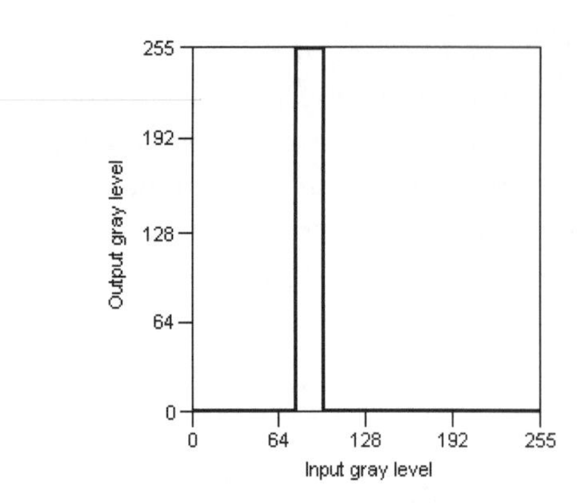

Figure 4.6c *Brightness-slicing mapping function.*

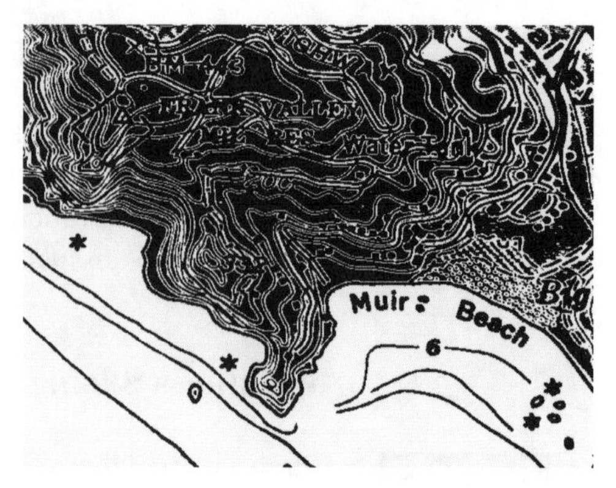

Figure 4.6d *Image after brightness slicing, showing only line boundaries and ocean as white (255) and everything else as black (0).*

Photometric Correction

In any image acquisition or display system, we encounter certain degradations that are caused by the equipment used. These degradations come primarily in the form of photometric and geometric distortions. *Photometric distortions* relate to the light-response incongruities of light-sensitive materials or sensor devices, instabilities in lighting subsystems, and nonlinearities of optical lens arrangements. *Geometric distortions* are spatially related degradations and will be discussed in a later section. Ideally, the imaging system should produce an image identical to the original scene with no added degradations. However, using real-world equipment, some distortions will occur, although they may be insignificant to the application. When the distortions are significant to the application, it is important to remove them.

Often imaging equipment is purposely not optimized for low-distortion characteristics because of size, weight, and cost considerations. This situation arises in virtually any commercial product such as video cameras, image scanners, darkroom enlargers, copy machines, facsimile (fax) machines, and so on. It also becomes very significant in spacecraft and other such restrictive applications. In such cases, the degradations of the system can be characterized before use and then corrected out of the images later. This type of image-distortion correction is generally based on factual, measured knowledge of the degradations, and is therefore a form of image restoration, not enhancement.

An example of photometric correction can be applied to solid-state image sensors, such as those used in most spacecraft imaging systems and virtually all video cameras. These photosensor devices convert light intensities to a voltage signal. The voltage signal, in turn, represents image brightnesses. The devices do not, however, respond linearly to light intensity. This means that the level of their output voltage signal is not exactly proportional to the amount of light hitting the

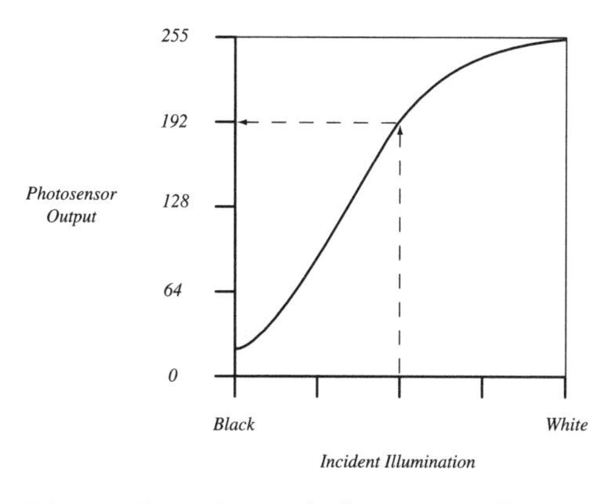

Figure 4.7a *Typical photosensor illumination response curve.*

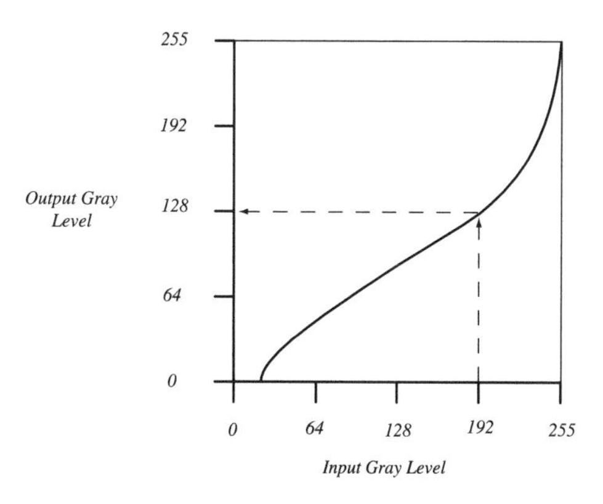

Figure 4.7b *Mapping function to inversely correct the photosensor response nonlinearity.*

surface of the device. A typical solid-state photosensor device response is similar to the one depicted in Figure 4.7. Looking at this response curve, we can see nonlinearities where the response deviates from a straight line. To correct this distortion, we can use a pixel process mapping function that inversely maps the sensor's response back to a linear response. The resulting image will look identical to the scene's actual brightnesses, as if the sensor had no photometric distortion. Incidentally, photographic film also has a similar response as that shown in Figure 4.7. When a film image is over- or underexposed, its dark and bright brightnesses can enter the nonlinear portion of the film's response. This distortion can be corrected using the techniques in the photosensor case described above.

The same operation can be used to precorrect an image that will be displayed on a cathode-ray tube (CRT) display device, to compensate for a display distortion known as the *gamma response characteristic.* Because of the nonlinear response properties of CRT phosphors to an electron beam, a video display monitor does not display brightnesses that are perfectly proportional to the signal voltage representing them. Because the gamma property of a particular video display monitor can be characterized, an image can be precorrected to inversely compensate for the effect. When the image is then displayed, it will appear without the distortions induced by the gamma response characteristic of the phosphors.

Pixel Point Processing—Multiple Image

We have seen how pixel point operations work on single images. Now let's apply this technique to multiple input images. Instead of mapping pixel brightness from one image to an output image, we can map two or more pixel brightnesses, one from each input image, to an output image.

The most common multiple-image point processes involve two input images. These processes are called *dual-image point processes*, and they are the primary focus of this discussion. The mapping function for dual images is more involved than for single images. With 8-bit input pixels, each can take on one of 256 different brightnesses. Because each pixel in an input pair is independent, we have a total of 256 × 256 = 65,536 different possible input combinations being mapped into 256 possible output gray levels. The map for this type of function is usually displayed in three dimensions and, unlike the single-image point process map, is not easily interpreted. For this reason, we generally work in terms of a *combination function*. This name is appropriate because it refers to the way in which each pixel is combined from the two input images. Our dual-image point process equation is

$$O(x,y) = I_1(x,y) \# I_2(x,y)$$

where $I_1(x,y)$ and $I_2(x,y)$ represent the two input images. The symbol # is used here to denote the combination function. As in the single image case, it is implied that each pixel of the input images is processed to create the output image.

Combination functions include mathematical and logical operators such as +, −, ×, ÷, AND, OR, and EXclusive-OR. With these functions, we have the ability to mix the attributes of multiple images, opening an additional world of point processing.

We can lump multiple-image point processes into two general classes—*image combination operations* and *image compositing operations*. Image combination operations mix similar images of essentially the same scene to create a resulting image that better conveys some attribute of the scene. Image compositing operations mix unrelated images, or portions of images, to create an entirely new scene.

Image Combining

Image combination operations merge multiple images that are related to one another in some way. Usually, the images are of identical scenes, but may be acquired at different times or through different spectral filters. Let's look at some of the most common image combination operations.

Differencing

Differences in similar images can often be very subtle. These differences can be made up of things like slight shifts in the location of objects or object brightness variations. When two images are essentially the same, but with subtle variations, image differencing can instantly expose these discrepancies. This is illustrated in Figure 4.8. By subtracting one image from another, pixel by pixel, the pixels in

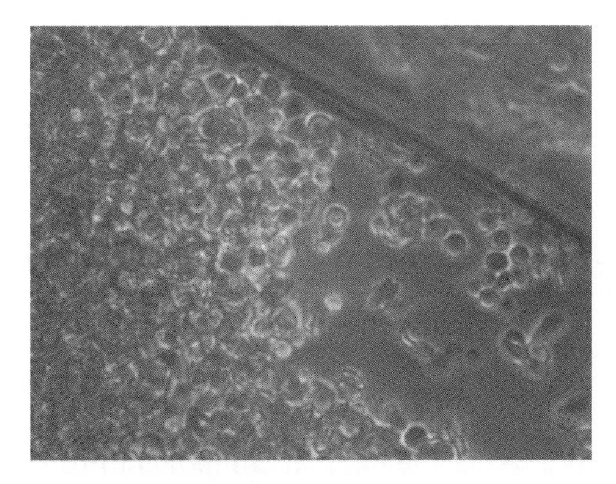

Figure 4.8a *Original microscope specimen image 1.*

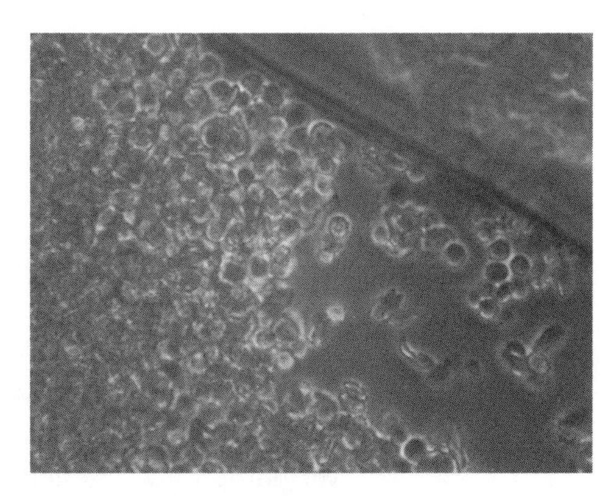

Figure 4.8b *Original image 2 acquired a fraction of a second after the first image.*

Figure 4.8c *The difference image showing only the changes between the two input images—while most of the objects show a little motion, the particle to the lower right of center shows relatively more motion than the other objects.*

the images that are identical will subtract out to black (0). Portions of the images that are different, however, will yield a brightness other than zero. Because the eye can have great difficulty in discerning minor differences, the image differencing operation can be an efficient way of finding scene variations.

Image differencing combines two images using the following equation:

$$O(x,y) = I_1(x,y) - I_2(x,y)$$

where it is implied that each pixel in the input images is processed through the combination function to create the output image.

Image differencing is often used to discern movement of an object between two images of the same scene that were acquired at different times. This technique is essential in satellite reconnaissance image interpretation. The differencing operation can quickly uncover military equipment and troop movement, as well as forest fire progression and crop yields from year to year. Further, by knowing the time interval between two images, we can also calculate the speed of an object's movement. Other primary uses for image differencing include *background removal* and *illumination equalization*.

Spectral Ratioing

Earth resources satellites acquire images of the Earth in several different spectral bands. Like the red, green, and blue components of a color image, multispectral images also include other bands such as infrared (heat). Each image of a particular scene is composed of several spectral component images. Because objects on the Earth's surface reflect light in differing proportions and spectrums, each spectral component image sees the objects differently, depending on their composition. For instance, live vegetation reflects a lot of infrared energy and very little red light energy. Dead vegetation acts oppositely: It reflects a lot of red light energy and very little infrared energy. As a result, the infrared component image will show live vegetation as a bright object and dead vegetation as a dark one. In the red component image, however, the opposite occurs—live vegetation appears as a dark object and dead vegetation as a bright object.

In analyzing multispectral images, we often need to determine the type and abundance of the groundcover in an image. For instance, we may want to know how much of the groundcover in a particular region is living vegetation and how much of it is dead. By knowing the reflectance characteristics of live vegetative groundcover, it is possible to answer this question.

To do this, we might look at the infrared image for bright regions or the red image for dark regions. Unfortunately, other objects also appear in these images. For instance, "warm" objects other than live vegetation also appear bright in the infrared image, and objects other than live vegetation lacking red energy also appear dark in the red image. Therefore, neither the infrared nor red component images can be used alone to show live vegetation accurately. By combining the two images using spectral ratioing, however, accurate determination can be accomplished.

The technique of *spectral ratioing* simply divides one spectral component image by another, pixel by pixel, using a dual-image point process. This is called an *image ratioing* operation. Image ratioing combines two images using the following equation:

$$O(x,y) = \frac{I_1(x,y)}{I_2(x,y)}$$

where it is implied that each pixel in the input images is processed through the combination function to create the output image.

This creates an image that highlights certain objects while obscuring others. For instance, dividing the infrared component image by the red image creates an image where live vegetation is highly accentuated, as shown in Figure 4.9. This is because the bright (large) vegetation brightnesses in the infrared image divided by the dark (small) vegetation brightnesses in the red image create very large resulting brightnesses. Similarly, dividing the red image by the infrared image accentuates the dead vegetation.

Figure 4.9a *Original satellite image showing Morro Bay, California through an infrared filter.*

Figure 4.9b *Original image showing the same region through a red filter.*

Figure 4.9c *Original infrared image divided by original red image with the resulting brightness scaled down not to exceed 255. This image shows the live vegetation significantly highlighted. (Images courtesy of Earth Observation Satellite Co.)*

Temporal Noise Reduction

Time-varying, or temporal, random noise in an image can be reduced if several images of the identical scene are available. Each image must be acquired at different times. As a result, they will each have different random noise patterns. By adding them together, pixel by pixel, and then dividing each resulting pixel brightness by the number of images in the summation, an average of all the image frames can be created. The resulting averaged image will contain less noise. This is because the averaging process will reinforce the parts of the images that do not change from image to image. The parts that do change, the random noise, will be averaged out. If enough images of the exact same scene are averaged together, the random noise can be eliminated entirely.

Temporal image averaging combines two images using the following equation:

$$O(x,y) = \frac{I_1(x,y) + I_2(x,y)}{2}$$

where it is implied that each pixel in the input images is processed through the combination function to create the output image.

Image Compositing

Image compositing operations merge unrelated objects from multiple images. The result is a new scene that may never have existed physically. Image compositing has gained widespread use in the graphic and photographic art fields. As a result, these fields have changed dramatically since the advent of digital image processing. Image compositing operations, previously applied in the darkroom, can be done quickly, repeatedly, and without chemicals, greatly increasing flexibility and quality while reducing costs.

To composite a part of one image into another image, we must first create a *mask image*, often referred to as a *digital matte*. This is done by outlining the object of interest in the first image, then setting the pixels inside the outline to 0 and those outside to 255. The mask image is then subtracted from the first image, using a dual-image point process, creating an image that shows just the outlined object with black everywhere else. A second mask is created that is the exact opposite of the first—255 inside the outlined object and 0 everywhere else. The second mask is subtracted from the second image, yielding a black hole in the image exactly the size of the object to be inserted from the first image. The two images are then summed, again using a dual-image point process, producing a final image with the object of the first image composited into the second. The compositing operation is shown in Figure 4.10.

Many techniques have been developed to make the compositing process appear flawless. For instance, instead of using a sharp 0-to-255 transition at the mask boundaries, a smooth slope from 0 to 255 creates a softer transition between the object and the image into which it is inserted. Because the soft transition is closer to how the eye sees an object in front of a background, the results appear to be more natural.

Figure 4.10a *Original source image of San Francisco—a portion of the image is to be composited into a destination image.*

Figure 4.10b *The mask image outlining the objects to be cut out.*

Figure 4.10c *The difference of the source image and mask image.*

Figure 4.10d *The destination image of an open field.*

Figure 4.10e *The inverse mask image.*

Figure 4.10f *The difference of the destination image with the inverse mask image.*

Figure 4.10g *The summation of the masked source image and masked destination image, creating the final composite image.*

Pixel Group Processing

As we have seen in the previous section, pixel point processing provides image combinations and image gray-scale alterations and corrections—all of which are fundamentally important digital image processing tools. However, point operations cannot provide the ability to alter spatial scene details within an image. This is because point processes act pixel by pixel by mapping a single input pixel to a single corresponding output pixel. The point process does not consider neighboring input pixels in its processing. *Pixel group processing* operates on a group of input pixels surrounding a center pixel. The adjoining pixels provide valuable information about brightness trends in the area being processed. Using these brightness trends open the doors to the world of *spatial filtering*.

We discussed the concept of spatial frequency in Chapter 3. An image is composed of basic frequency components, ranging from low frequencies to high frequencies. Where rapid brightness transitions are prevalent, there are high spatial frequencies. Slowly changing brightness transitions represent low spatial frequencies. The highest frequencies in an image are found wherever sharp edges or points are present—like a transition from white to black within a one- or two-pixel distance.

An image can be filtered to accentuate or remove a band of spatial frequencies, such as the high frequencies or low frequencies. These digital image processing operations are known as spatial filtering operations. Other spatial filtering operations make it possible to highlight only the sharp transitions in the image, such as the edges of objects. These operations are a subset of spatial filtering operations known as edge enhancement operations.

Spatial filters are implemented through a process called *spatial convolution*. Spatial convolution is the method used to calculate what is going on with the pixel brightnesses around the pixel being processed. It is a mathematical method used in signal processing and analysis, and although the operation is mathematically complex, we can study its action in an intuitive, pictorial manner. The spatial convolution process is also referred to as a *finite impulse response (FIR) filter*.

As in point processing, the spatial convolution process moves across the image, pixel by pixel, placing resulting pixels in the output image. Each output pixel's brightness is dependent on a group of input pixels surrounding the pixel being processed. By using the brightness information of the center pixel's neighbors, spatial convolution calculates spatial frequency activity in the area, and is therefore capable of filtering based on the area's spatial frequency content.

The spatial convolution process uses a *weighted average* of the input pixel and its immediate neighbors to calculate the output pixel brightness value. The group of pixels used in the weighted average calculation is called the *kernel*. Kernel dimensions are generally that of a square with an odd number of mask values in each dimension. The kernel can have the dimensions of 1×1, which is the trivial case of simply a point process, 3×3, 5×5, and so on. The larger the size of the kernel of pixels used in the calculation, the greater the *degrees of freedom* of the spatial filter. This means that the flexibility and precision of the spatial filter are increased when more neighboring pixels are taken into account in the calculation. In practice, 3 × 3 and 5 × 5 kernels are used in most spatial filtering operations.

A weighted average calculation is called a *linear process* because it involves the summation of "elements" multiplied by "constant" values. The "elements" are the pixel brightnesses in the kernel and the "constant" values are the weights, or *convolution coefficients*. In the simple case where the weights are each equal to 1/number-of-elements in the kernel, we have a conventional averaging process. We are left with the average brightness of the kernel's pixels. If we alter the weights, certain pixels in the kernel will have more or less influence on the overall average. In fact, the selection of these weights directly determines the spatial filtering action, such as high-pass, low-pass, or edge enhancement filtering.

The mechanics of spatial convolution are straightforward. In carrying out a 3 × 3 kernel convolution, nine convolution coefficients are defined and labeled, as seen below:

$$\begin{matrix} a & b & c \\ d & e & f \\ g & h & i \end{matrix}$$

This array of coefficients is called the *convolution mask*. Every pixel in the input image is evaluated with its eight neighbors, using this mask to produce an output pixel value. This process is illustrated in Figure 4.11. We can visualize that the mask is placed over an input pixel. The pixel and its eight neighbors are multiplied by their respective convolution coefficients and the multiplicands are summed. The result is placed in the output image at the same center pixel location. This process occurs pixel by pixel for each pixel in the input image. The equation for the spatial convolution process is

$$O(x,y) = aI(x-1,y-1) + bIi(x,y-1) + cI(x+1,y-1) + dI(x-1,y) + eI(x,y) + fI(x+1,y) + gI(x-1,y+1) + hI(x,y+1) + iI(x+1,y+1)$$

where it is implied that every input pixel is processed through the equation, creating a corresponding output pixel value.

Convolution mask values can generally take on any numeric value. It is important, however, that when the convolution process is executed, the final resulting value be between 0 and 255 (for an 8-bit output image). This is typically handled by clipping resulting values that are greater than 255 to 255, and clipping values that are less than 0 to 0.

In the case of a 640 pixel × 480 line image, the spatial filtering operation requires the pixel-by-pixel weighted average process to occur 640 × 480 = 307,200 times. Each weighted average process requires nine multiplications and nine additions per input image pixel. This means that a spatial filtering operation applied to an entire 640 × 480 image requires nearly three million multiplications and additions.

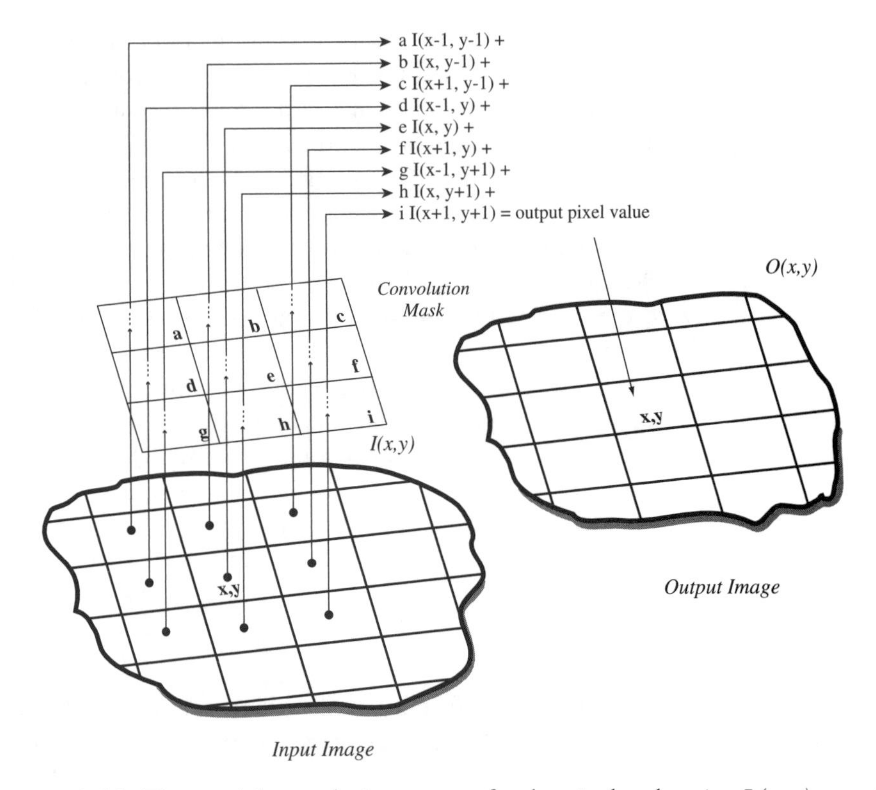

Figure 4.11 *The spatial convolution process for the pixel at location* I(x,y)*, creating the output pixel at location* O(x,y)*.*

Let's look at several commonly used convolution masks and discuss how they work to effect important digital image processing operations.

Spatial Filtering

Common *spatial filtering* operations include high-pass, low-pass, and edge enhancement filters. A high-pass filter accentuates the high-frequency details of an image and attenuates the low-frequency details. A low-pass filter has the inverse effect. Edge enhancement filters detect and enhance image edge details. These filters are covered separately in the following sections.

Low-Pass Spatial Filter

A spatial *low-pass filter* has the effect of passing, or leaving untouched, the low spatial frequency components of an image. High-frequency components are attenu-

ated and are virtually absent in the output image. A common low-pass convolution mask is composed of all nine coefficients having the value of 1/9:

$$\begin{matrix} \frac{1}{9} & \frac{1}{9} & \frac{1}{9} \\ \frac{1}{9} & \frac{1}{9} & \frac{1}{9} \\ \frac{1}{9} & \frac{1}{9} & \frac{1}{9} \end{matrix}$$

This mask carries out a straight pixel brightness averaging process, as described earlier. It is often referred to as a *box filter*. Two aspects are immediately evident—that the coefficients sum to 1 ($9 \times 1/9 = 1$) and that they are all positive numbers. These two facts hold true for all low-pass filter masks. Figure 4.12 illustrates a low-pass filtered image.

To gain some insight into how the low-pass filter works, let's discuss the convolution output values as the mask is passed over regions of an image having different spatial frequency characteristics. First, we can look at the low spatial frequency case where each pixel in a 3×3 group has the same brightness value. The result is calculated as the sum of the nine constant brightness values, each multiplied by 1/9. This is, of course, simply the constant brightness itself, as shown in Figure 4.13. The same result occurs as the mask moves over the other pixel groups in the region that have constant brightnesses. In other words, the output brightness in a region of constant brightness pixels is the same as the input brightness. This correlates with the fact that there is no spatial activity in the region (as evidenced by a lack of gray-level changes), which means that there is a spatial frequency of zero. Zero is the lowest possible frequency and, of course, would be expected to be passed, unchanged, by the low-pass filter.

Looking at the high spatial frequency case, pixel brightnesses in the group change rapidly from white to black and back to white, every other location. Again, the result is calculated as the sum of the nine white and black brightnesses,

Figure 4.12a *Original Space Shuttle image. (Image courtesy of NASA.)*

Figure 4.12b *Low-pass filtered image.*

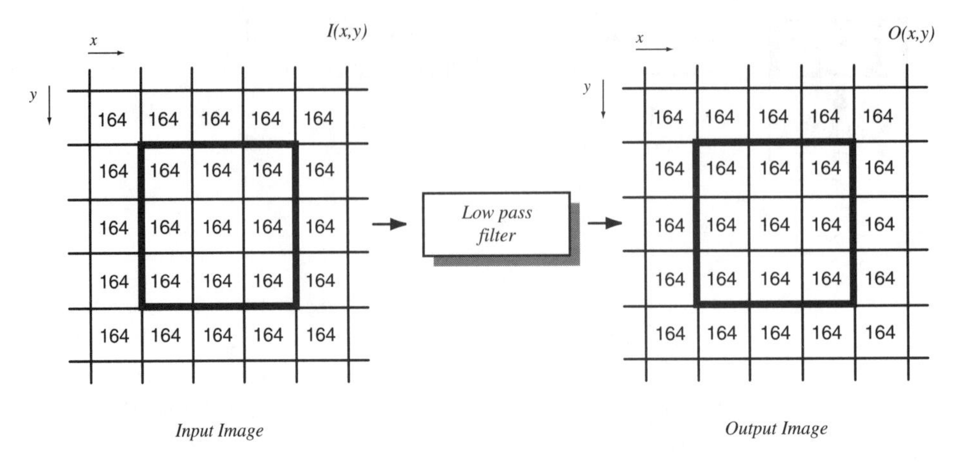

164	164	164	164	164
164	164	164	164	164
164	164	164	164	164
164	164	164	164	164
164	164	164	164	164

164	164	164	164	164
164	164	164	164	164
164	164	164	164	164
164	164	164	164	164
164	164	164	164	164

Figure 4.13 *The low-pass filter process over an image area of constant brightness. The constant brightness is passed untouched.*

each multiplied by ⅑. The result is a middle gray value between the blacks and whites in the group, as shown in Figure 4.14. As the mask moves over the pixel groups in the high-frequency region, similar middle gray values result. This produces an output image that is composed of slightly varying middle gray brightnesses. The high-frequency black-to-white transitions of the input image have been attenuated to transitions of minimal gray-level values. This is exactly the high frequency attenuation that we expect in a low-pass filter.

The visual effect of a low-pass filter is image blurring. This is because the sharp brightness transitions that we perceive as edges become attenuated to small brightness transitions. The small brightness transitions appear to have less detail and look fuzzy or blurred.

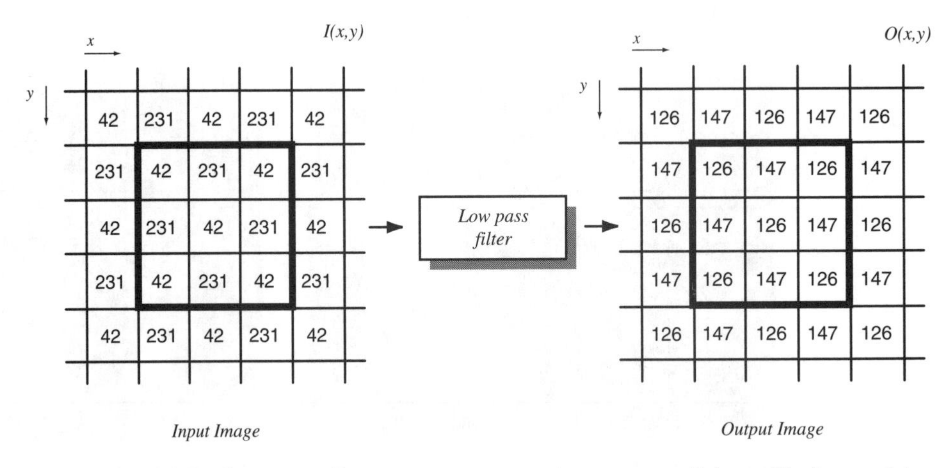

42	231	42	231	42
231	42	231	42	231
42	231	42	231	42
231	42	231	42	231
42	231	42	231	42

126	147	126	147	126
147	126	147	126	147
126	147	126	147	126
147	126	147	126	147
126	147	126	147	126

Figure 4.14 *The low-pass filter process over an image area of sharp black-to-white transitions. The brightness range becomes greatly attenuated.*

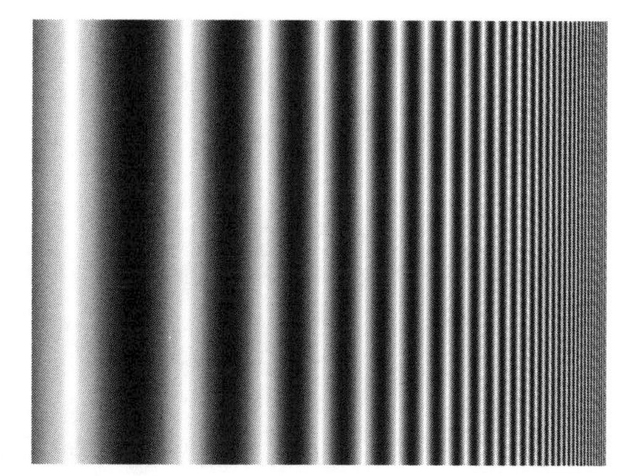

Figure 4.15a *Original image of ascending spatial frequencies.*

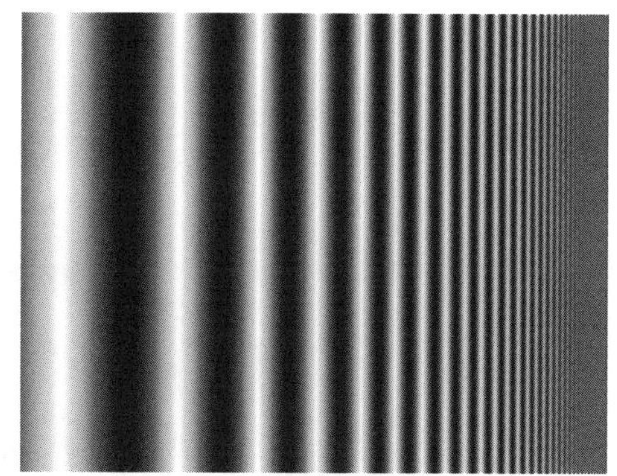

Figure 4.15b *Low-passed image of ascending frequencies. The higher spatial frequencies have been attenuated.*

We can see that the output image is related to the spatial frequency content of the input image—slow-changing areas are left unchanged, or are changed slightly, and fast-changing areas are averaged to yield only the slow-changing aspects. Figure 4.15 shows the effect of the above low-pass filter on a pattern of ascending spatial frequencies.

If we were to pass the low-pass mask over an area with a single-pixel-width line having a constant background pixel brightness, we would expect the line to be blurred. This is because the line represents a high spatial frequency content. In fact, a sharply defined line is actually composed of frequency components spanning a wide spectrum of low to high frequencies. As we move the mask over the line, pixels are replaced by the average of the bright line pixels and the constant background pixels. The resulting values are somewhere between the two. The line blurs into the background. Only the low-frequency components of the line remain at the end of the entire image convolution.

The low-pass filter's blurring effect can provide for better analysis of an image's low-frequency details by removing visually disruptive high-frequency edges and patterns. By subtracting a low-pass filtered image from its original, a sharpened image can by created. This operation is known as *unsharp masking enhancement.*

High-Pass Spatial Filter

The *high-pass filter* has the opposite effect of the low-pass filter. It accentuates high-frequency spatial components while leaving low-frequency components untouched. A common high-pass mask is composed of a 9 in the center location with −1s in the surrounding locations:

$$\begin{matrix} -1 & -1 & -1 \\ -1 & 9 & -1 \\ -1 & -1 & -1 \end{matrix}$$

Figure 4.16a Original Space Shuttle image.

Figure 4.16b High-pass filtered image.

We can see that the coefficients add to 1, and that smaller coefficients surround the large positive center coefficient. Figure 4.16 illustrates a high-passed image.

The fact that the high-pass mask contains a large positive coefficient in the center surrounded by smaller coefficients gives us a clue about its operation. It tells us that the center pixel in the group of input pixels being processed carries a high influence, whereas the surrounding pixels act to oppose it. If the center pixel possesses a brightness that is vastly different from that of its immediate neighbors, the surrounding pixel effect becomes negligible and the output value becomes an accentuated version of the original center pixel. The large difference indicates a sharp transition in gray level, which in turn indicates the presence of high-frequency components. We would therefore expect the transition to be accentuated in the output image. On the other hand, if the surrounding pixel brightnesses are large enough to counteract the center pixel's weight, the ultimate result is based more on an average of all pixels involved.

It is interesting to note that if all pixel brightnesses in a 3×3 group are equal, the result is simply the same value. This is equivalent to the low-pass filter's response over constant regions. This means that the high-pass filter does not attenuate low-frequency spatial components. Rather, it emphasizes high-frequency components while leaving low-frequency components untouched.

Because high-pass filtering accentuates the high-frequency details in an image, its effect is a sharper image, which is usually more pleasing to view than a blurred image. Additionally, high-pass filtering can render more visible any details that are obscured by haziness and poor focus in the original image.

High-pass and low-pass filters form the basis for most spatial filtering operations. When combined with point processes, they provide the adaptive part of the adaptive thresholding operation discussed earlier. This is because spatial filters can evaluate the pixel brightnesses of a group of pixels, and can therefore calculate local brightness trends. This trend information is then used to determine the appropriate thresholding value.

Edge Detection/Enhancement

Image *edge enhancement* reduces an image to show only its edge details. The edges appear as the outlines of objects within the image. The edge outlines can be used in subsequent image analysis operations for feature or object recognition.

Edge enhancements are implemented through spatial filters, as discussed in the previous section. Three particularly useful filters are quite common in many image processing tasks. They are known as shift and difference, Prewitt gradient, and Laplacian edge filters. All three enhancements are based on the *pixel brightness slope* occurring within a group of pixels. Additionally, numerous other edge enhancing filters have been developed, such as the *Sobel*, *Kirsch*, and *Robinson* filters. Each depends on spatial filters for its implementation.

To further define the term "slope" in an image context, think of the brightness of each pixel as being represented by a height coming out from the page toward an observer, as shown in Figure 4.17. We see an image mound rather than the standard gray-level representation; the brighter the pixel, the higher it appears. By measuring the slope of the mound within any given pixel group, we have a value for how steep the incline is. A large value corresponds to a steep slope and means a large change in gray level. A small value indicates small slope, which is equivalent to a small change in gray level. Because edges are, by definition, sharp brightness changes, a large slope indicates the presence of an edge.

The simplest edge enhancement operation is the *shift and difference* method. This procedure can enhance horizontal or vertical edge information. Shifting an image to the left by one pixel and then subtracting it from the original image will make vertical edges apparent. This is because each input pixel brightness value is subtracted from its horizontal neighbor, giving a value of their brightness difference, or slope. If two adjacent pixels have greatly differing brightnesses (an edge), a large brightness difference results. If two adjacent pixels have similar brightnesses (no

Figure 4.17a *Original image of text characters.*

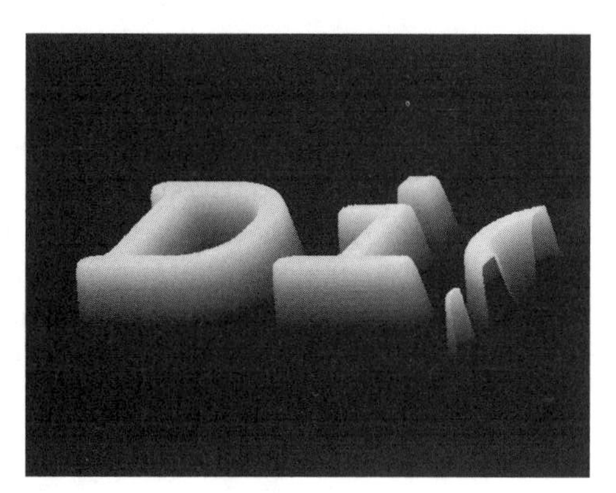

Figure 4.17b *Oblique view of image with brightnesses represented by the heights of pixels.*

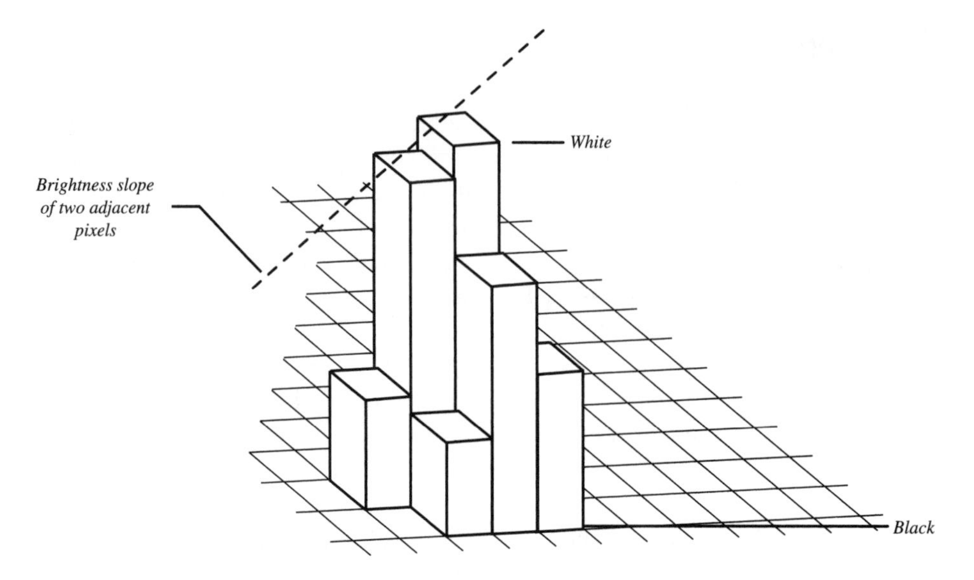

Figure 4.17c *Close-up view of a few pixel heights, illustrating the concept of brightness slope.*

edge), a small brightness difference results. The effect is an image that appears with directional outlines, as shown in Figure 4.19.

The analogous horizontal edge enhancement is implemented by shifting the image upward by one pixel and carrying out the subtraction. The shift and difference operation can be carried out using either a dual-image subtraction point process or a group process with these masks:

$$\begin{matrix} 0 & 0 & 0 \\ -1 & 1 & 0 \\ 0 & 0 & 0 \end{matrix} \qquad \begin{matrix} 0 & -1 & 0 \\ 0 & 1 & 0 \\ 0 & 0 & 0 \end{matrix}$$

Vertical *Horizontal*

Note that the coefficients add to 0. This means that as the mask passes over a region of the image having a constant brightness (no edges), a result of 0 is produced. Of course, this represents a brightness slope of 0, which is exactly what a region of constant brightness has.

The *Prewitt gradient operation* forms a directional edge enhancement. Using a 3 × 3 kernel, eight gradient images may be generated from an original. Each highlights edges oriented in one of the eight compass directions—N, NE, E, SE, S, SW, W, and NW. The mask for the northwest direction is

$$\begin{matrix} 1 & 1 & 1 \\ 1 & -2 & -1 \\ 1 & -1 & -1 \end{matrix}$$

Again, the coefficients add to 0. As the mask passes over a region of the image having a constant brightness, a result of 0 is produced, representing a brightness slope of 0.

Using the east mask, a transition from dark to light, going from left to right, will be accentuated. This is because a positive east brightness slope exists. Brightness slopes in other directions sum to a negative value, which is set to 0, or black. Figure 4.18 shows the response of the Prewitt gradient operation to a one-dimensional edge. Where the gradient generates negative results, the output value

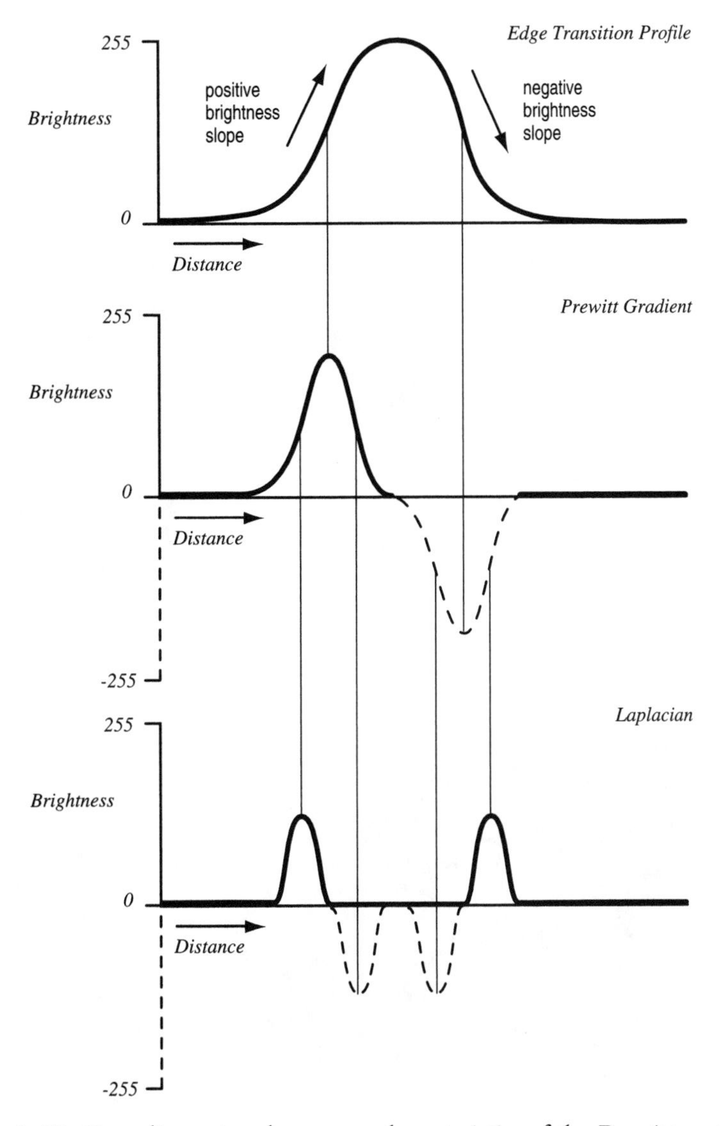

Figure 4.18 *One-dimensional response characteristics of the Prewitt gradient and Laplacian edge enhancement operations.*

is set to 0, because negative brightnesses are undefined. The gradient image appears as black wherever the original image brightnesses are constant. Edges with the correct directional orientation in the original image are seen as white, as shown in Figure 4.19.

Laplacian edge enhancement is an omnidirectional operation that highlights all edges in an image, regardless of their orientation. This operation is based on the rate of brightness slope change within a 3 × 3 pixel kernel. The common Laplacian mask is composed of an 8 in the center location with –1s in the surrounding locations:

$$\begin{matrix} -1 & -1 & -1 \\ -1 & 8 & -1 \\ -1 & -1 & -1 \end{matrix}$$

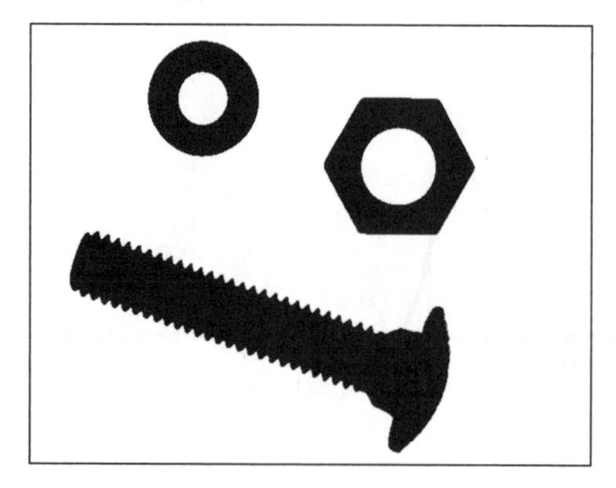

Figure 4.19a *Original image of a washer, nut, and bolt.*

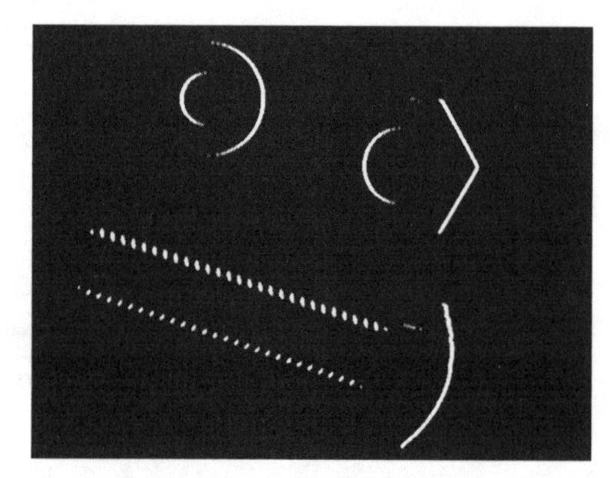

Figure 4.19b *Vertical shift and difference edge enhancement.*

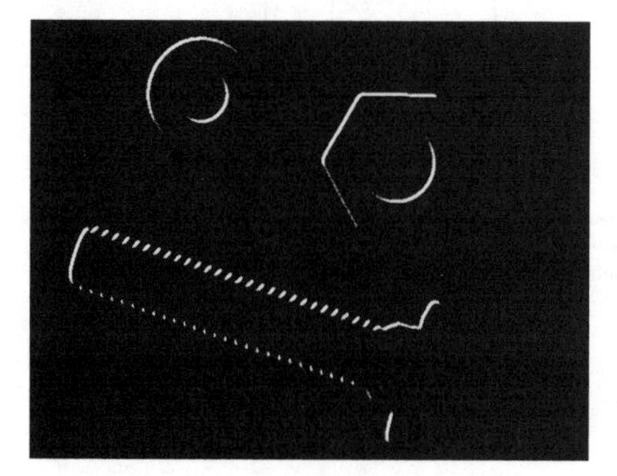

Figure 4.19c *Northwest direction Prewitt gradient edge enhancement.*

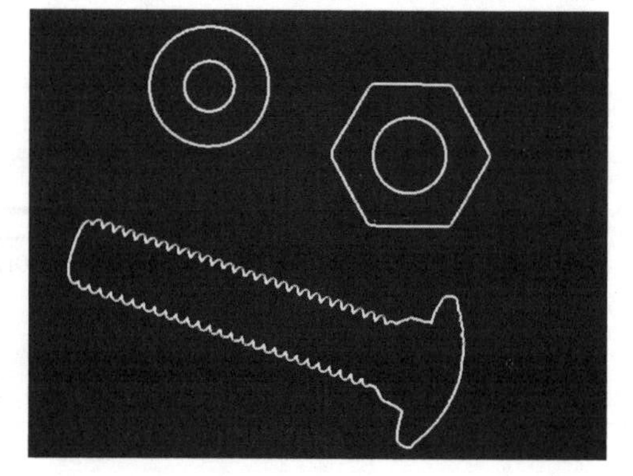

Figure 4.19d *Laplacian omnidirectional edge enhancement.*

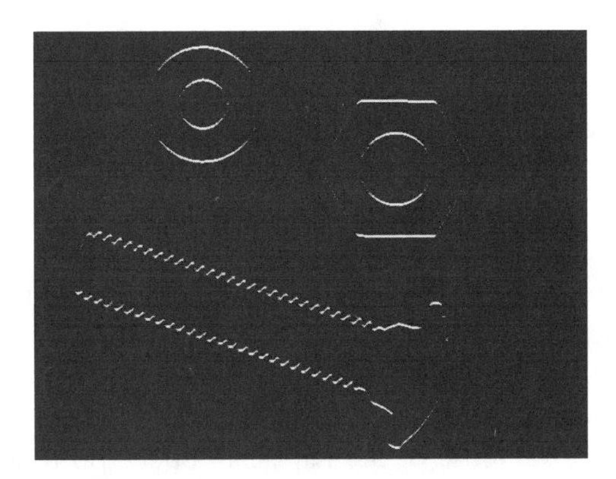

Figure 4.19e *Horizontal line-segment enhancement.*

Figure 4.19f *Original Space Shuttle image.*

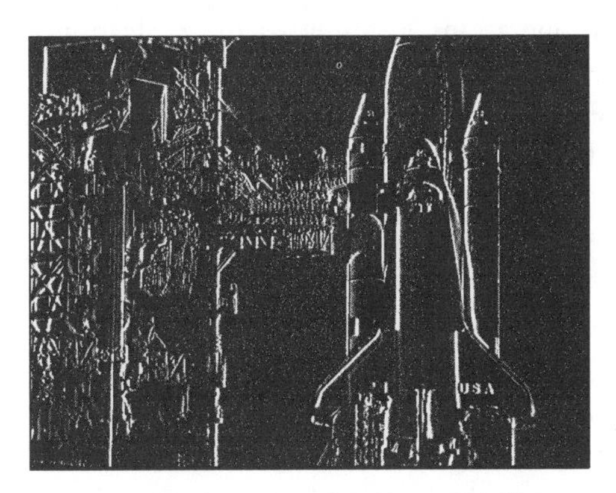

Figure 4.19g *Vertical shift and difference edge enhancement.*

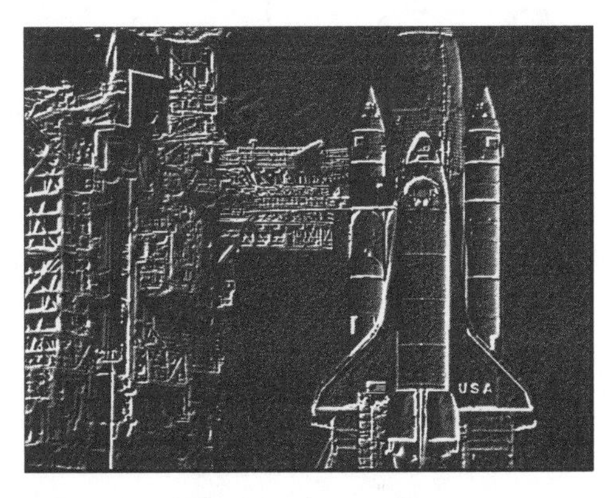

Figure 4.19h *Northwest direction Prewitt gradient edge enhancement.*

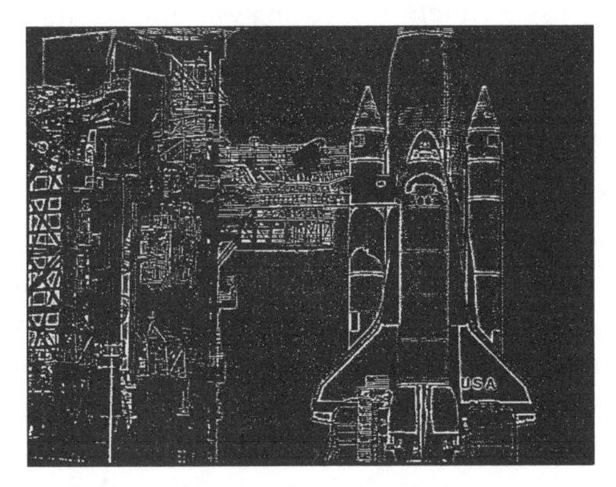

Figure 4.19i *Laplacian omnidirectional edge enhancement.*

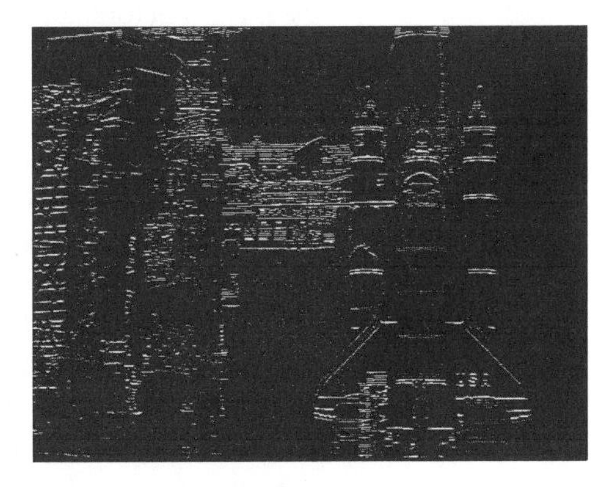

Figure 4.19j *Horizontal line-segment enhancement.*

The coefficients add to 0, and as in the high-pass filter mask, negatively valued coefficients surround the large positive center coefficient.

The Laplacian enhancement operation generates sharper peaks at edges than does the gradient operation. Any brightness slope, whether positive or negative, is accentuated, giving the Laplacian its omnidirectional quality, as shown in Figure 4.19. In the human visual system, the eye–brain system applies a Laplacian-like enhancement to everything it views. We can simulate the human visual system by adding a Laplacian enhanced image to the original image using a dual-image point process. This operation often results in an image that appears naturally sharper with subjectively pleasing qualities.

Figure 4.18 illustrates the response of the Laplacian operation to a one-dimensional edge. The Laplacian image is black wherever the original brightnesses are constant or linearly changing. Edges made of nonlinear brightness transitions are highlighted as white.

Line segment enhancement operations are often useful to clean up the edges of an image following an edge enhancement operation. These operations emphasize line segments within an image, as shown in Figure 4.19. The mask for the horizontal line segment enhancement operation is

$$\begin{matrix} -1 & -1 & -1 \\ 2 & 2 & 2 \\ -1 & -1 & -1 \end{matrix}$$

Again, the coefficients add to 0. This means that constant brightness regions of the original image will become black when processed. Only line segments will remain highlighted.

The above methods for edge enhancements play a major role in machine vision applications. Whether the application is automated, assembly-line material inspection or object recognition, these processes are usually used first to condition the raw images before applying a subsequent image analysis operation.

Nonlinear Spatial Filters

There are other spatial filters that are not computed as a linear summation of elements (pixel brightnesses) multiplied by constant weights (mask values). These filters are referred to as *nonlinear spatial filters*. They, too, use a pixel group process to operate on a kernel of input pixels surrounding a center pixel. But rather than using a weighted average, they use various other techniques to combine the group's input pixel brightnesses.

As an example of a nonlinear spatial filter, let's look at the *median filter*. The median filter is well suited for removing impulse noise from images. It works by evaluating the pixel brightnesses in the kernel and determining which pixel

Figure 4.20a *Original Space Shuttle image corrupted with impulse noise.*

Figure 4.20b *Median-filtered image.*

brightness value is the median value of all pixels. The median value is determined by placing the pixel brightnesses in ascending order and selecting the center value so that an equal number of pixel brightnesses are less than, and greater than, the center value.

The median filters cleans up images with bright noise spikes, because the bright pixels tend to end up at the top of the ascending order of pixels in each pixel group. As a result, the bright spikes are replaced by the median values of the group. Figure 4.20 shows the effect of the median filter on an image corrupted by impulse noise.

Another very important set of nonlinear spatial filters are image morphological operations. These operations will be discussed in depth in the following chapter.

Frequency Domain Processing

We introduced the spatial frequency transform as an image quality measurement tool in Chapter 3. Now, we will fully examine its ability to perform image frequency filtering and enhancement.

To reiterate earlier discussions, images are composed of spatial details that are seen as brightness transitions cycling from dark to light, and back to dark. The rate at which these transitions cycle is their spatial frequency. Spatial frequencies can be oriented horizontally, vertically, or at any diagonal in between. An image is generally composed of many spatial frequencies that, when combined in the correct magnitude and phase, form the complex details of the image.

A *frequency transform* decomposes an image from its spatial-domain form of brightnesses, into a frequency-domain form of fundamental frequency components. Each frequency component has a magnitude and phase value. Similarly, an *inverse frequency transform* converts an image from its frequency form back to a spatial form. The frequency form of an image is also depicted as an image, where brightnesses represent the magnitudes of the various fundamental frequency components.

Numerous frequency transforms exist, and each has an inverse transform to convert a frequency image back to its original spatial form of brightnesses. The most common transform is the *Fourier transform*. The type of Fourier transform used to process digital images is called the *discrete Fourier transform* (*DFT*). A faster, more computationally efficient form of the DFT is the *fast Fourier transform* (*FFT*). Generally, any Fourier transform implemented by digital techniques, such as a computer, uses the FFT.

There are other transforms, such as the *Hadamard*, *Haar*, *slant*, *Karhunen-Loeve*, *sine*, and *cosine transforms*. Each transform has certain unique characteristics. The *frequency component functions* (or *basis functions*) into which a transform decomposes an image are different for each transform. For instance, the Fourier transform decomposes an image into pure, sinusoidal frequency component functions, which can be considered the most fundamental frequency domain form. In this section, we will concentrate on the Fourier transform because of its fundamental nature. In Chapter 6, we will discuss the discrete cosine transform in conjunction with image compression operations.

Conversion to the Frequency Domain

The Fourier transform of an image is a two-dimensional process. First, each row of pixels is processed, followed by each column. The result is a two-dimensional array of values, each having two parts—a magnitude part and a phase part. Each value represents a distinct spatial frequency component. There are the same number of values in the frequency transform image as there were pixels in the original image. The magnitude portion of the values can be displayed as an image, visually showing the frequency components in the original image, as shown in Figure 4.21. The inverse Fourier transform can be applied to the values (both magnitude and phase parts) to convert the frequency image back to a spatial image.

When the magnitude of the frequency image is displayed, it appears to be symmetrical about the center of the image. The center is the zero-frequency point of the display. Two axes exist through the zero-frequency point; one is horizontal and one is vertical. The horizontal axis defines horizontal frequency and the vertical axis defines vertical frequency. Frequencies begin at 0 (or "DC" for direct current) at the center of the image and progress outward. The outermost frequency is the Nyquist rate, which is one-half the sampling rate (as described in Chapter 3). The magnitude of each frequency component is indicated by the brightness of the pixel at each location in the image. The brighter the pixel, the greater the magnitude of

Figure 4.21a *Original Space Shuttle image.*

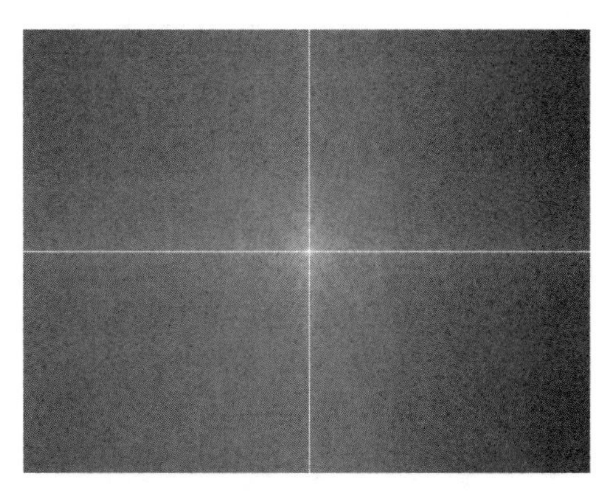

Figure 4.21b *Fourier-transform image. The origin of 0 horizontal frequency and 0 vertical frequency is at the center. Horizontal frequency ascends to the right and vertical frequency ascends up. The brightness of the image at a location indicates the magnitude of the corresponding frequency content.*

the corresponding spatial frequency. The negative frequencies in the frequency image are symmetric mirrors of the positive frequency quadrant; although necessary for the inverse Fourier transform, they visually convey only redundant information.

Figure 4.22 shows several single- and multiple-frequency images and their respective Fourier transform images. Because the single-frequency images contain only one spatial frequency component, their corresponding frequency images appear as a single point of brightness with their associated negative-frequency mirrors.

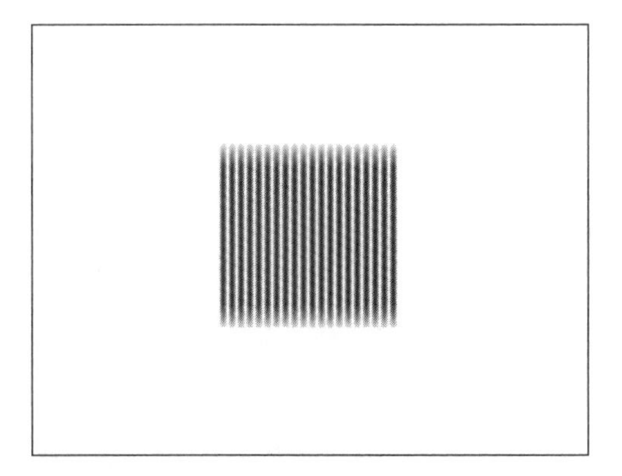

Figure 4.22a *Original image of single, horizontal high frequency. Although the lines are vertical, their frequency is horizontal because that is the direction of brightness change.*

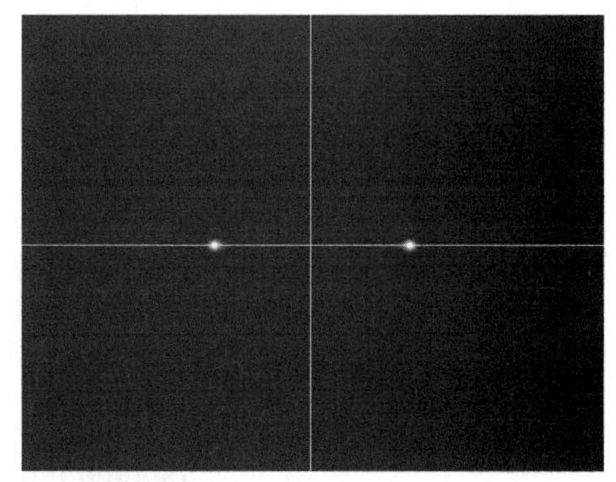

Figure 4.22b *Fourier-transform image showing a single point (and its negative-frequency mirror) representing the single pure frequency.*

Figure 4.22c *Original image of single, vertical low frequency.*

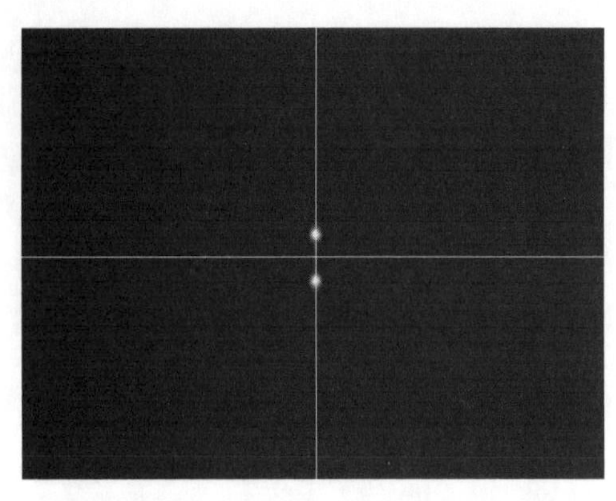

Figure 4.22d *Fourier-transform image also showing a single point and its negative-frequency mirror.*

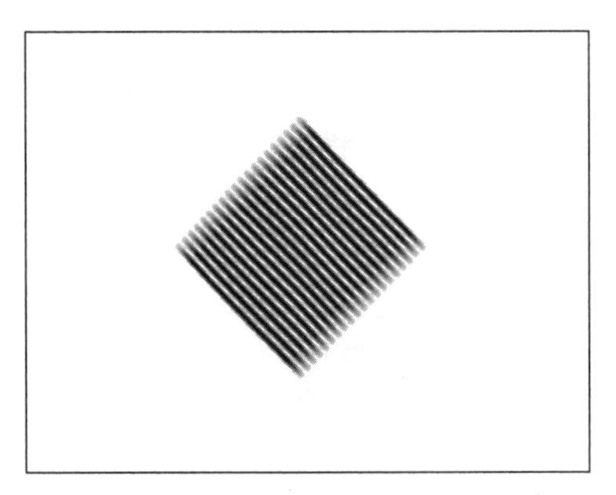

Figure 4.22e *Original image of single, diagonal high frequency.*

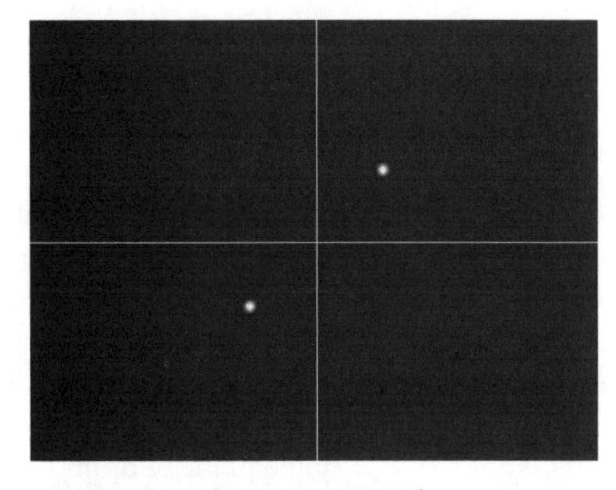

Figure 4.22f *Fourier-transform image also showing a single point and its negative-frequency mirror.*

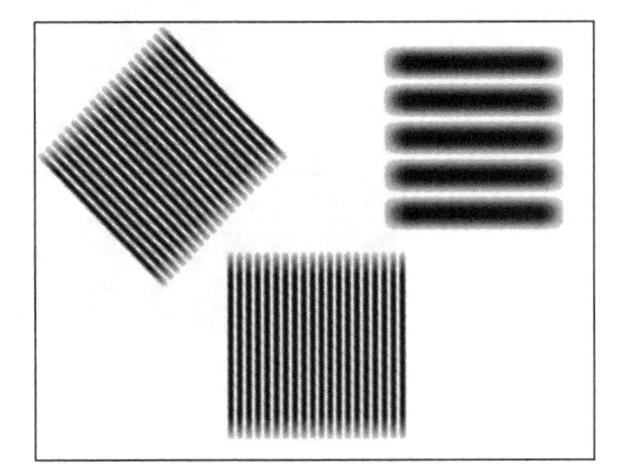

Figure 4.22g *Original image of several pure frequencies.*

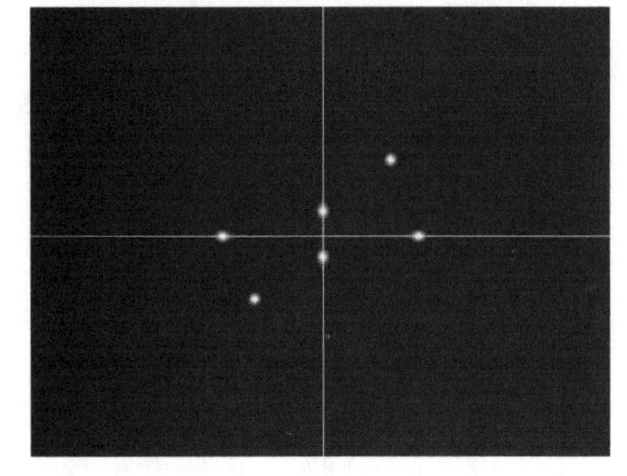

Figure 4.22h *Fourier-transform image showing several points representing the multiple pure frequencies.*

Spatial Filtering in the Frequency Domain

Earlier, we discussed spatial filtering and the use of a pixel group process to carry out spatial convolution. Numerous convolution masks were identified to create low-pass, high-pass, and edge enhancement filters. Identical filtering operations can also be carried out on a frequency domain image. In fact, theoretically, the process of spatial convolution between a spatial image and a convolution mask is identical to the multiplication of a frequency image and a *frequency mask*. Filtering in the frequency domain, however, can be very selective, removing only specific frequency components or bands of frequency components.

Frequency domain filtering is carried out as illustrated in Figure 4.23. First, the original spatial image is transformed to a frequency image using a frequency transform like the Fourier transform. The frequency image is multiplied by a frequency mask image, using a dual-image point process. The frequency mask image is 0 wherever we want to eliminate a frequency and 1 everywhere else. Wherever the frequency image is multiplied by zero, it will become 0 as well; wherever it is multiplied by one, it will be unchanged. The frequency image is then inverse Fourier-transformed back to the spatial domain. The resulting image will be devoid of the frequencies that were multiplied by 0.

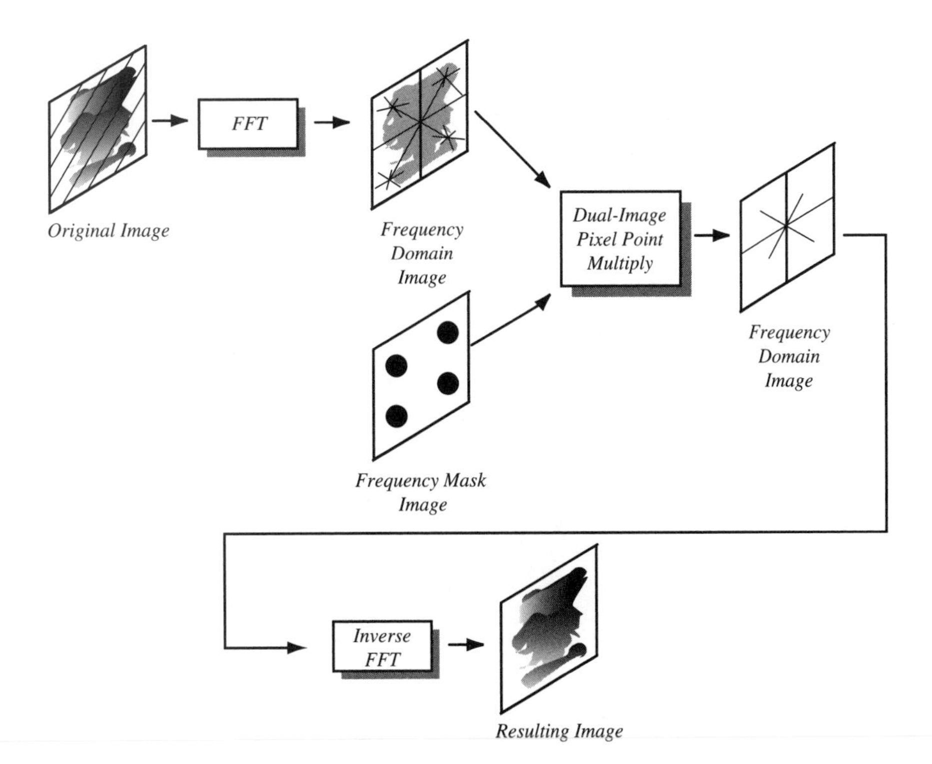

Figure 4.23 *The spatial filtering process using the Fourier transform.*

In the case of an image corrupted with periodic noise, as shown in Figure 4.24, the frequency image will contain obvious bright spots representing the spatial frequency of the noise bands. We can multiply the frequency image by 0 in the area of the spots to eliminate them. This creates a band-reject filter that attenuates only the frequency of interest. The inverse Fourier-transformed image will no longer contain the repetitive noise. Of course, any legitimate spatial details in the image that had the same frequency will also be gone, perhaps creating a visible degradation. A low-pass filter can be implemented by multiplying the high-frequency portions of a frequency image by 0, as seen in Figure 4.25. Similarly, high-pass and band-pass filters can be implemented by multiplying the appropriate frequencies by 0. Alternately, mutliplication by values other than 0 can provide an attenuation or accentuation of the corresponding frequencies.

Figure 4.24a *Original jet fighter image corrupted with a diagonal periodic noise pattern.*

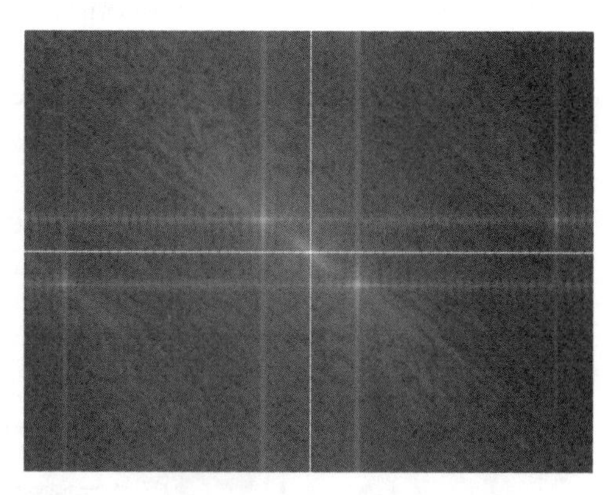

Figure 4.24b *Fourier-transform image.*

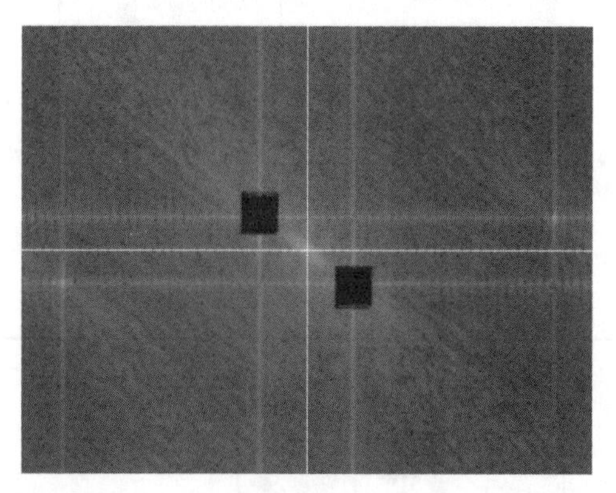

Figure 4.24c *Fourier-transform image with the noise frequencies zeroed.*

Figure 4.24d *Inverse Fourier-transform image with most of the periodic noise removed—visibility of the aircraft is greatly improved.*

Figure 4.25a *Original Space Shuttle image.*

Figure 4.25b *Fourier-transform image.*

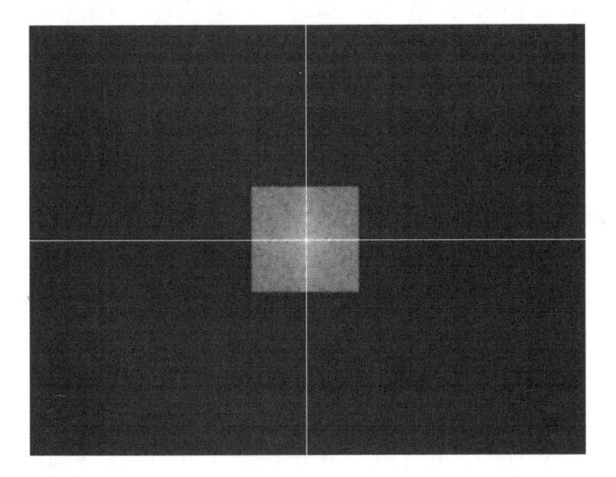

Figure 4.25c *Fourier-transform image with the high frequencies zeroed.*

Figure 4.25d *Inverse Fourier-transform image showing a low-pass filtered version of the original image.*

Image Windowing

It is important to understand the Fourier transform's behavioral quirks before attempting to use it. In particular, it models its frequency representation of the original spatial image as though it were periodic. This means that from the perspective of the Fourier transform, the image frame wraps from its right edge around to its left edge and from its bottom edge around to its top edge. If significant differences exist between the right and left edges of the image, or between the top and bottom edges, the Fourier transform sees an abrupt spatial discontinuity in the image. This is shown in Figure 4.26.

Spatial discontinuities generally contain a wide band of spatial frequency components. As a result, the frequency image will contain erroneous spatial frequency components caused by the Fourier transform's attempt to model the dis-

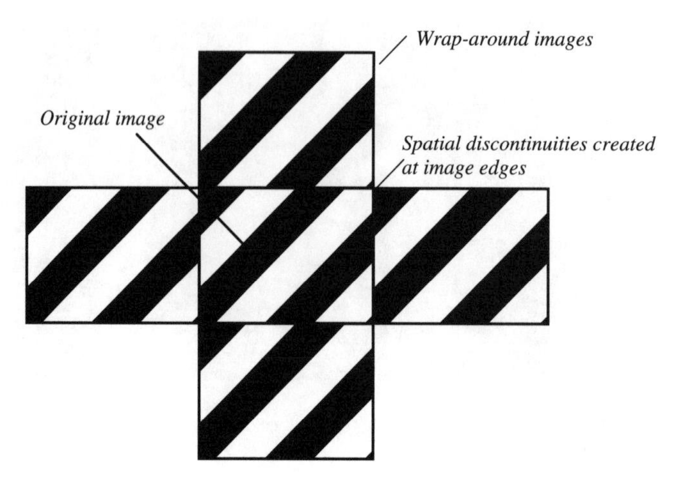

Figure 4.26 *The Fourier transform sees the image as though it were periodic. Usually, discontinuities are introduced at the edges of the image, causing erroneous frequency components to appear in the Fourier-transformed image.*

continuity. When the frequency image is transformed back to the spatial domain, it will have spatial distortion artifacts, because the inverse Fourier transform will try to recreate the edge discontinuities that did not really exist in the original image.

A *windowing function* can be used to make the edges of the original spatial image match with the same brightness value. A windowing function works by slowly reducing image brightnesses to 0 as the edges of the image are approached. Because all edges are then at the same 0 brightness, the spatial discontinuity is eliminated. The resulting frequency image will no longer contain the erroneous spatial frequency components.

It is also helpful to use a windowing function when creating a frequency mask image. Instead of just setting unwanted frequencies to 0, causing a spatial discontinuity, a smooth transition to 0 is used. By applying a windowing function to the frequency mask image, spatial distortion artifacts are minimized in the inverse Fourier-transformed image. Figure 4.27 illustrates the windowing process on a frequency-masked image.

Several windowing functions are commonly used. The *triangular window* linearly transitions the frequency image from 1 to 0 in equal steps as the edges of the image frame are approached. The *Gaussian*, *Hamming*, and *Von Hann* (also referred to as *Hanning*) windows perform smoother transitions that gradually roll off from 1 down to 0. One-dimensional versions of these windowing functions are illustrated in Figure 4.28. Of course, for images, two-dimensional windows must be used to roll off in both the x and y directions.

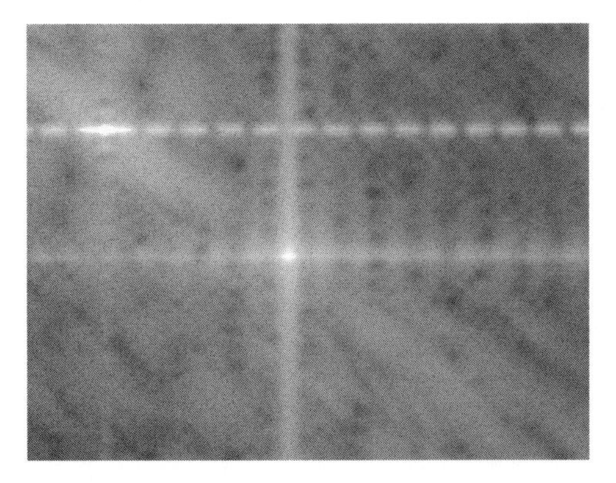

Figure 4.27a *Close-up of a portion of a Fourier-transformed image.*

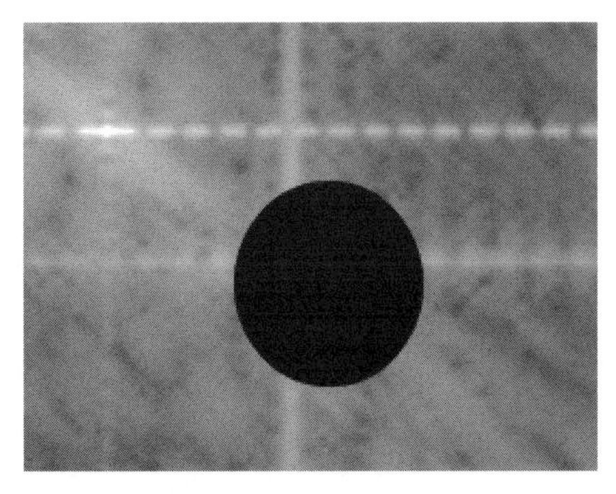

Figure 4.27b *The Fourier-transform image with a frequency mask using no windowing function. Notice the sharp, discontinuous transition from image to masked area.*

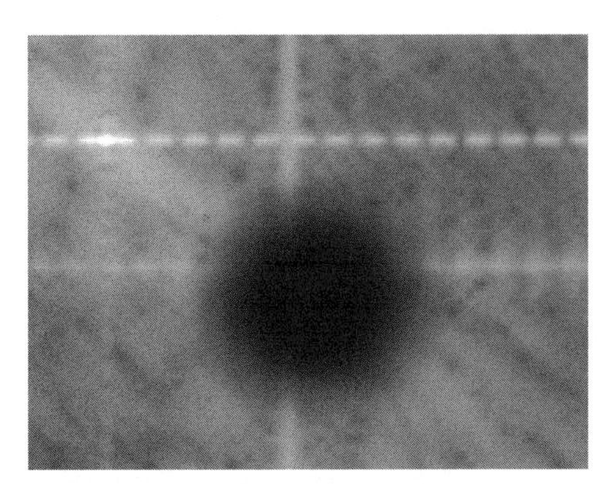

Figure 4.27c *The Fourier-transform image with a frequency mask using a triangular windowing function. The transition from image to masked area is smooth and continuous.*

Deconvolution

Often, we can inverse-filter an image to counteract a degradation previously encountered. In particular, blurring caused by lens misfocus and motion are good candidates for deconvolution operations. Simply, *deconvolution* inverts the action of an earlier convolution operation—one caused by physical means, such as a lens.

When a camera's lens is misfocused, it is essentially applying a low-pass filter to the image being acquired. This is much like the digital low-pass filter operation that we discussed earlier. In this case, though, it is caused by the physical position of the lens. If we know the equivalent convolution mask for the filter function caused by the lens, we can often create an inverse-filter convolution mask to reverse the effect. By applying the inverse filter with a spatial filtering operation, we can usually restore the original image to some extent.

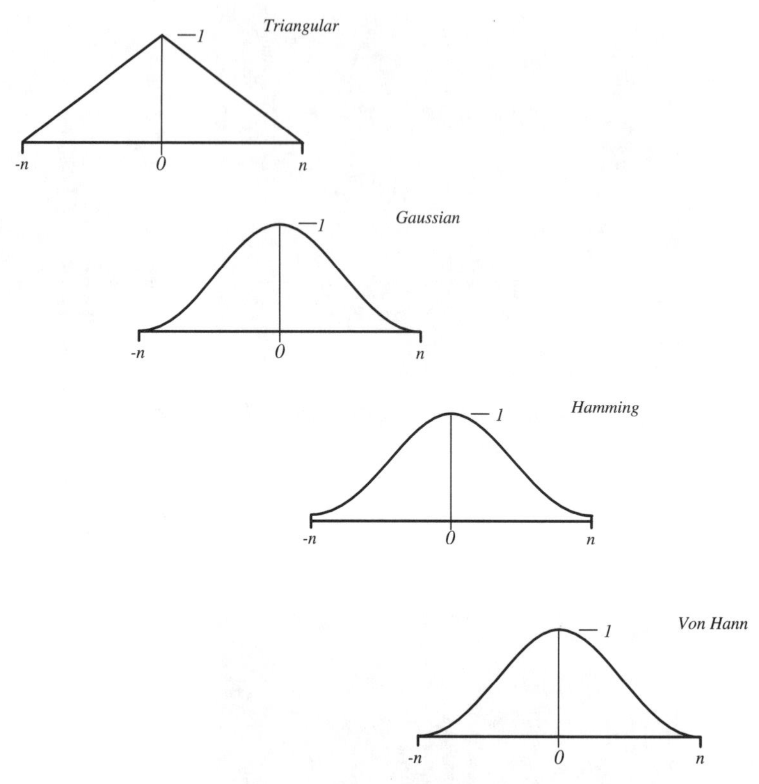

Figure 4.28 *One-dimensional windowing function curves for the triangular, Gaussian, Hamming, and Von Hann windows. These windows smoothly transition the image edges to zero brightness over* n *samples.*

Similarly, we can apply the inverse filter in the frequency domain. First, we Fourier-transform the original spatial image to a frequency image. Then the frequencies attenuated by the misfocus are multiplied by corresponding inverse values. The frequency image is then inverse-Fourier-transformed back to the spatial domain. The resulting image will appear with improved spatial details where the misfocus had previously attenuated them.

Hidden Clues for Deconvolution

Sometimes we do not know the filter function that an image encountered due to misfocusing or another process. In these cases, we can often glean important clues by examining the image itself. In particular, edges or points of light can yield excellent clues. If the image contains what was a sharp brightness transition edge in the original scene, then it will appear blurred in the image. The amount of blurring that it shows directly relates to how out-of-focus the lens was. Likewise, if the image has motion blur, the edge will be streaked in some direction. The streaking tells us how much motion occurred and in what direction it happened.

In both of these cases, a telltale edge or point of light can be used to create an inverse filter. Alternately, it can indicate which frequencies in the frequency image need to be attenuated or accentuated to reverse the original blurring process.

Geometric Transformation Processing

Geometric transformations provide the ability to reposition pixels within an image. Using a mathematical transformation, pixels are relocated from their (x,y) spatial coordinates in the input image to new coordinates in the output image. Geometric transformations are used to move, spin, size, and arbitrarily contort the geometry of an image. These transformations are used for correcting geometric distortions in an image, as well as for adding visual effects such as a perspective foreshortening.

There are two primary forms of geometric transformation operations. First, there are *linear geometric operations*, which include translation, rotation, and scaling. These operations, sometimes referred to as *affine transformations*, do not introduce any curvature to the processed image, hence the term *linear*. Second, there are *nonlinear geometric operations*, known as *warping transformations*. Warping transformations can introduce localized curvatures as well as overall bends and contortions to the processed image. Geometric transformations can be used to register multiple images and restore various geometric distortions caused by lens aberrations and viewing geometry.

All geometric operations are performed by moving pixel brightnesses from their original spatial coordinates in the input image to new coordinates in the output image. The general equation for these operations is

$$I(x,y) \rightarrow O(x',y')$$

where (x',y') are the transformed coordinates of the pixel brightness originally located at coordinates (x,y). It is implied that every input pixel location is processed through this transformation, creating a geometrically transformed output pixel location. Each geometric operation is therefore defined by a coordinate transformation equation that defines the new x' and y' output coordinates in terms of the input pixel at coordinates (x,y).

The process of transforming pixel locations from the input image to the output image is called *source-to-target mapping*. Each pixel of the input image (source) is transformed, pixel by pixel, to its new location in the output image (target). In practice, though, this transformation method does not work well. As the input pixels are transformed, some output pixel locations may be missed because no input pixels were transformed there. The missed locations will be devoid of any brightness and will appear black. The black holes create a poor resulting output image. This phenomenon occurs especially when a scaling or warping operation is involved.

By using a reverse transformation, known as *target-to-source mapping*, the transformation mapping can be reversed, avoiding black holes in the output image. The general equation for target-to-source geometric transformations is

$$O(x,y) \leftarrow I(x',y')$$

where (x',y') are the transformed coordinates of the input image pixel brightness to be relocated to coordinates (x,y) in the output image. For target-to-source mapping, it is implied that every output pixel location is processed through this transformation, creating a geometrically transformed input pixel location. With the target-to-source form, each geometric operation is defined by a coordinate transformation equation that defines the x' and y' input coordinates in terms of the output pixel at coordinates (x,y).

This transformation method maps target (output) pixel locations to source (input) pixel locations. By processing each output image pixel location through the transformation, an input pixel location is identified. This input pixel location's brightness is then relocated to the new output pixel location. By stepping through each output pixel location and filling it with its transformed input pixel brightness, it is impossible to miss any output pixel locations. The geometric transformation result is identical to the source-to-target mapping result, except that there are no black holes in the output image. Figure 4.29 shows the source-to-target and target-to-source mappings.

In the following discussions, both the source-to-target and target-to-source transformation equations will be given for the various geometric transformation operations. In practice, we will generally implement the target-to-source transformations to avoid the black hole problem. But because the source-to-target mapping is easier to envision, our examples will use the source-to-target mapping convention.

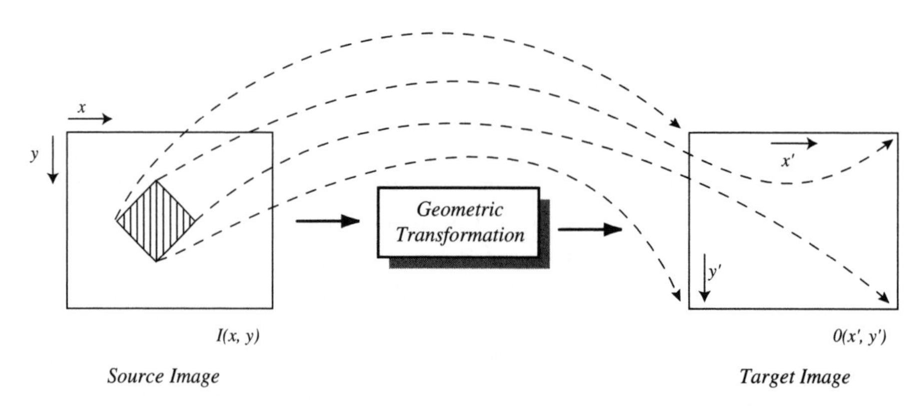

Figure 4.29a *The source-to-target mapping. The transformation sequences, pixel by pixel, through the input pixel locations and places their brightnesses into resulting transformed output pixel locations.*

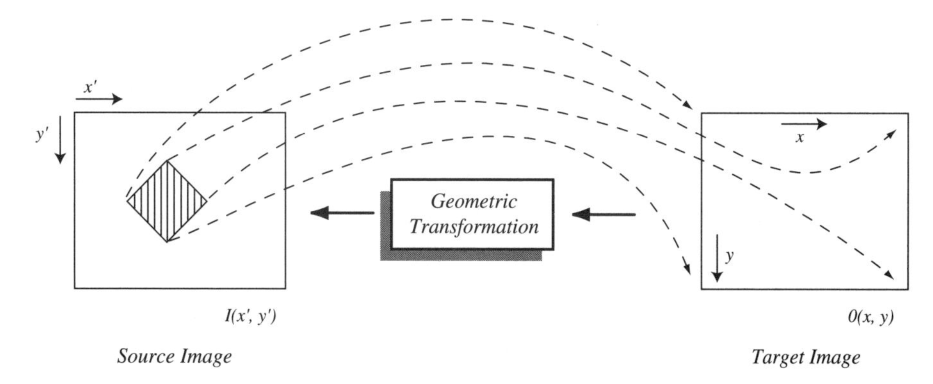

Figure 4.29b *The target-to-source mapping. The transformation sequences, pixel by pixel, through the output pixel locations and fetches transformed input pixel location brightnesses to fill them.*

Translation, Rotation, and Scaling

The linear geometric transformations can spin an image, enlarge and shrink it, and move it left, right, up, and down. These operations cannot introduce curvatures, so straight lines will remain straight in the resulting output image. These transformations are shown in Figure 4.30.

Image translation slides an image left, right, up, or down. An x-value defines the amount of left or right slide, while a y-value defines the amount of up or down slide. The coordinate transformation equations for image translation are

Source-to-target mapping

$x' = x + T_x$

and

$y' = y + T_y$

where x and y comprise the input pixel coordinates, x' and y' are the output coordinates, and T_x and T_y define the amount of translation in the x and y directions.

Target-to-source mapping

$x' = x - T_x$

and

$y' = y - T_y$

where x and y comprise the output pixel coordinates, x' and y' are the transformed input coordinates, and T_x and T_y define the amount of translation in the x and y directions.

As an example, using the source-to-target equations with translation values of $T_x = -100$ and $T_y = -50$, we get the following transformation equations:

$x' = x - 100$

and

$y' = y - 50$

The pixel at location (126,68) will be mapped to the location (26,18) in the output image. When applied to all pixels, this transformation shifts the input image to the left by 100 pixels and up by 50 pixels.

In the case where the T_x or T_y translation values are not integers, the transformation will translate the image by some amount that does not fall directly onto a new output pixel location. In this case, we need some sort of pixel interpolation scheme to estimate the resulting pixel brightness at integer pixel locations in the output image. We will discuss various pixel interpolation methods later in this chapter.

IOS 19 Rotation Transformation

Image rotation spins images about a center point. The coordinate transformation equations for image rotation are

Source-to-target mapping

$x' = x\cos\theta + y\sin\theta$

Figure 4.30a *Original Space Shuttle image.*

Figure 4.30b *Scaled image where* $S_x = 0.5$ *and* $S_y = 0.5$.

Figure 4.30c *Rotated (and scaled) image where* $\theta = 19°$.

Figure 4.30d *Translated (and scaled and rotated) image where* $T_x = -100$ *and* $T_y = -50$.

and

$y' = -x\sin\theta + y\cos\theta$

where x and y comprise the input pixel coordinates, x' and y' are the output coordinates, and T defines the angle of clockwise rotation of the image about the (0,0) pixel location.

Target-to-source mapping

$x' = x\cos\theta - y\sin\theta$

and

$y' = x\sin\theta + y\cos\theta$

where x and y comprise the output pixel coordinates, x' and y' are the transformed input coordinates, and T defines the angle of clockwise rotation of the image about the (0,0) pixel location. Any rotation angle between 0 and 360 degrees may be specified.

As an example, using the source-to-target equations with the rotation angle of $\theta = 19°$, we get the following transformation equations:

$$x' = x\cos\theta + y\sin\theta$$
$$= x\cos(19°) + y\sin(19°)$$
$$= x(0.946) + y(0.326)$$

and

$$y' = -x\sin\theta + y\cos\theta$$
$$= -x\sin(19°) + y\cos(19°)$$
$$= -x(0.326) + y(0.946)$$

The pixel at location (126,68) will be transformed to the location (141.4,23.3) in the output image. By applying the rotation transformation to all pixels in the input image, we create an output image that is rotated by 19 degrees.

When rotation angles are selected that are not multiples of 90 degrees, pixels will often get transformed to noninteger pixel locations in the output image, as in the above example. Like the translation transformation, we must use a pixel interpolation scheme to estimate the pixel brightnesses at the integer pixel locations in the output image.

Image scaling enlarges and shrinks an image. An x-value defines the amount of x-direction scaling, while a y-value defines the amount of y-direction scaling. The coordinate transform equations for image scaling are given by the equations

Source-to-target mapping

$x' = xS_x$

and

$y' = yS_y$

where x and y comprise the input pixel coordinates, x' and y' are the output coordinates, and S_x and S_y define the amount of scaling in the x and y directions.

Target-to-source mapping

$x' = x/S_x$

and

$y' = y/S_y$

where x and y comprise the output pixel coordinates, x' and y' are the transformed input coordinates, and S_x and S_y define the amount of scaling in the x and y directions.

As an example, using the source-to-target equations with scaling factors of S_x = 0.5 and S_y = 0.5, we get the following transformation equations:

$x' = x(0.5)$

and

$y' = y(0.5)$

The pixel at location (126,68) will be transformed to the location (63,34) in the output image. By applying the scaling transformation to all pixels in the input image, we create an output image that is half the size in both the x and y dimensions. Like the other transformations, if the scaling operation transforms input pixels to noninteger output pixel locations, we need a pixel interpolation scheme to estimate pixel brightnesses at integer locations.

Pixel interpolation is especially important for the scaling transformation when the scaling factor is greater than 1. This is because the input image becomes enlarged and pixels are stretched out. Say, for instance, we use a scaling factor of 2 for both S_x and S_y. The pixel at location (126, 68) will be transformed to the location (252,136). Notice, though, that the neighboring input pixel at location (127,68) is transformed to location (254,136), as Figure 4.31 illustrates. The obvious question is, what happens to the pixel at location (253,136) in the output image? In fact, with the scaling factors of 2, every odd pixel and line location in the output image will have nothing transformed to it. Pixel interpolation schemes are used to fill in these pixels with estimates of the brightnesses that should be there.

Further, let's look back to the scaling example where both S_x and S_y are 0.5. In this case, the pixel at location (126,68) is transformed to the location (63,34). The pixel at (127,68) is transformed to (63.5,34). But, because the output image only has integer pixel locations, the pixel is extraneous—there are more pixels in the

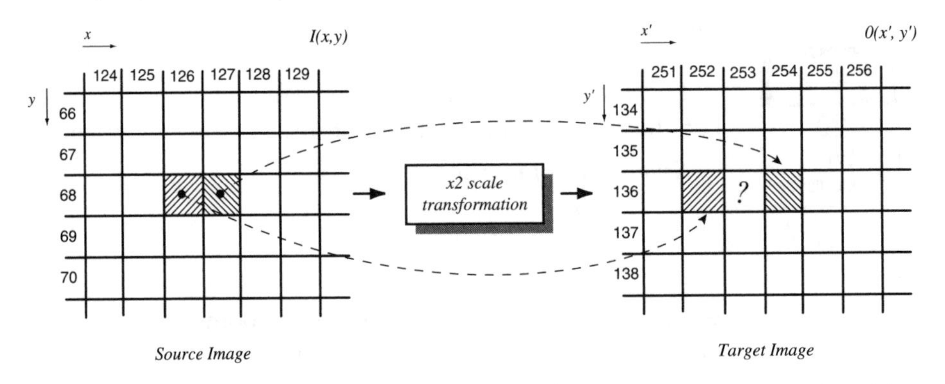

Figure 4.31 *The need for pixel interpolation—the source-to-target mapping skips every other pixel in the output image, leaving distracting black holes.*

Figure 4.32a *Original Space Shuttle image.*

Figure 4.32b *Perspective distortion foreshortens image geometry based on the relative distance of parts of the image from the viewer.*

input image than necessary. The good part is that there is an abundance of input pixels, and therefore pixel interpolation is not of prime importance. The bad part, however, is that spatial aliasing artifacts will appear because the extra pixels must be eliminated. This aliasing effect can be eliminated by using a low-pass filtering operation, as discussed later in the section on downsampling.

Perspective distortions in images caused by the camera-target viewing geometry can be restored by scaling operations. As Figure 4.32 shows, perspective distortion appears as the reduction in scale of an object as it recedes from the foreground of an image to the background. Try holding a book at arm's length and slowly tilting the top away from you. When the book is almost flat, line up the top edge (which is farthest from you) to the bottom edge (which is closest to you). The top edge will appear to be smaller than the bottom edge. Perspective distortion has reduced the scale of the part of the book that is farthest away, relative to that which is closest. By applying two scaling transformations, one that stretches the y axis and one that stretches the x axis with a descending scale factor, we can remove perspective distortion and restore the original flat geometry.

Warping

As mentioned before, linear geometric transformations cannot introduce curvature in their mapping process. Sometimes the ability to introduce curvature is important when an image has been distorted through lens aberrations and other nonlinear processes. *Warping transformations*, often called *rubber sheet transformations*, can arbitrarily stretch and pull the image about defined points, yielding a resulting image that conforms to particular geometric requirements.

Looking back at the source-to-target equations for translation, rotation, and scaling transformations, we can combine them into the following equation:

$$x' = (x\cos\theta + y\sin\theta)S_x + T_x$$
$$= (S_x\cos\theta)x + (S_x\sin\theta)y + T_x$$

which can be generalized to the form

$$x' = a_2x + a_1y + a_0$$

where the coefficients a_2, a_1, and a_0 are constants with the values $S_x\cos\theta$, $S_x\sin\theta$, and T_x, respectively, and

$$y' = (-x\sin\theta + y\cos\theta)S_y + T_y$$
$$= (-S_y\sin\theta)x + (S_y\cos\theta)y + T_y$$

which can be generalized to the form

$$y' = b_2x + b_1y + b_0$$

where the coefficients b_2, b_1, and b_0 are constants with the values $-S_y\sin\theta$, $S_y\cos\theta$, and T_y, respectively.

The source-to-target linear geometric transformations of translation, rotation, and scaling reduce to the following generalized equations, called *polynomials*:

$$x' = a_2x + a_1y + a_0$$

and

$$y' = b_2x + b_1y + b_0$$

Warping transformations are simply these generalized polynomial equations with the addition of higher-order terms, such as x^2, y^2, x^3, y^3, and so on. The warp is said to have an *order*, dependent on the highest exponential term of x or y appearing in the polynomial. A first-order warp means that only the x and y terms exist, which is the linear case of translation, rotation, and scaling. A second-order warp includes the x^2 and/or y^2 terms. A third-order warp includes the x^3 and/or y^3 terms, and so on. The higher the order of a warp, the more complex geometric warping it can offer.

The choice of how high an order warp to perform is totally dependent upon the requirements of the application. The computational time for a warping transformation increases with its order; therefore, performing the minimum order warp is always desired. Also, as in the earlier linear transformations, pixel interpolation is required in higher order transformations, because often input pixel locations will not be transformed to integer output pixel locations.

Controlling the Warp Transformation

Warping transformations can be conveniently discussed using a rubber sheet analogy. We can think of a warping process as operating on an input image printed on a rubber sheet. The rubber sheet can be stretched at arbitrary pixel locations and pinned down to maintain the desired geometric effect. The pins are called *control points*, as shown in Figure 4.33. In practice, we are often interested in aligning two

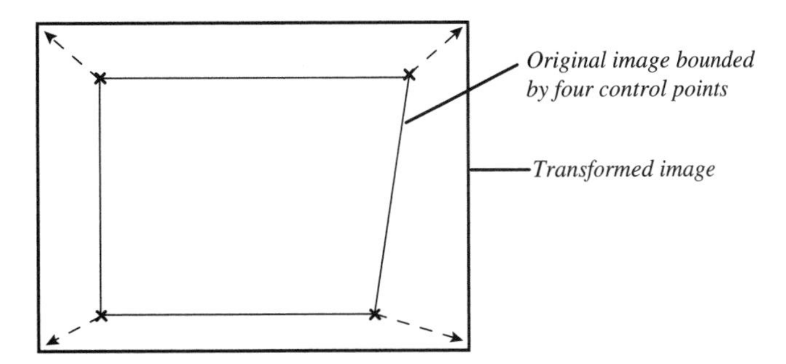

Figure 4.33 *Control points give the warping transformation the necessary information to control the warp parameters.*

images that must be joined together along a seam or overlapped for a subsequent operation. In these cases, we choose control points on each image that correspond to pixel locations that must align together.

The number of control points necessary to control the alignment of an image relates directly to the order of warp required to carry out the transformation. Three control points corresponds to a first-order warp. Six control points corresponds to a second-order warp; ten to a third-order warp, and so on.

The *pincushion lens distortion*, noted in Chapter 2, and its close relative, *barrel distortion*, can be removed from an image by using a third-order warping transformation. The equations for this warp are as follows:

$$x' = a_9x^3 + a_8y^3 + a_7x^2y + a_6y^2x + a_5x^2 + a_4y^2 + a_3x + a_2y + a_1xy + a_0$$

and

$$y' = b_9x^3 + b_8y^3 + b_7x^2y + b_6y^2x + b_5x^2 + b_4y^2 + b_3x + b_2y + b_1xy + b_0$$

Resampling

Whenever we apply a geometric transformation to an image, a *resampling* process occurs. This means that the original sample rate and orientation used to acquire the image change. Resampling occurs in the form of downsampling or upsampling. The processes of low-pass filtering and pixel interpolation can be used to estimate new or intermediate pixel brightnesses so that the resampling process more closely approximates the image appearance, as if it had been originally sampled with the transformed geometry.

Low-Pass Filtering for Downsampling

Downsampling is a spatial resolution reduction that is present whenever a geometric transformation operation transforms an image, or a portion of an image, to a size smaller than the original image. Because the resulting image becomes smaller, it has fewer pixels defining it than the original had.

As a result of downsampling, an image has a reduced spatial frequency content capability. Remember, the sampling theorem says that the highest spatial frequency in an image can be no greater than one-half its sampling rate. So, when we shrink an image, high-frequency components in the original image may be aliased in the transformation process, and thus appear as aliasing artifacts in the resulting image.

To keep spatial aliasing from occurring, we can first reduce the frequency content of the image, using a low-pass filtering operation, so that it doesn't exceed the limit of the new sampling rate. Then, we scale the image down to the new size. As an example, if the image is to be scaled by the factors $S_x = 0.5$ and $S_y = 0.5$, then the resulting image size will be one-half that of the original in both the x and y directions. This means that the new sampling rate of the resulting image will also be one-half that of the original image. By first applying a low-pass filtering operation, we can remove all the high frequencies from the original image that exceed the new one-half resampling rate. The scaling transformation can then be applied without causing aliasing artifacts in the resulting image.

In practice, the artifacts of the aliasing phenomenon may have only minimal impact on the visual quality of the resulting image. The worst aliasing artifacts occur when repetitive patterns of high spatial frequency are present in the original image or when a motion sequence of images is involved. When these are not the case, we can often avoid the low-pass filtering operation without significantly sacrificing image quality.

Pixel Interpolation for Upsampling

Upsampling is the process of increasing the spatial resolution of an image. It is present whenever a geometric transformation operation transforms a pixel to, or from, a noninteger location. In the source-to-target transformation, this occurs when the output pixel location is not an integer location. In the target-to-source transformation, this happens when the input pixel location is not an integer location, as shown in Figure 4.34.

Whenever a transformed pixel does not fall directly on an integer pixel location, a form of pixel interpolation is used. Because the interpolation process is an estimation process that determines the pixel brightness that would exist between integer pixel locations, many differing schemes can be used, depending on the application's accuracy requirements. The most rudimentary form of interpolation is that of *nearest neighbor interpolation,* or *zero-order interpolation.* Nearest neighbor interpolation simply determines which pixel location is closest to the one desired and uses it, as shown in Figure 4.35. For instance, in scaling an image to twice its original size in both the x and y directions, the new output pixels created between the original pixels will have brightnesses equal to those of their adjacent neighbors. The visual

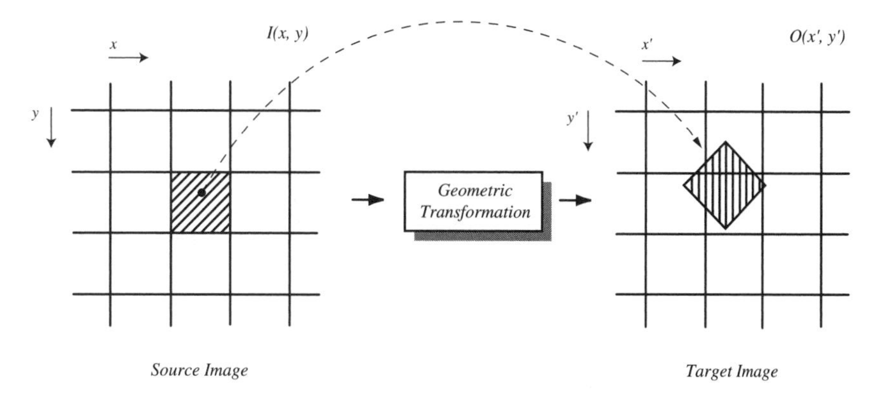

Figure 4.34a *The source-to-target transformation when output pixel locations are not integers.*

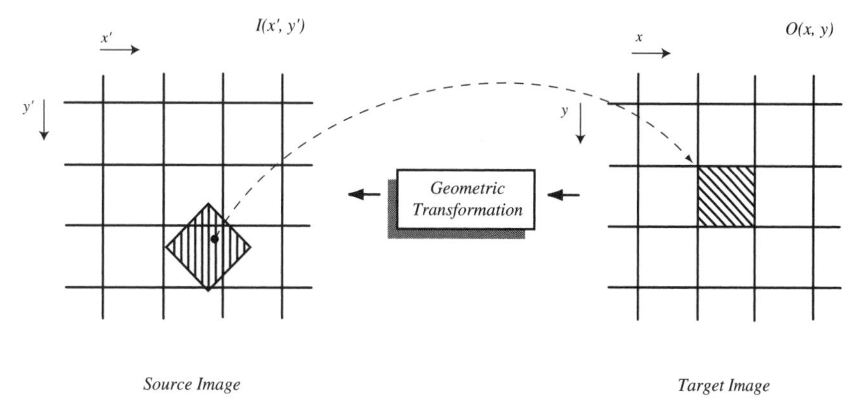

Figure 4.34b *The target-to-source transformation when input pixel locations are not integers.*

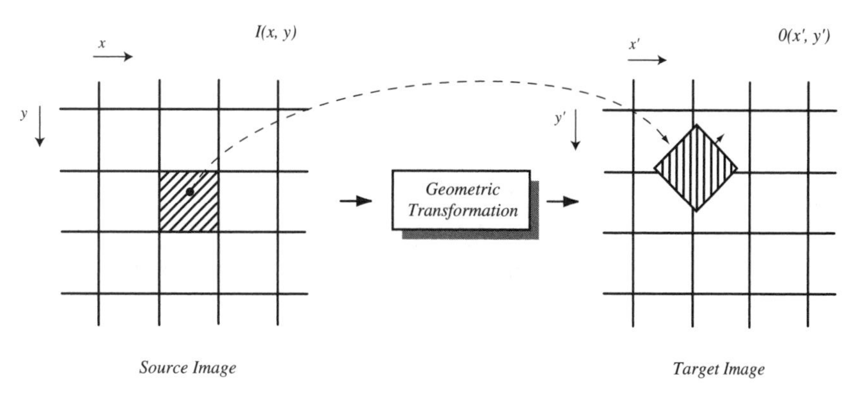

Figure 4.35a *Nearest neighbor interpolation applied to source-to-target transformations.*

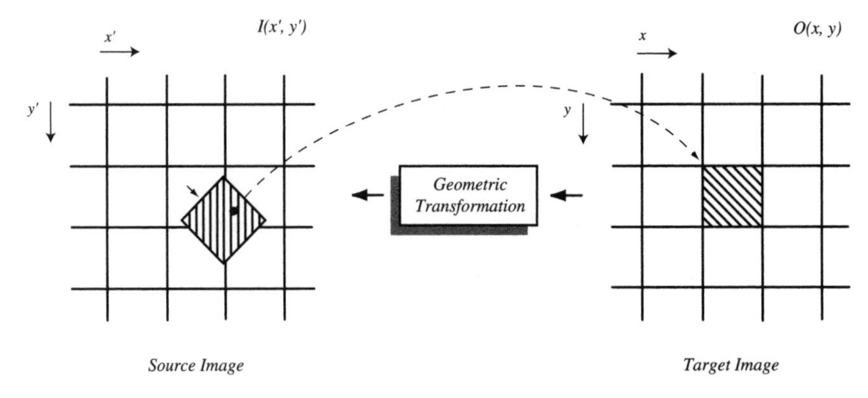

Figure 4.35b *Nearest neighbor interpolation applied to target-to-source transformations.*

effect of pixel blocking can become noticeable, however, and can degrade the resulting image.

A more accurate interpolation scheme is *bilinear interpolation,* or *first-order interpolation.* Bilinear interpolation takes a weighted average of the brightnesses of the four pixels surrounding the pixel location of interest, as shown in Figure 4.36. In this way, some of the brightness trends of the pixel neighborhood are used to estimate the pixel of interest. The weights of the average calculation are proportional to how close each of the neighboring pixels are to the desired pixel location. This is given by the equation

$$O(x,y) = (1-a)(1-b)I(x',y') + (1-a)bI(x',y'+1) + a(1-b)I(x'+1,y') + abI(x'+1,y'+1)$$

where the output pixel at $O(x,y)$ is surrounded by the mapped pixels from locations $I(x',y')$, $I(x'+1,y')$, $I(x',y'+1)$, and $I(x'+1,y'+1)$ for a target-to-source mapping.

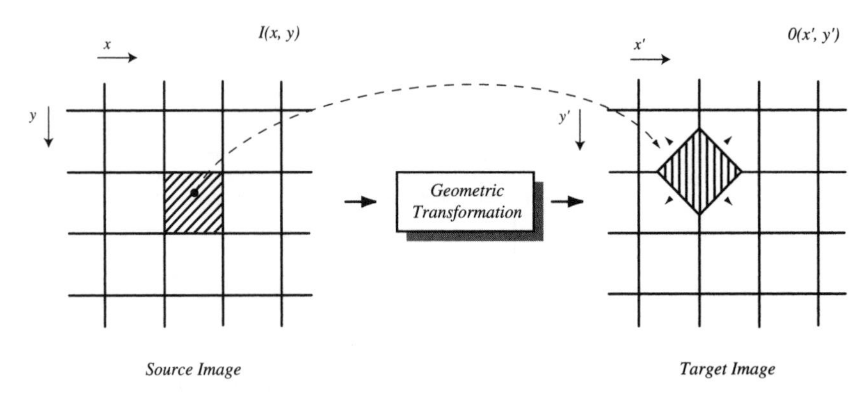

Figure 4.36a *Bilinear interpolation applied to source-to-target transformations.*

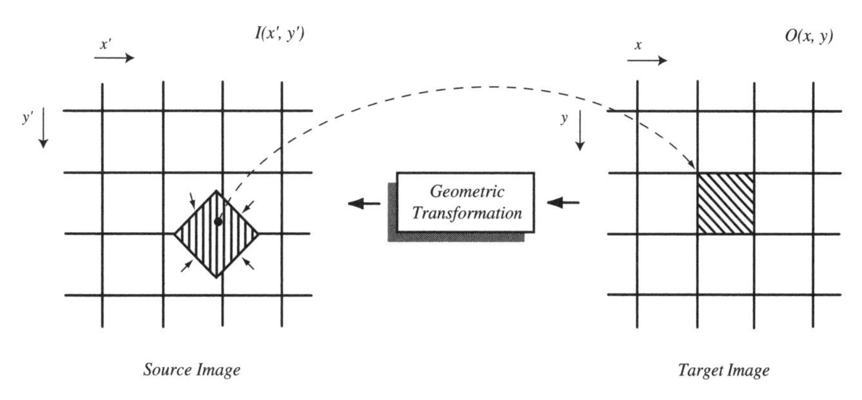

Figure 4.36b *Bilinear interpolation applied to target-to-source transformations.*

If the desired pixel is very close to one of the four neighboring pixels, its brightness will be most influenced by that neighbor. If the desired pixel is perfectly in the center of the four neighbors, each neighbor will have equal weight in the average. Bilinear interpolation produces resampled images that appear smoother and considerably more appealing than those using the nearest neighbor approach, as shown in Figure 4.37.

Other, higher-order interpolation schemes essentially factor more neighboring pixels into the weighted average. The more neighbors involved, the better the brightness estimate of a fractional pixel location. Commonly, the *second-order interpolation* scheme, known as *cubic convolution*, is used when more accuracy is needed. This scheme generally uses a neighborhood of 16 pixels in its computation. In extreme cases, up to 64 neighboring pixels can be used, yielding an interpolated pixel brightness that is virtually a perfect estimate. This type of pixel interpolation is usually reserved for use only on motion sequences where the absolute temporal smoothness of the geometric transformation is paramount. For still images, the

Figure 4.37a *Original Space Shuttle image.*

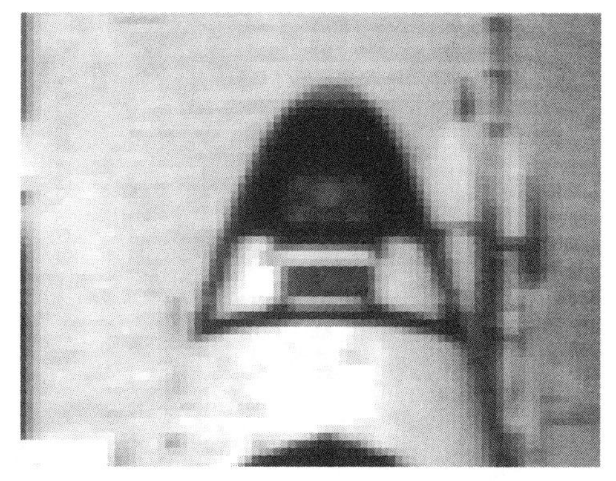

Figure 4.37b *Image scaled by a factor of eight using nearest neighbor interpolation.*

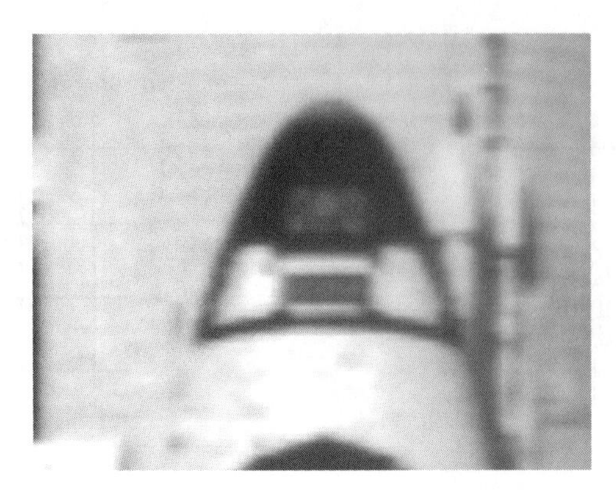

Figure 4.37c *Image scaled by a factor of eight using bilinear interpolation.*

differences between bilinear and larger interpolation neighborhoods is usually undetectable. If an application requires the use of higher order schemes, a greater computational price must be paid for such accuracy.

5 *Image Analysis*

Image analysis operations are used in applications that require the measurement and classification of image information. They are different from all other digital image processing operations because they almost always produce nonpictorial results. The mission of image analysis operations is to understand an image by quantifying its elements. The elements of interest are generally objects in an image, such as cells, bolts, aircraft, or characters on a page of text. Their quantification includes such things as measures of size, indicators of shape, and descriptions of outlines. Other elements of interest can include attributes such as brightness, color, and texture. Image analysis operations play a major role in automated machine vision and image interpretation applications of digital image processing.

An image exists as a large array of pixels with associated brightnesses. This native image form contains no inherent intelligence about the contents of the scene. That interpretation is left entirely to the observer. Image analysis operations seek to break an image into a concise list of individual objects meeting specific measures and/or descriptions. These operations convert an image from a visual form to a descriptive form. The result is a drastically reduced form of the image providing object recognition and identification.

We have already studied the image histogram, which is probably the most fundamental and often used image analysis operation. As discussed in Chapter 3, the histogram accumulates image brightness distribution information and presents it with either a graphical or tabular representation. With the histogram's measure of an image's brightness attributes, overall image quality can be assessed. Further, the histogram can be applied only to the pixels in a particular object in an image. This way, the object's brightness attributes can be independently analyzed.

In this chapter, we study the sequence of image analysis—breaking an image into individual objects, measuring the objects, and classifying the objects based on their measurements. These processes are generally referred to as image segmentation, feature extraction, and object classification, respectively. This process flow is illustrated in Figure 5.1.

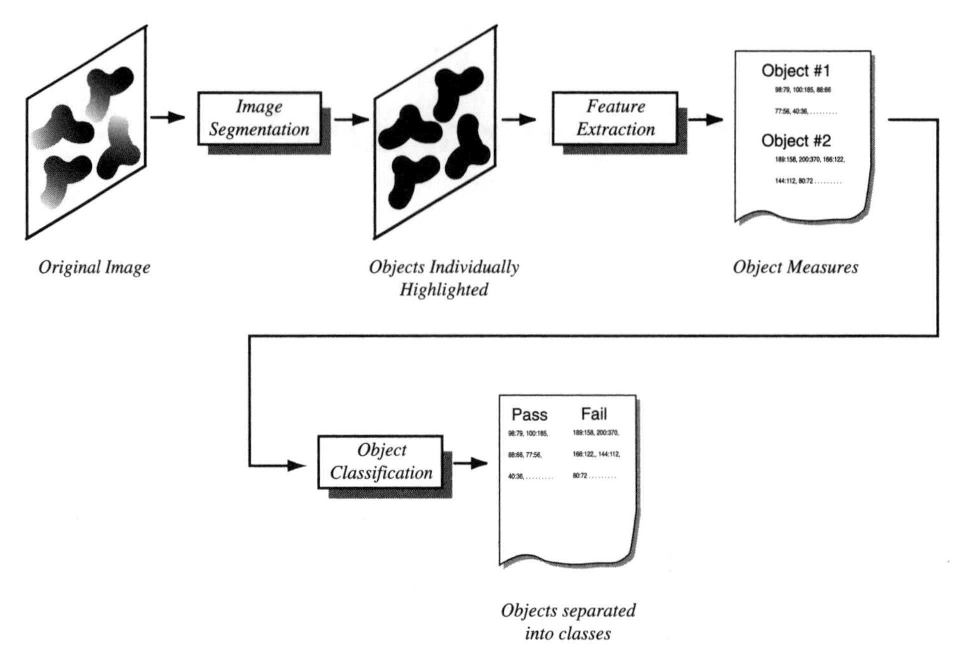

Figure 5.1 *The flow diagram for image analysis operations.*

Image Segmentation

The first step in any image analysis endeavor is to simplify the image, reducing it to its basic component elements, or *objects*. This is the domain of *image segmentation* operations. A segmentation operation is any operation that highlights, or in some way isolates, individual objects within an image. The goal of segmentation operations is to simplify the image without discarding important image features. The definition of "important image features" generally depends on an application's particular requirements.

We will break the image segmentation process into three stages. The first is *image preprocessing*. In this stage, the image is visually improved to make it as free of zero-information-carrying clutter as possible. The second stage is *initial object discrimination*, where objects are grossly separated into groups with similar attributes. The third stage is *object boundary cleanup*, where object boundaries are reduced to single-pixel widths. In this final stage, noise clutter and other artifacts are removed from the image. Figure 5.2 illustrates the image segmentation process flow.

The techniques to implement the first and second stages of the image segmentation process have been discussed in Chapter 4. These techniques generally come from the image-enhancement class of digital image processing operations. In the following sections, we revisit these operations with the goals of image analysis in mind. Then, we explore the techniques of image morphological processing that are frequently used for stage-three boundary cleanup operations.

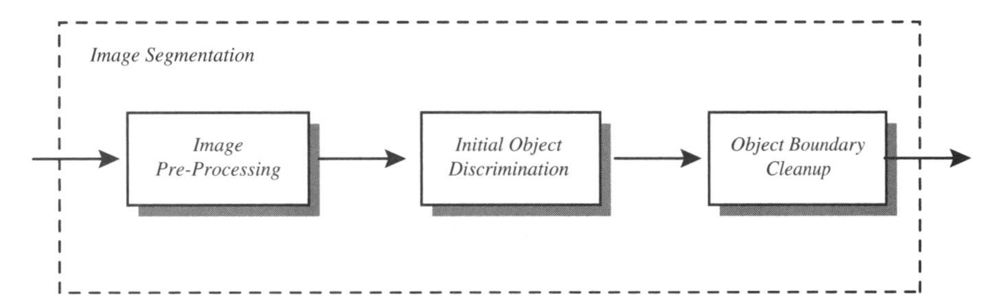

Figure 5.2 *The image segmentation process.*

Image Preprocessing

The goal of the first image segmentation stage is to remove distracting and useless information from the image. It is important, however, to make sure that this process does not remove significant information about the objects being analyzed. Preprocessing operations must not degrade the image in ways that will interfere with the overall image analysis goals of the application.

In Chapter 4, we covered several image enhancement operations that can be appropriate preprocessing operations for many image analysis applications. In particular, contrast balancing operations, using point processes, can improve the distribution of gray levels within an image. These operations are usually applied based on the appearance of the image histogram. Further, the background subtraction operation is very useful in removing illumination nonuniformities and static background clutter that have no value in the image analysis process. This subtraction operation is shown in Figure 5.3. Also, spatial filtering and other group processes can be used to reduce image noise prior to further segmentation operations.

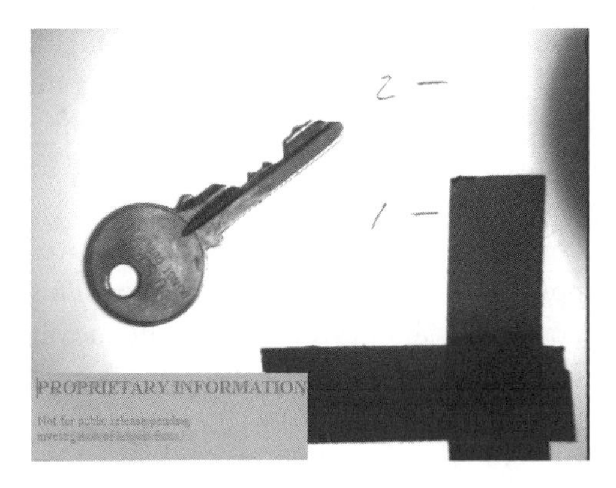

Figure 5.3a *Original image of key with background clutter.*

Figure 5.3b *Background image.*

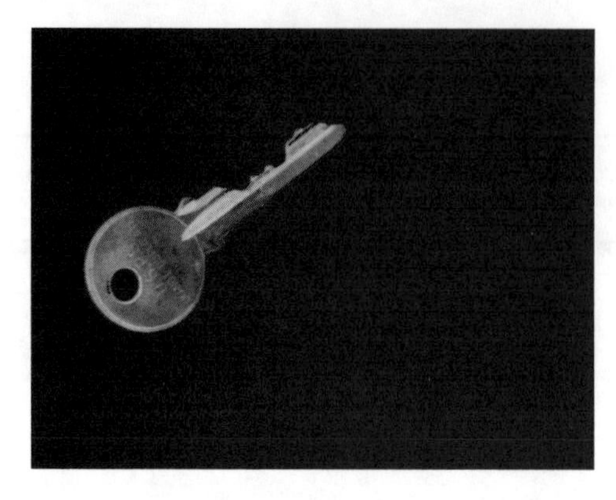

Figure 5.3c *Difference of original and background image yielding an image of only the key.*

Basically, any image enhancement operation can be used in image preprocessing as long as we keep in mind the requirements of the subsequent image analysis tasks. Choosing appropriate preprocessing operations can reduce image degradations, leading to better object detection and improved image analysis performance.

Initial Object Discrimination

The second stage of the image segmentation process separates image objects into rough groups with like characteristics. These characteristics are defined by the image analysis goals of the particular application. As in the preprocessing activity, it is important not to remove significant information about the objects being analyzed. Otherwise, subsequent processing may be impaired.

The image enhancement operations covered in Chapter 4 are also relevant to object discrimination operations. Binary contrast enhancement, brightness slicing, and all of the edge enhancement operations are of the greatest use. Each of these operations works to isolate objects in an image, either by highlighting similar objects with a common brightness or by accentuating their boundaries (edges). Figure 5.4 illustrates how these three operations can provide a simplified resulting image. Of course, if inappropriately applied, important object information can end up being degraded, or worse, entirely discarded.

The results of preprocessing and initial object discrimination are usually objects with ragged edges, overlapping portions, and noise clutter. This object image debris can disrupt the accuracy and abilities of subsequent measurement operations. Therefore, we need to implement the third stage of image segmentation—object boundary cleanup. An important group of image analysis operations, called *image morphological operations*, are often called upon for this task.

Figure 5.4a *Original image of printed text on a semiconductor package.*

Figure 5.4b *Binary contrast enhanced image.*

Figure 5.4c *Sobel edge-enhanced image.*

Binary Morphological Processing

Image morphology pertains to the study of the structure of objects within an image. Following image preprocessing and initial object discrimination, morphological operations work to clarify the underlying structure of objects. This is done by further simplifying object boundaries to their most rudimentary single-pixel-wide outlines or skeletons. These outline and skeletal forms yield an object's most primitive essence. Subsequent measurement operations can then more easily quantify object shape measures for use in classification operations.

There are two forms of image morphological processing—binary and gray-scale. *Binary morphological processes* operate upon binary images, such as those created using a binary contrast enhancement operation. The binary form of morphological processing establishes the premises and methodology of image morphological processing. *Gray-scale morphological processes* are a continuation of the binary techniques and include operations that directly process gray-scale images. First, we will focus on binary morphological processes.

Binary image morphological processes work much like the spatial convolution group process discussed in Chapter 4. The spatial convolution process computes a resulting pixel's brightness value based on the spatial frequency activity of neighboring pixels. Morphological processes, on the other hand, logically combine pixel brightnesses with a structuring element, looking for specific patterns.

Instead of multiplying the pixel brightnesses by weights and summing the results, like the spatial convolution process, morphological processes use set theory operations, such as intersection (AND), union (OR), and complement (NOT), to combine the pixels logically into a resulting pixel value. Among other things, this means that morphological processes are considered to be nonlinear group processes. This is because the process does not involve the summation of "elements" multiplied by "constant" values—as in the definition of a linear process (see Chapter 4).

In the case of binary image morphological processing, input images are assumed to be composed of pixels that have one of two brightness values, either black (0) or white (255). The resulting image will also be of a binary form. When discussing the binary morphological process, we refer to pixels as having logical values rather than brightness values. A brightness of black (0) is called an "off" or "0-state" pixel, whereas a white pixel (255) is an "on" or "1-state" pixel. In this way, a pixel is considered either off or on, which is black or white, respectively.

Like spatial convolution, the morphological process moves across the input image, pixel by pixel, placing resulting pixels in the output image. At each input pixel location, the pixel and its neighbors are logically compared against a *structuring element*, or *morphological mask*, to determine the output pixel's logical value. A resulting output pixel of logical value 0 appears as black (0), while a pixel of logical value 1 appears as white (255).

The structuring element is analogous to the convolution mask used in spatial convolution. Instead of weighting values, though, it is an array of logical values. The structuring element is generally composed of square dimensions of size 3 × 3, 5 × 5, and sometimes greater, depending upon the application. Each logical value can take on the value of 0 or 1 (off or on), or a third state of X which is the "don't-care" state.

The mechanics of the binary morphological process are as straightforward as the spatial convolution process. For the 3 × 3 structuring element case, nine logical values are defined and labeled, as seen below:

$$\begin{matrix} X_3 & X_2 & X_1 \\ X_4 & X & X_0 \\ X_5 & X_6 & X_7 \end{matrix}$$

Every pixel in the input image is evaluated with its eight neighbors to produce a resulting output pixel value. We can think of the evaluation as placing the mask over an input pixel. The input pixel and its eight neighbors are logically compared to the values of the mask. Where a don't-care state exists in the mask, no comparison is made, because the mask doesn't care what value the corresponding input pixel is. In the case where all nine input pixels (or fewer, if don't-care states are present) are identical to their respective mask values, the resulting pixel value is set to a predefined logical value (either 0 or 1). Where one or more input pixels do not match their respective mask values, the resulting value is set to the opposite state. This process repeatedly occurs, pixel by pixel, for each pixel in the input image. A simplified equation for the binary morphological process is

$$
\begin{aligned}
\mathrm{O}(x,y) = 0 \text{ or } 1 \text{ (predefined)} \quad \text{if } \; & X = I(x,y) \text{ AND} \\
& X_0 = I(x+1,y) \text{ AND} \\
& X_1 = I(x+1,y-1) \text{ AND} \\
& X_2 = I(x,y-1) \text{ AND} \\
& X_3 = I(x-1,y-1) \text{ AND} \\
& X_4 = I(x-1,y) \text{ AND} \\
& X_5 = I(x-1,y+1) \text{ AND} \\
& X_6 = I(x,y+1) \text{ AND} \\
& X_7 = I(x+1,y+1)
\end{aligned}
$$

otherwise,

$\mathrm{O}(x,y) =$ *opposite state*

where each input and output pixel value is either logical value 0 (off) or 1 (on), and it is implied that every input pixel is processed through the equation to create a corresponding output pixel value.

This generalized implementation of the binary morphological process is commonly referred to as the *hit or miss transform*. When the mask values match their respective input pixel values, we call the evaluation a "hit." Otherwise, it is a "miss." The hit or miss transform, as we will see, provides a convenient way to define numerous morphological operations. Figure 5.5 illustrates the binary morphological process.

Erosion and Dilation

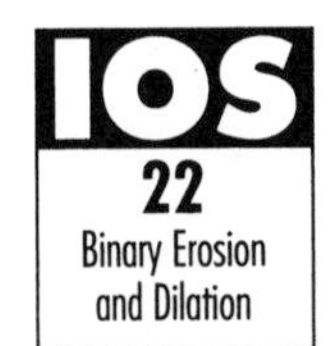

The two most fundamental morphological operations are *erosion* and *dilation*. The erosion operation uniformly reduces the size of objects in relation to their background. The dilation operation—the inverse of the erosion operation—uniformly expands the size of objects. Erosion and dilation operations are used to eliminate small-image object features, such as noise spikes and ragged edges. Various forms of these operations provide the basis for many additional operations.

First, let's define an "object" in a binary image as having white (1) pixels, and the "background" as having black (0) pixels. With this convention, the generalized erosion mask is as follows:

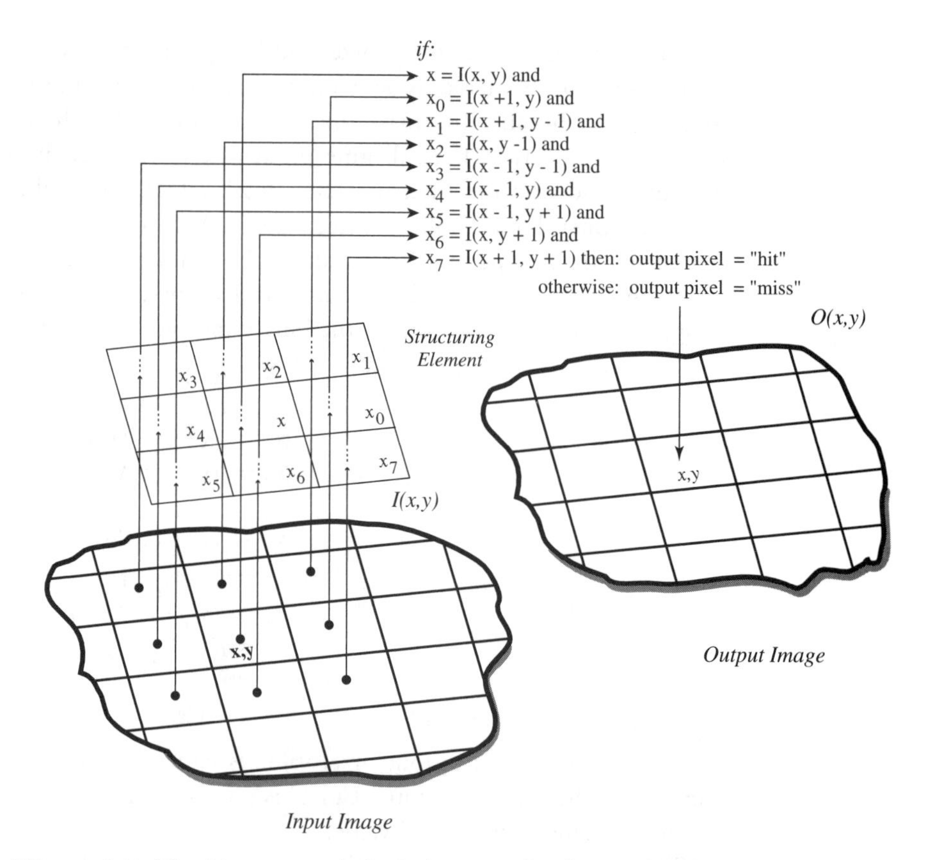

Figure 5.5 *The binary morphological process for the pixel at location* I*(*x,y*), creating the output pixel at location* O*(*x,y*).*

$$\begin{matrix} 1 & 1 & 1 \\ 1 & 1 & 1 \\ 1 & 1 & 1 \end{matrix}$$

where $O(x,y) = 1$ (white) for a hit

$= 0$ (black) for a miss

This mask has the effect of removing a single pixel from the perimeter of a white object. Figure 5.6 illustrates an eroded binary image.

We can see how the erosion operation works by looking at the effect of its mask as it passes over image regions having various black and white pixel combinations. There are four distinct cases to consider. In each case, the morphological process compares each pixel's brightness against its corresponding mask value. When all nine pixels match their respective mask values, a hit occurs and the out-

Figure 5.6a *Original binary "text" image*

Figure 5.6b *Eroded binary image showing the shrinkage of white objects.*

put pixel value is set to 1 (white). When at least one pixel does not match its mask value, a miss occurs and the output pixel value is set to 0 (black).

The four input pixel cases are illustrated in Figure 5.7 and are described as follows:

Input pixel = 1, neighbors = 1 → hit; $O(x,y) = 1$
Input pixel = 0, neighbors = 0 → miss; $O(x,y) = 0$
Input pixel = 0, neighbors = mix of 0 and 1 → miss; $O(x,y) = 0$
Input pixel = 1, neighbors = mix of 0 and 1 → miss; $O(x,y) = 0$

In the first case, the input pixel has a value of 1 and its eight neighbors in the group are also equal to 1. This is the hit case, and the output pixel value becomes 1. In the second case, the input pixel has a value of 0 and its eight neighbors are also equal to 0. This is a miss case, so the output pixel value is set

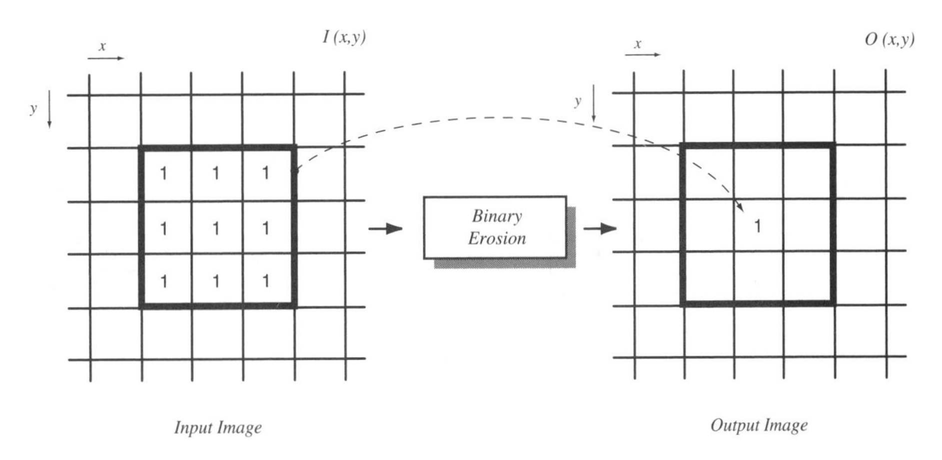

Figure 5.7a *The four input pixel cases of binary erosion. Case 1—input pixel equals 1, neighboring pixels equal 1.*

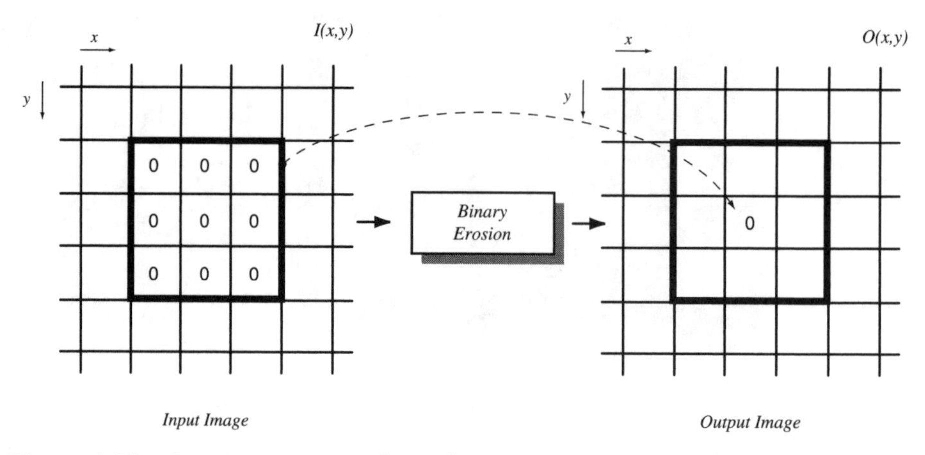

Figure 5.7b *Case 2—input pixel equals 0, neighboring pixels equal 0.*

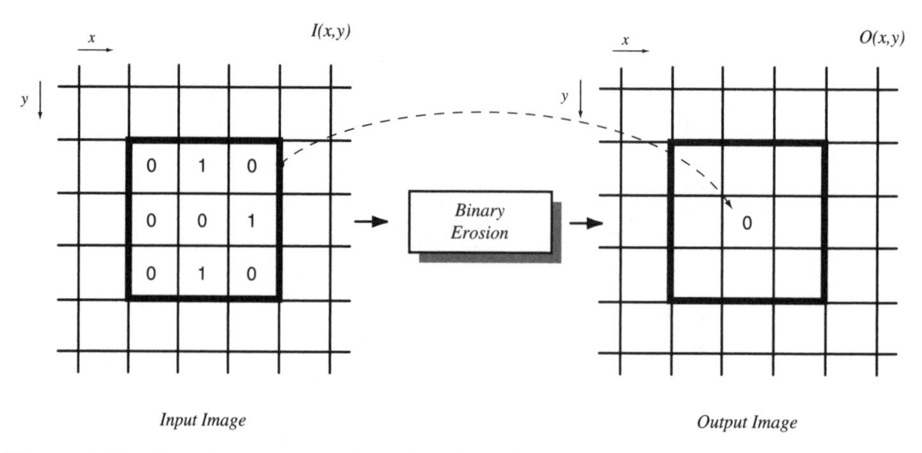

Figure 5.7c *Case 3—input pixel equals 0, neighboring pixels equal a mix of 0s and 1s.*

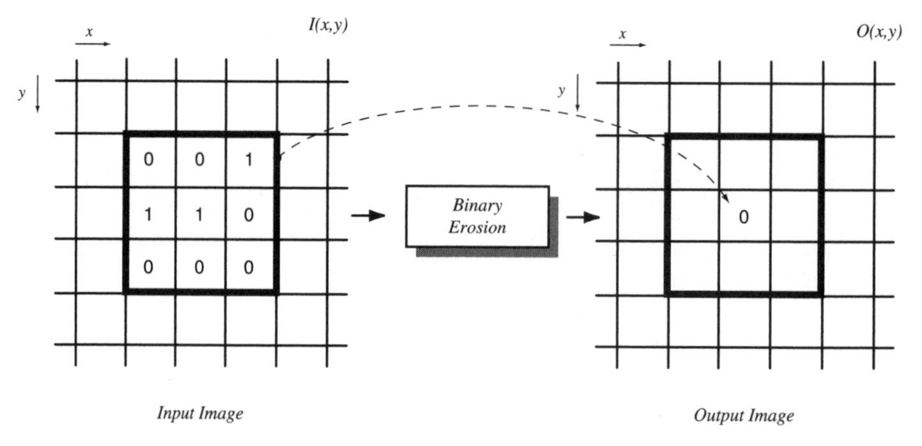

Figure 5.7d *Case 4—input pixel equals 1, neighboring pixels equal a mix of 0s and 1s.*

to 0. The third case is where the input pixel value is equal to 0 and its eight neighbors are a mix of 0 and 1 values. This is also a miss case, so the output pixel value is set to 0. In all three of these cases, the output pixel value is the same as the input pixel value. The fourth case is different, though. In the fourth case, the input pixel value is equal to 1 and its eight neighbors are a mix of 0 and 1 values. This is also a miss case, so the output pixel value is set to 0. In this fourth case, however, the output pixel value is changed from the input pixel's original value of 1 to a value of 0.

In evaluating these cases, we can see that in constant regions, where all pixel values in the input pixel group are equal (all 0s or all 1s), the output pixel value is unchanged from the input pixel value. Also, whenever the input pixel value is 0 (black), regardless of its neighboring pixels, the output value is also unchanged from the input pixel value. But, when the input pixel is 1 (white) and at least one neighboring pixel is 0 (black), the output pixel changes to 0 (black). The white input pixel erodes away from the white object and becomes part of the black background (remember, white pixels were defined as "objects" and black pixels as "background"). The erosion of white pixels occurs only in neighborhoods where a mix of black and white pixels exists in the input image. Essentially, if any pixel value in the input pixel group is equal to 0, the output pixel value is set to 0; otherwise, the output value is set to 1.

When we perform an erosion operation on a binary image, the white objects shrink in size. Very small features, one pixel in size, will disappear entirely. If we continue to apply the erosion operation, white objects will continue to shrink and will ultimately disappear entirely. The erosion operation is useful in separating objects that are touching each other. This is because often each object shrinks independently and finally pulls away from other objects.

Dilation is the inverse operation of erosion. The generalized dilation mask is as follows:

$$\begin{matrix} 0 & 0 & 0 \\ 0 & 0 & 0 \\ 0 & 0 & 0 \end{matrix}$$

$$\text{where } O(x,y) = 0 \text{ (black) for a hit}$$
$$= 1 \text{ (white) for a miss}$$

This mask has the effect of adding a single pixel to the perimeter of a white object. Figure 5.8 illustrates a dilated binary image.

Like the erosion operation, the dilation operation has four distinct input pixel cases to consider. These cases are described as follows:

Input pixel = 0, neighbors = 0 → hit; $O(x,y) = 0$
Input pixel = 1, neighbors = 1 → miss; $O(x,y) = 1$
Input pixel = 1, neighbors = mix of 0 and 1 → miss; $O(x,y) = 1$
Input pixel = 0, neighbors = mix of 0 and 1 → miss; $O(x,y) = 1$

Figure 5.8a Original binary "text" image.

Figure 5.8b Dilated binary image showing the expansion of white objects.

Again, as in the erosion operation, neighborhoods of constant valued pixels, all either 0 or 1, result in an output pixel value that is unchanged from the input pixel value. The difference is in the other two cases—they are the reverse of the erosion operation. When the input pixel has a value of 1, with neighbors that are a mix of 0s and 1s, the output pixel is set to 1, which is unchanged from the input pixel value. When the input pixel has a value of 0, with at least one neighbor that is 1, the output pixel changes from the input value of 0 to a value of 1. The black pixel dilates to white, increasing the size of the white object and reducing the black background. The dilation of black pixels occurs only in neighborhoods where a mix of black and white pixels exists in the input image. Essentially, if any pixel value in the input pixel group is equal to 1, the output pixel value is set to 1; otherwise, the output value is set to 0.

When we perform a dilation operation on a binary image, the white objects grow in size. Very small features will enlarge, exaggerating their shape. If we continue to apply the dilation operation, white objects will continue to expand and will ultimately fill the image frame with white. This operation is useful in combining objects that are divided because of clutter and debris in an image. This type of operation can be used when the analysis goal is to recreate the general shapes of objects in a scene.

Figure 5.9 illustrates the effects of erosion and dilation operations upon a one-dimensional profile of pixels with binary values. Where white objects are surrounded by black background, the erosion operation reduces the width of the object. The dilation operation does just the opposite: It expands the white object.

The effects of erosion and dilation exactly reverse when an image has objects that are black (0) and a background that is white (1). For such a reversed image, the erosion operation, as we have defined it, becomes the dilation operation, and the dilation operation becomes the erosion operation.

Other masks can be applied to binary erosion and dilation. Instead of the generalized masks defined above, hit patterns that include only a few of the nine input

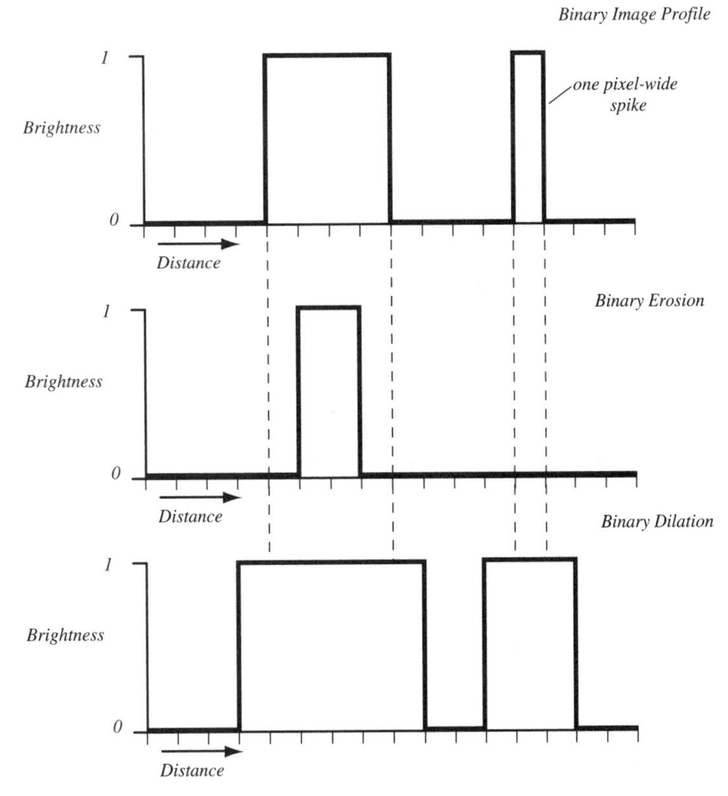

Figure 5.9 *One-dimensional effect of the binary erosion and dilation operations on a binary profile.*

group pixels can be defined, leaving the others as don't-care (*X*) values. In these cases, only the pixels in the mask that are 0s or 1s will be used in the hit or miss evaluation. For instance, the masks for diagonal erosion and dilation are as follows:

$$\begin{matrix} X & X & 1 \\ X & 1 & X \\ 1 & X & X \end{matrix}$$

where $O(x,y) = 1$ (white) for a hit
$= 0$ (black) for a miss

Diagonal Erosion

$$\begin{matrix} X & X & 0 \\ X & 0 & X \\ 0 & X & X \end{matrix}$$

where $O(x,y) = 0$ (black) for a hit
$= 1$ (white) for a miss

Diagonal Dilation

These masks erode and dilate only in the diagonal direction. Their effect is to shrink and expand features oriented diagonally. Masks can also be created for noise spike, and vertical and horizontal erosion and dilation operations.

Opening and Closing

Erosion and dilation operations are considered the primary morphological operations. The operations of *opening* and *closing* are secondary operations and are implemented using erosion and dilation operations.

The opening operation is simply an erosion operation followed by a dilation operation. Like erosion, the effect is to remove single-pixel object anomalies such as small spurs and single-pixel noise spikes. As a result, object contours are smoothed. Unlike using just the erosion operation, opening tends to maintain the original shapes and sizes of objects in the image. This is because, although the erosion operation shrinks the objects by one pixel, the dilation operation expands them back to their original size. Figure 5.10 shows an image opening operation.

The opening operation, like erosion and dilation, can be applied numerous times to achieve the necessary effect. When it is applied multiple times, it effects multiple applications of the erosion operation first, followed by the same number of dilation operations. Multiple opening operations remove larger object anomalies.

The closing operation is the opposite of the opening operation. It is a dilation operation followed by an erosion operation. Closing fills in single-pixel object anomalies, such as small holes and gaps. As with opening operations, closing smooths object contours. Also, the closing operation maintains shapes and sizes of objects. Figure 5.11 shows an image closing operation.

Figure 5.10a *Original binary "text" image.*

Figure 5.10b *Opened binary image showing original sized objects with small spurs and noise spikes removed.*

Figure 5.11a *Original binary "text" image.*

Figure 5.11b *Closed binary image showing original sized objects with small gaps and holes removed.*

The closing operation can also be applied multiple times in a manner similar to the opening operation. In this way, larger object anomalies can be filled.

Outlining

The erosion operation can be used to create outlines of objects in an image. The *outlining* operation is similar to the edge enhancement operations discussed in Chapter 4. Generally, the outlining operation produces one-pixel-wide outlines and tends to be more immune to image noise than most edge enhancement operations.

The outlining operation is implemented by first applying an erosion operation to an image. As described earlier, the objects in the image will shrink by one pixel around their perimeters. We then use a dual-image point process to subtract the eroded image from the original image. The result is an image showing only the outlines of the objects, as seen in Figure 5.12. Figure 5.13 illustrates the outlining operation as applied to a one-dimensional profile of binary-valued pixels.

We can use the outline of an object in subsequent feature extraction operations to measure object size, shape, and orientation. These measurements can then be used in object classification operations to determine whether the object meets particular desired criteria.

Skeletonization

Another useful application of erosion operations is *object skeletonization*. This operation uses numerous different erosion masks, oriented in various directions, to whittle away at the objects in an image. Ultimately, the skeleton of each object is all that remains in the image.

Figure 5.12a Original binary "text" image.

Figure 5.12b Eroded binary image.

Figure 5.12c Outline binary image created by subtracting the eroded image from the original image.

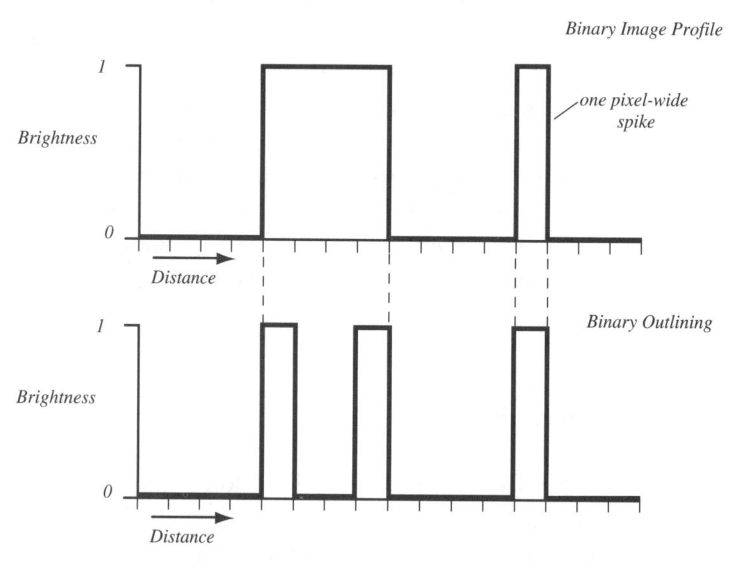

Figure 5.13 *One-dimensional effect of the binary outlining operation on a binary profile.*

Figure 5.14a *Original binary "text" image.*

Figure 5.14b *Cleaned up "text" image using opening and closing operations.*

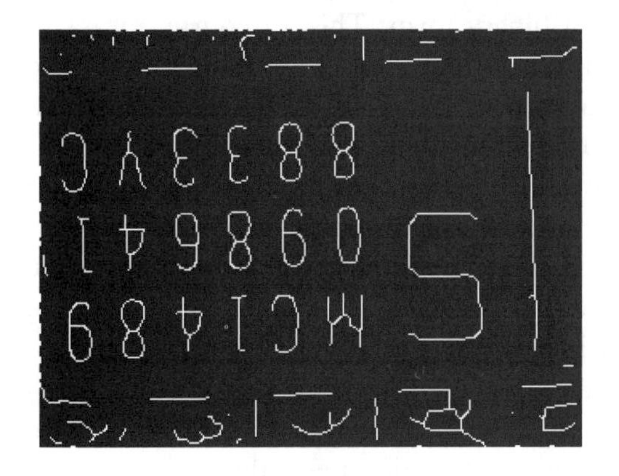

Figure 5.14c *Binary skeletonized image showing text as stick-figures.*

The classical analogy of the creation of an object skeleton begins by setting fire to the entire perimeter of the object, all at once. As the fire burns inward, toward the center of the object, it meets up with adjacent fires and they extinguish themselves along a particular path. This path is the skeleton of the object, as seen in Figure 5.14.

The skeleton of an object is an important measure because it represents the object's underlying structure. We can measure and compare this structure to other known structures to make decisions about the object's characteristics.

Gray-Scale Morphological Processing

The techniques of binary morphological processing can be extended beyond the world of binary images to gray-scale images. Although the implementation is a little different for *gray-scale morphological processes*, many analogous operations exist, corresponding to the binary processes discussed earlier. The gray-scale versions of these operations can be used in cases where we don't intend to first perform an initial object discrimination operation like binary contrast enhancement. These cases include the processing of images where these initial operations do not work well or, perhaps, significantly degrade an image. This occurs when gray levels vary widely within the object or background portions of an image.

Gray-scale morphological processes do not generally produce images with single-pixel-wide object boundaries or skeletons. They are, therefore, usually followed by some binarization operation (like a binary contrast enhancement), or perhaps even a binary morphological process. This way, we can accomplish the final image segmentation goal of yielding an object's underlying structure—where object boundaries are reduced to their most rudimentary single-pixel-wide outlines or skeletons. As in binary morphological processes, subsequent feature-extraction measurement operations can more easily quantify object shape measures for subsequent classification operations.

In gray-scale morphological processing, input and output images are of a gray-scale form, rather than of a binary form. This, of course, means that both input and output pixel values can range from 0 (black) to 255 (white). Also, the way that we evaluate a group of input pixel values to create an output pixel value differs from the evaluation method used in a binary morphological process.

Like spatial convolution and binary morphological processes, the gray-scale morphological process moves across the input image, pixel by pixel, placing resulting pixels in the output image. At each input pixel location, the input pixel and its neighbors are combined using a structuring element (or morphological mask) to determine the output pixel's brightness value.

The structuring element for gray-scale morphological processes is an array of values ranging from –255 to 255. Alternately, a value can be assigned the don't-care state, meaning that it does not participate in the evaluation at all. The structuring element is generally composed of square dimensions, size 3 × 3, 5 × 5, or larger, depending upon the application. The nine values of the structuring element are labeled and arranged just like those for binary morphological processing:

$$\begin{matrix} X_3 & X_2 & X_1 \\ X_4 & X & X_0 \\ X_5 & X_6 & X_7 \end{matrix}$$

The mechanics of the gray-scale morphological process are similar to, but different from, those of binary processing. Every pixel in the input image is evaluated with its eight neighbors to produce a resulting pixel value. Two different evaluation methods can be followed, one for each of the two fundamental morphological operations—gray-scale erosion and gray-scale dilation. We can think of the each evaluation method as placing the mask over an input pixel. For the erosion case, mask values can range from –255 to 0, but will generally be 0. The center mask value and its eight neighbors are each added to the corresponding input pixel and its eight neighbors. Where don't-care states exist in the mask, no addition is made. The output value is determined as the *minimum* value of all nine addends. This process repeats, pixel by pixel, for each pixel in the input image. The equation for the gray-scale erosion process is as follows:

$$\begin{aligned} O(x,y) = \min\{ & X + I(x,y), \\ & X_0 + I(x+1,y), \\ & X_1 + I(x+1,y-1), \\ & X_2 + I(x,y-1), \\ & X_3 + I(x-1,y-1), \\ & X_4 + I(x-1,y), \\ & X_5 + I(x-1,y+1), \\ & X_6 + I(x,y+1), \\ & X_7 + i(x+1,y+1)\} \end{aligned}$$

where it is implied that every input pixel is processed through the equation to create a corresponding output pixel value. The gray-scale erosion morphological process is illustrated in Figure 5.15.

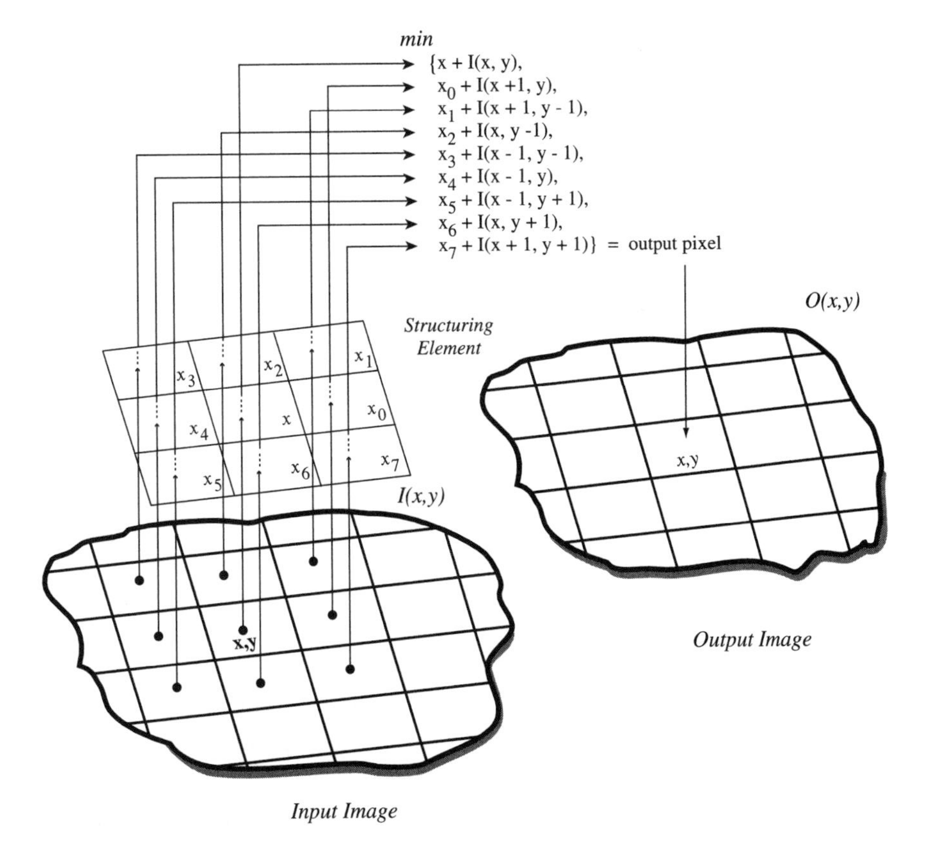

Figure 5.15 *The gray-scale erosion morphological process for the pixel at location I(x,y), creating the output pixel at location O(x,y).*

For the dilation case, mask values can range from 0 to 255, but will generally be 0. The center mask value and its eight neighbors are each added to the corresponding input pixel and its eight neighbors. Where don't-care states exist in the mask, no addition is made. For dilation, the output value is determined as the *maximum* value of all nine addends. This process repeats, pixel by pixel, for each pixel in the input image. The equation for the gray-scale dilation process is as follows:

$$\begin{aligned} O(x,y) = \max\{&X + I(x,y), \\ &X_0 + I(x+1,y), \\ &X_1 + I(x+1,y-1), \\ &X_2 + I(x,y-1), \\ &X_3 + I(x-1,y-1), \\ &X_4 + I(x-1,y), \\ &X_5 + I(x-1,y+1), \\ &X_6 + I(x,y+1), \\ &X_7 + I(x+1,y+1)\} \end{aligned}$$

where it is implied that every input pixel is processed through the equation, creating a corresponding output pixel value. The gray-scale dilation morphological process is illustrated in Figure 5.16.

The generalized implementation of the gray-scale morphological process is very similar to that of its binary counterpart. In fact, by applying a gray-scale morphological structuring element with mask values of 0 to a binary image, identical results are achieved as with the corresponding binary morphological process.

Erosion and Dilation

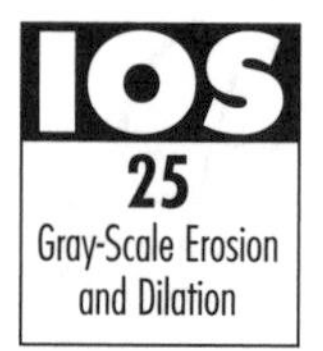

Like the similar binary operations, *gray-scale erosion* and *gray-scale dilation* operations are the two most fundamental gray-scale morphological operations. The erosion operation reduces the size of objects relative to their background. The dilation operation is the inverse operation: it expands the size of objects. Erosion and dilation operations are used to eliminate small image features such as noise spikes and ragged edges. Various forms of these operations provide the basis for many additional operations.

For gray-scale morphological operations, an "object" is composed of pixels with brightnesses that are greater than the brightnesses of "background" pixels. Gray-scale morphological operations work better when a greater number of gray levels separate the object and background brightnesses.

The generalized gray-scale erosion operation mask is as follows:

$$\begin{matrix} 0 & 0 & 0 \\ 0 & 0 & 0 \\ 0 & 0 & 0 \end{matrix}$$

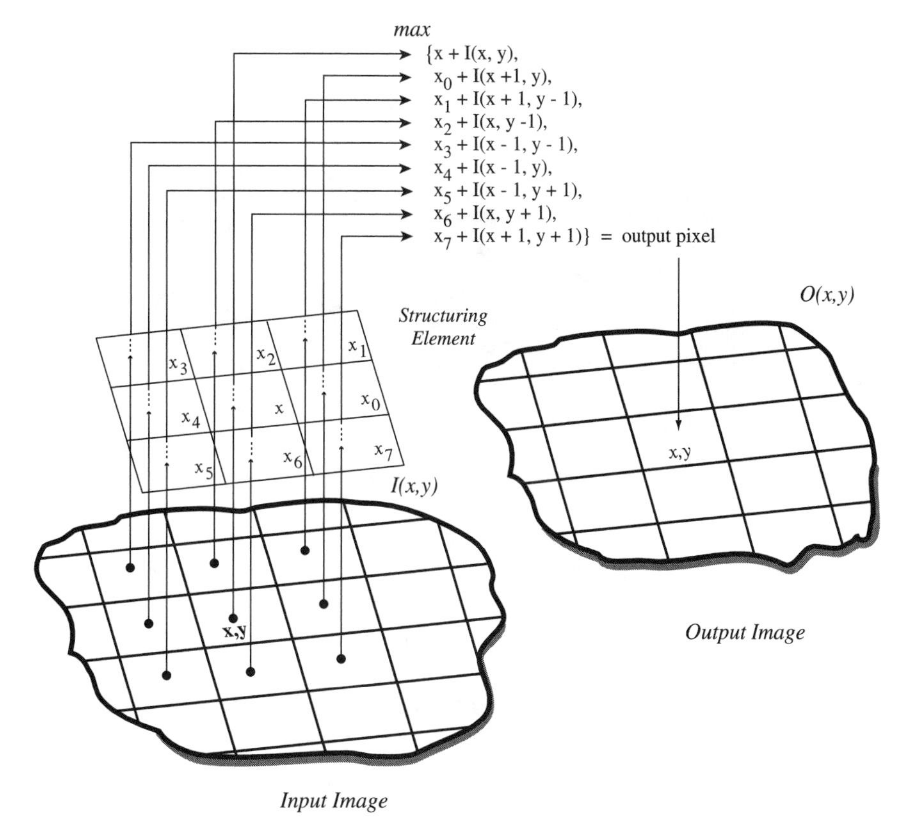

Figure 5.16 *The gray-scale dilation morphological process for the pixel at location* I*(*x,y*), creating the output pixel at location* O*(*x,y*).*

where the gray-scale erosion equation is used to extract the *minimum* value of the input pixel group once combined with their respective mask values.

This erosion mask has the effect of darkening bright objects, and thus making them appear smaller. The overall image brightness is reduced as well. Figure 5.17 illustrates the gray-scale erosion operation.

Looking at how this mask works, we first note that all mask values are equal to 0. This means that the gray-scale erosion equation adds 0 to the input pixel and its eight neighbors. Then, the minimum value of these nine addends becomes the output pixel value. So, the basic erosion operation creates an output pixel that has a brightness that is the smallest brightness of the input pixel and its eight neighbors.

Over regions of an image where the pixel brightnesses are constant, the output pixel value is unchanged from the input pixel value. In any other region, the output pixel value will assume the brightness of the darkest pixel in each input pixel group. Gray-scale erosion makes bright objects shrink in size. Very small bright objects will disappear entirely. When we continue to apply the gray-scale erosion operation, objects will shrink until they are ultimately eliminated.

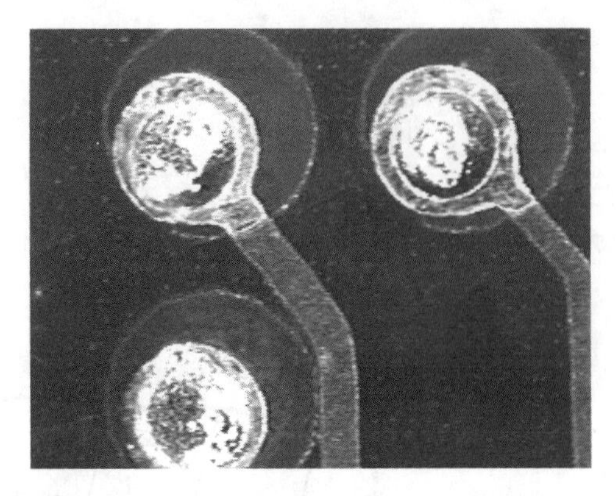

Figure 5.17a *Original image of printed circuit board pads.*

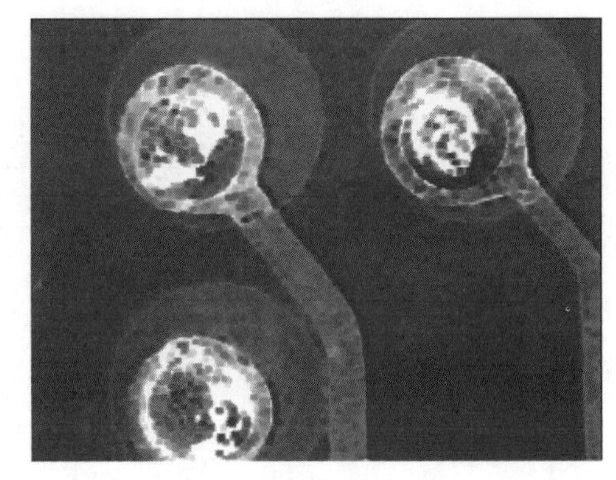

Figure 5.17b *Eroded gray-scale image showing the shrinkage of bright objects.*

By varying the erosion mask values between –255 and 0, we can achieve different effects. For instance, decreasing all values to a lower but equal value will produce a quicker erosion. Gray values will not only reduce to the minimum input pixel value, but will also decrease the additional amount of the mask values. This is because, before the minimum operator, the negative mask values are first added to the input group's pixel values. Other nonuniform mask patterns can provide the ability to erode gray levels based on how well the input group of pixels matches the mask pattern.

The gray-scale dilation operation is very similar to that of erosion. The generalized gray-scale dilation mask is identical to that of gray-scale erosion, as follows:

0	0	0
0	0	0
0	0	0

where the gray-scale dilation equation is used to extract the *maximum* value of the input pixel group once combined with their respective mask values.

This dilation mask has the effect of brightening bright objects, and thus making them appear larger. The overall image brightness also increases as a result. Figure 5.18 illustrates the gray-scale dilation operation.

Like the erosion mask, all values are equal to 0. This means that the gray-scale dilation equation adds 0 to the input pixel and its eight neighbors. Then, the maximum value of these nine addends becomes the output pixel value. So, the basic dilation operation creates an output pixel with a brightness that is the largest brightness of the input pixel and its eight neighbors.

Over regions of an image where the pixel brightnesses are constant, the output pixel value is unchanged from the input pixel value. In any other region, the output pixel value will assume the brightness of the brightest pixel in each input pixel group. Gray-scale dilation expands the size of bright objects. Very small bright objects become larger and their shapes become exaggerated. Repeated application

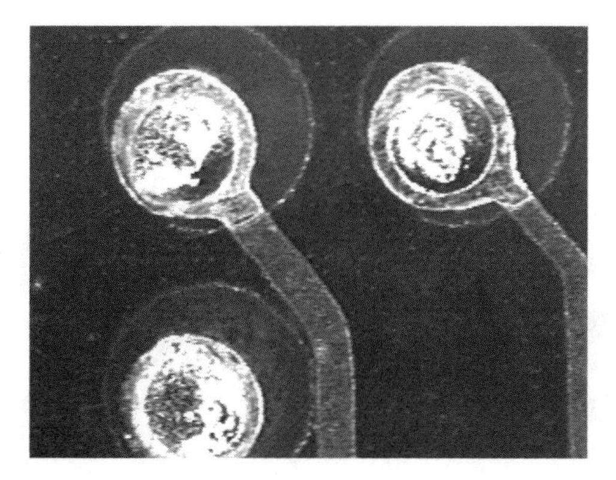

Figure 5.18a *Original "pads" image.*

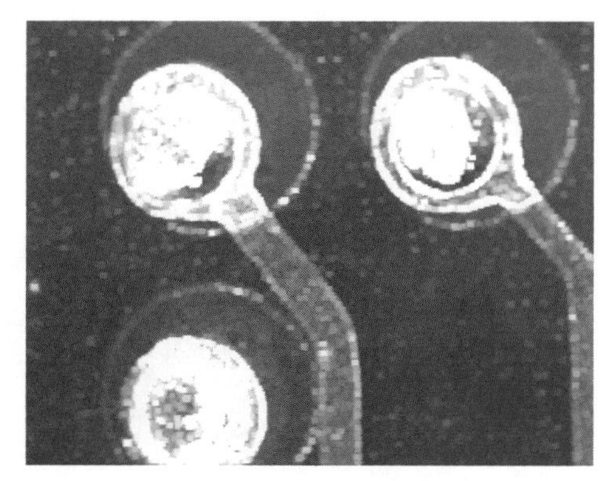

Figure 5.18b *Dilated gray-scale image showing the expansion of bright objects.*

of the gray-scale dilation operation will continue to expand objects until they fill the entire image frame with white.

By varying the dilation mask values between 0 and 255, we can achieve different effects. By increasing all values to a higher but equal value, we can produce a quicker dilation. Gray values will increase to the maximum input group value and will additionally increase by the amount of the mask values. This is because, before the maximum operator, the mask values are first added to the input group's pixel values. Other nonuniform mask patterns can give us the ability to dilate gray levels based on how well the input group of pixels match the mask pattern.

Opening and Closing

The *gray-scale opening* and *gray-scale closing* operations are implemented just like their binary counterparts discussed earlier. The gray-scale opening operation is a gray-scale erosion followed by gray-scale dilation. Like the binary version, gray-scale opening darkens small objects and entirely removes single-pixel objects like noise spikes and small spurs. Objects tend to retain their original shapes and sizes. Figure 5.19 shows the gray-scale opening operation.

The gray-scale opening operation can be applied numerous times to achieve the necessary effect. Like its binary counterpart, when it is applied multiple times, it effects multiple applications of the erosion operation first, followed by the same number of dilation operations. Using multiple gray-scale opening operations removes larger object anomalies.

The gray-scale closing operation is the opposite of the opening operation. It is a gray-scale dilation operation followed by a gray-scale erosion operation. Like the binary version, gray-scale closing brightens small objects and entirely fills in single-pixel objects like small holes and gaps. Objects tend to retain their original shapes and sizes. Figure 5.20 shows a gray-scale closing operation.

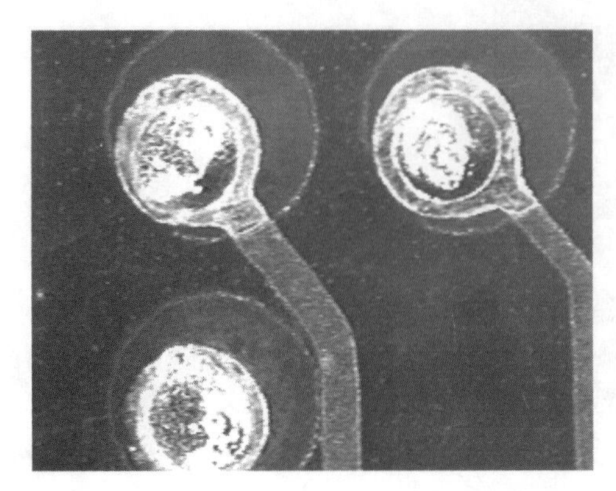

Figure 5.19a *Original "pads" image.*

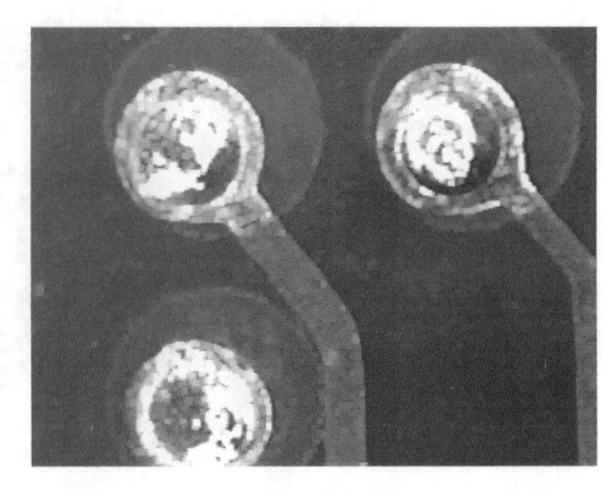

Figure 5.19b *Opened gray-scale image showing original sized objects with small spurs and noise spikes removed.*

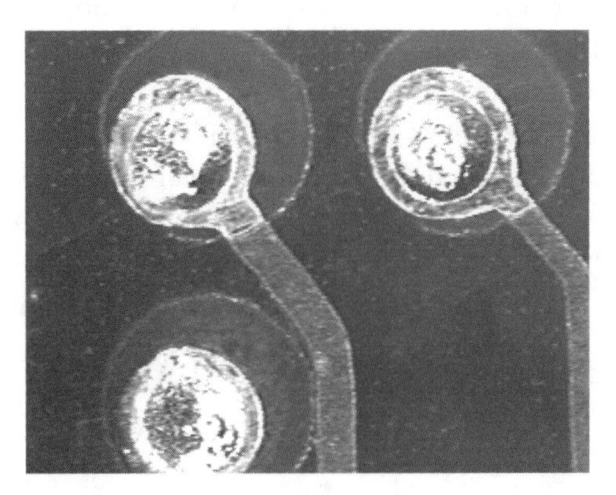

Figure 5.20a *Original "pads" image.*

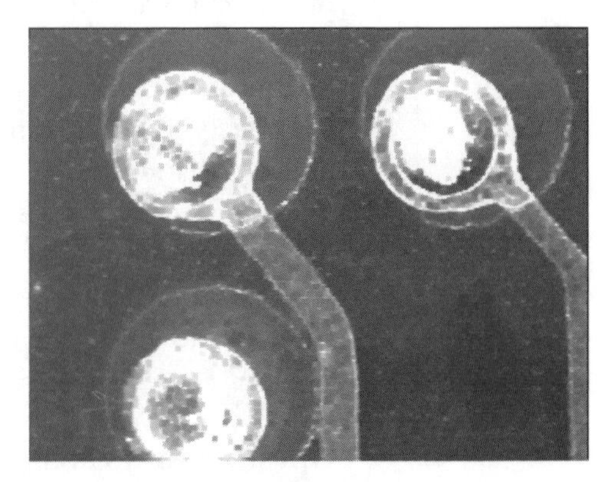

Figure 5.20b *Closed gray-scale image showing original sized objects with small gaps and holes removed.*

The gray-scale closing operation can also be applied multiple times in a manner similar to the opening operation. In this way, larger object anomalies are filled.

Top-Hat and Well Transformations

The *top-hat transformation* (or *peak detector*) and *well transformation* (or *valley detector*) are variants on the gray-scale opening and closing operations. The top-hat operation produces an output image showing only the bright peaks of an image. These peaks are generally small, bright features—like the bright tips of objects—within the image. We can create a top-hat image by first applying an opening operation (erosion followed by dilation) to an image. We then subtract the opened image from the original image using a dual-image point process. The result is an image in which only the bright peaks appear, as shown in Figure 5.21.

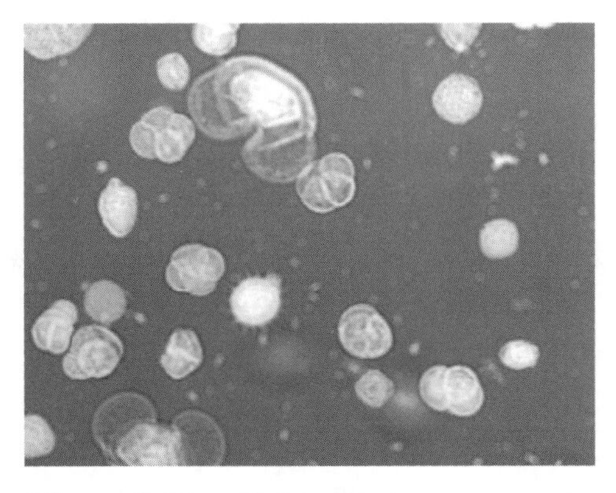

Figure 5.21a *Original microscope specimen image of various pollens. (Image courtesy of Data Translation, Inc.)*

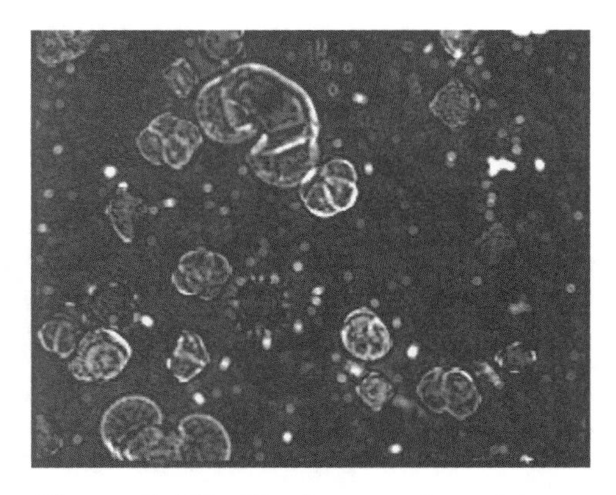

Figure 5.21b *Top-hat transformed image showing the brightest peaks of the objects.*

The effect of the top-hat transformation on a one-dimensional gray-level profile is shown in Figure 5.22. Top-hat images are useful for detecting small features with rising brightnesses.

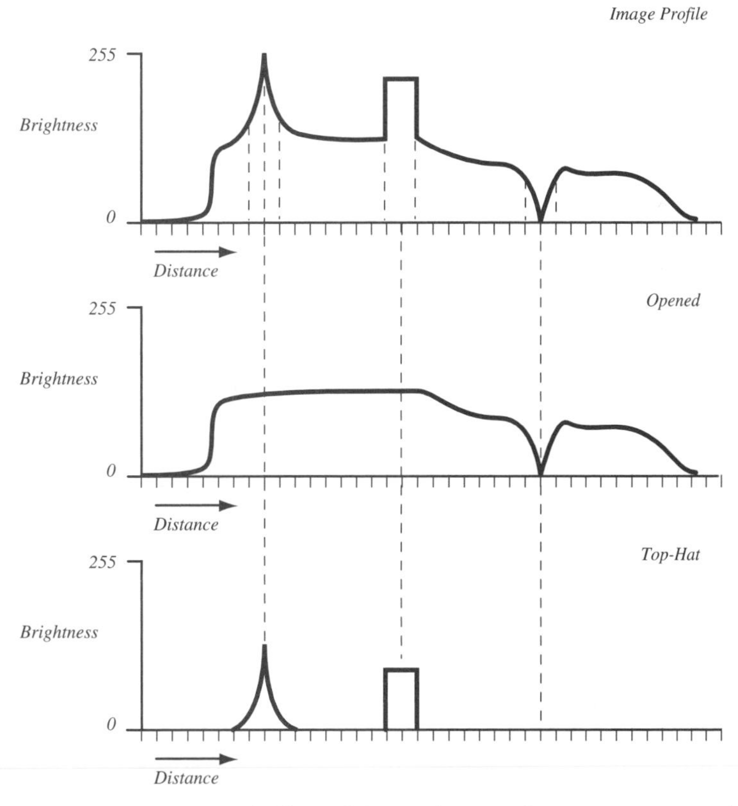

Figure 5.22 *One-dimensional effects of the top-hat transformation on a gray-level profile.*

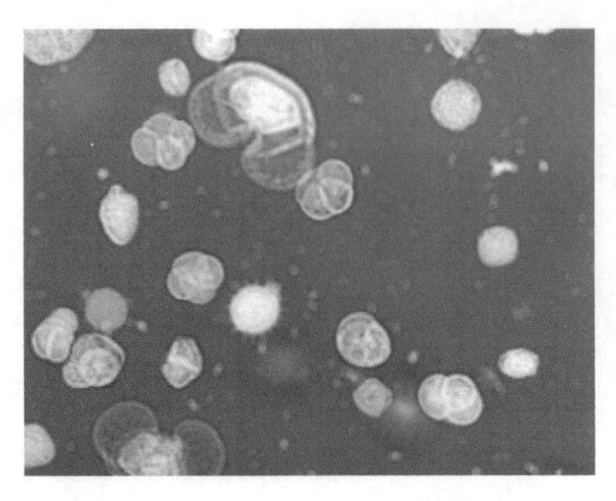

Figure 5.23a Original "pollen" image.

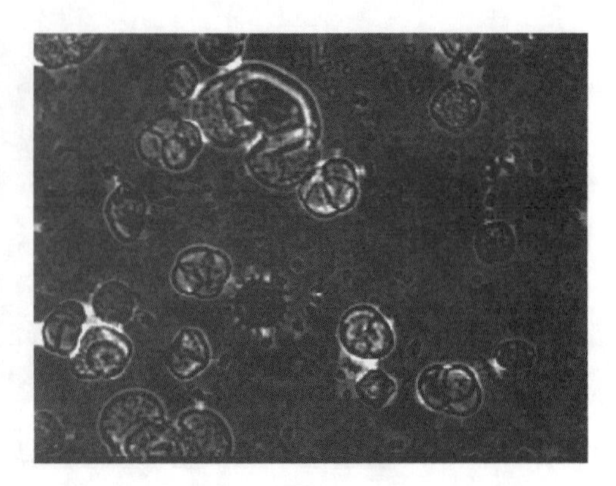

Figure 5.23b Well transformed image showing the darkest valleys of the background.

The well operation is the opposite of the top-hat operation. It produces an output image of only the dark valleys of an image. These valleys are generally small, dark features in the image. A well image is created by first applying a closing operation (dilation followed by erosion) to an image. We then subtract the closed image from the original image using a dual-image point process. The result is an image in which only the dark valleys appear, as shown in Figure 5.23.

The well transformation has the same action as the top-hat transformation, but it detects small features with falling brightnesses.

Morphological Gradient

The *morphological gradient* operation creates an edge-enhanced image of the boundaries of objects in an image, much like the binary outlining operation. The outlines represent the portions of the image where the steepest dark-to-bright and bright-to-dark transitions occur. A morphological gradient image is created by first making a copy of the original image. We then apply an erosion operation to the original image and a dilation operation to the copy of the original image. The eroded image is then subtracted from the dilated image using a dual-image point process. Because the dilated image slightly increases the size of objects and the eroded image slightly decreases the size of objects, their difference highlights the object boundaries. The result is an image showing only the object outlines, as shown in Figure 5.24. The effect of the morphological gradient operation on a one-dimensional gray-level pixel profile is shown in Figure 5.25.

Watershed Edge Detection

The *watershed edge detection* operation is another technique for enhancing edges between objects in an image. The object boundaries created by the watershed

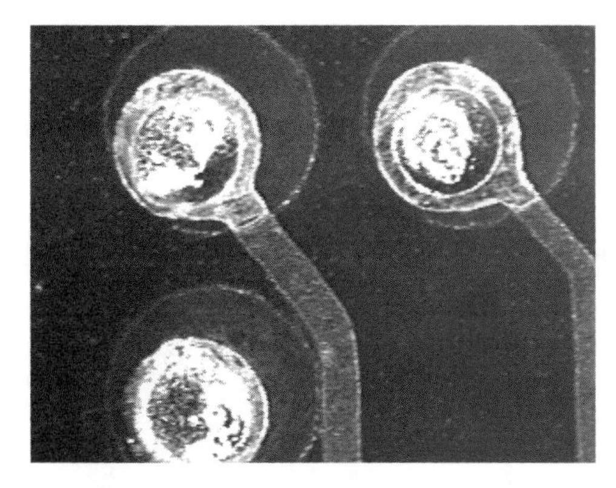

Figure 5.24a *Original "text" image.*

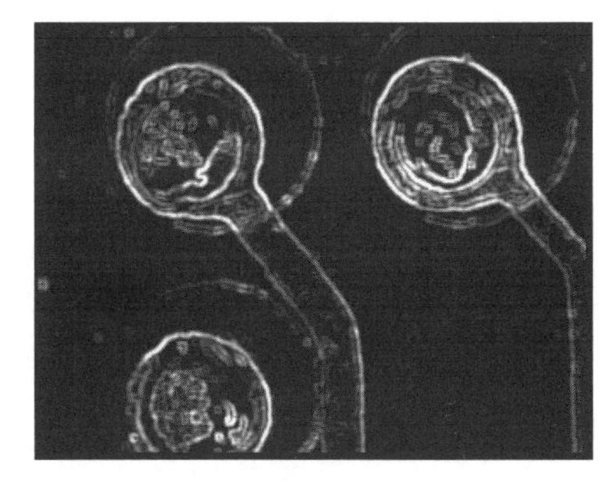

Figure 5.24b *Morphological gradient image showing object outlines.*

technique are especially good in areas of an image where objects are touching, or where few gray levels exist between different objects or an object and its background.

Referring to an earlier discussion in Chapter 3, we can think of an image in three dimensions, where the brightness of each pixel is represented by a height coming out from the page toward the observer. From this perspective, an image looks as shown in Figure 5.26, where bright objects are pixel mounds rising above the background. The question in edge detection is always the same: Where is the actual edge of the object? Is it near the top of the pixel mound or near the bottom? If we are measuring object sizes, these can be very important questions for the sake of accuracy. Also, where objects are close, the mounds may not reach back to the ground (background) level; instead, they may descend slightly and then begin to rise again. These separations between objects can be hard to detect using other object-outlining techniques. The watershed technique, however, can be effective with these in-between conditions.

The watershed edge detection operation can be applied to a gray-scale image or to a morphological gradient image. Applying the operation to a gray-scale image helps separate objects that are close together and hence have poor gray-level distinction between them. We can describe the way the watershed method works by referring to the three-dimensional image representation described above. First, we must complement the image, so that bright objects appear as pixel depressions in the image rather than pixel mounds. If we imagine flooding the image with water, the depressions (objects) initially fill with water, as shown in Figure 5.27. As the water level rises, neighboring objects collect water independently. The water fills the lowest portions of the image first—the water is said to have "shed" to these lowest points, just as it does in nature. As the water level rises, the operation begins to mark pixels where a single object pixel separates two water bodies. Figure 5.28 illustrates this process on a one-dimensional gray-level profile. The marked pixels

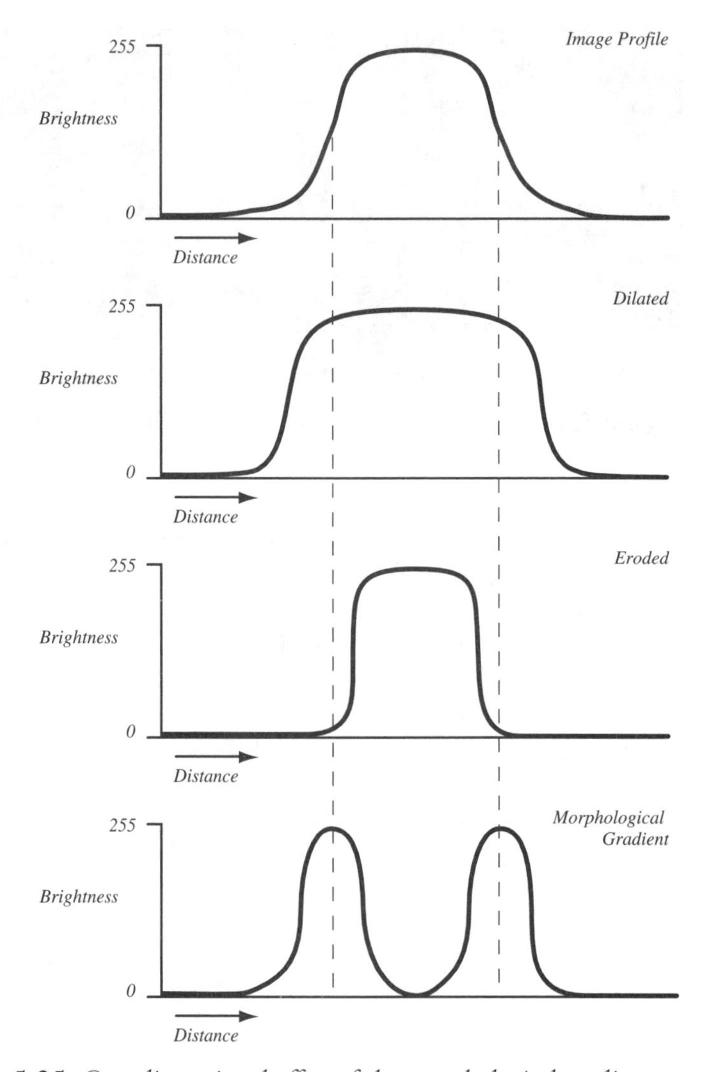

Figure 5.25 *One-dimensional effect of the morphological gradient operation on a gray-level profile.*

are the in-between points separating objects. At completion, the resulting image displays the marked points in the image that separated any two water bodies by a single-pixel-wide strip of dividing pixels, as shown in Figure 5.29. These pixels represent the dividing lines between two neighboring objects. The watershed operation is generally implemented using morphological processes that mark pixels as they detect the joining of water bodies within an input pixel group.

Applying the watershed edge detection operation to a morphological gradient image enhances edges by dividing the gray-level slope of an object at its center level. Instead of operating on an original gray-scale image, this method operates

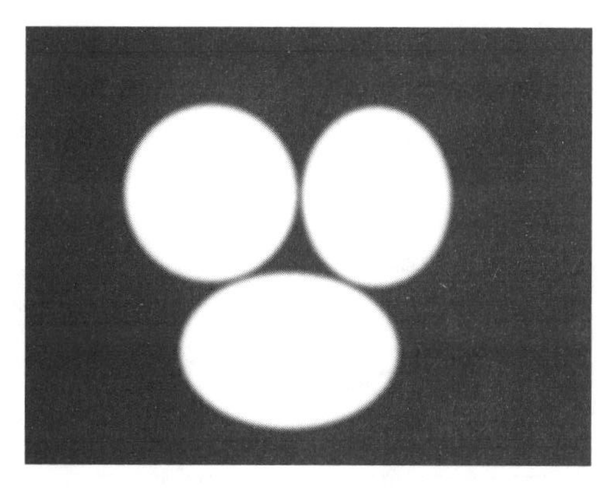

Figure 5.26a *A gray-scale image of three bright objects.*

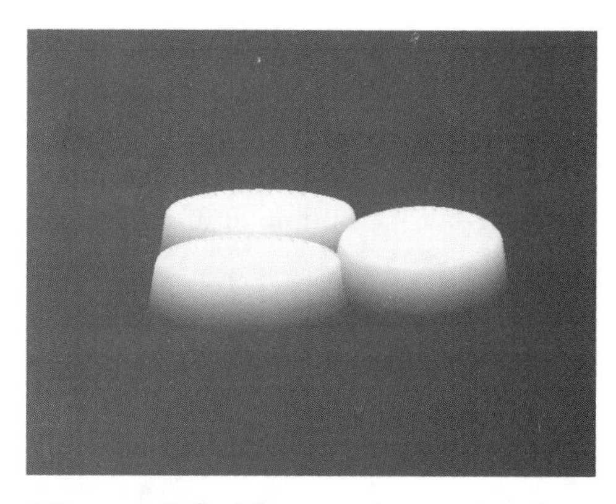

Figure 5.26b *The same image viewed with brightnesses portrayed as pixel heights.*

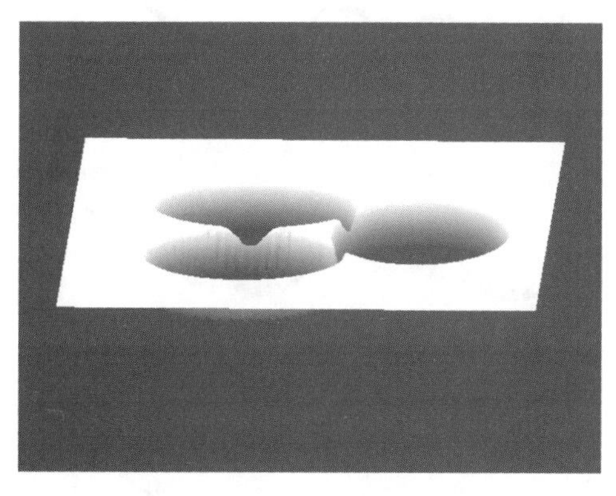

Figure 5.27a *The image is first complemented, making objects appear as depressions.*

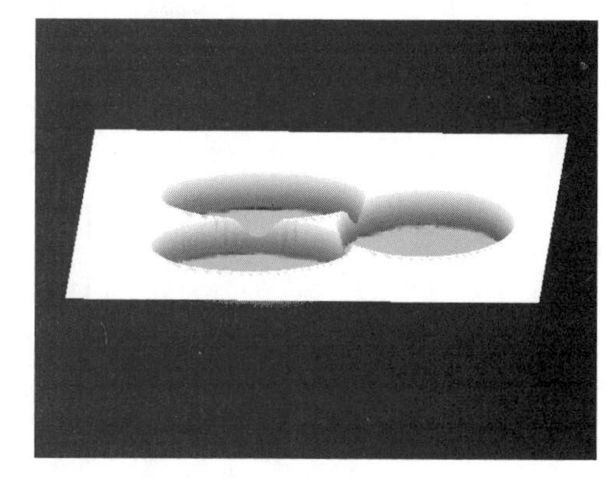

Figure 5.27b *The water begins to the fill the depressed objects.*

Figure 5.27c *As the water level rises higher, neighboring objects each independently collect water until adjoining pools meet.*

on a morphological gradient image that shows object edges as peaks. Each object appears to have a depression in its center bounded by two peaks that were the original object edges. Referring to our previous water analogy, as water fills the image to rising levels, the water bodies finally meet at the peaks of the object edges. The pixels of these peak edge boundaries are marked in the resulting output image. This way, the center of the object edges can be accurately determined. Figure 5.30 illustrates the watershed operation's action on a one-dimensional morphological gradient profile.

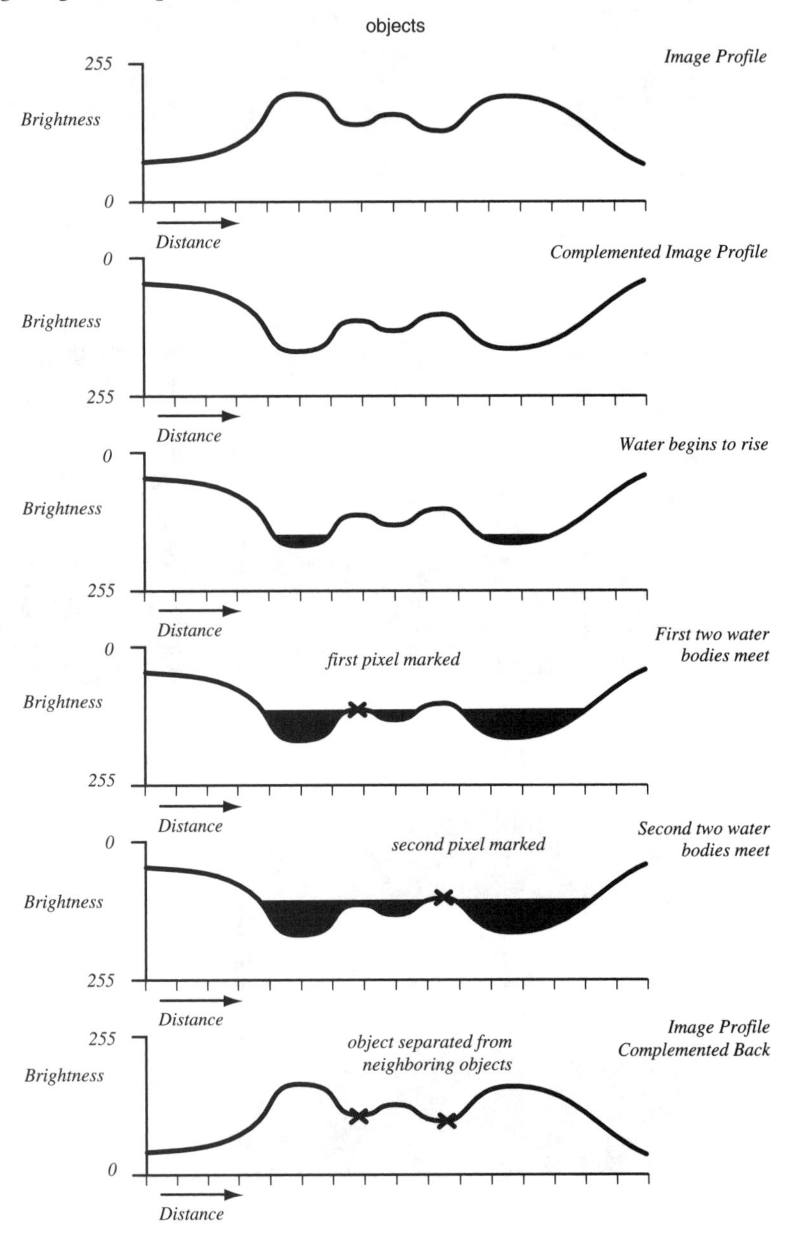

Figure 5.28 *One-dimensional effect of the watershed edge detector on a gray-level profile.*

Figure 5.29a *Original image of three bright objects.*

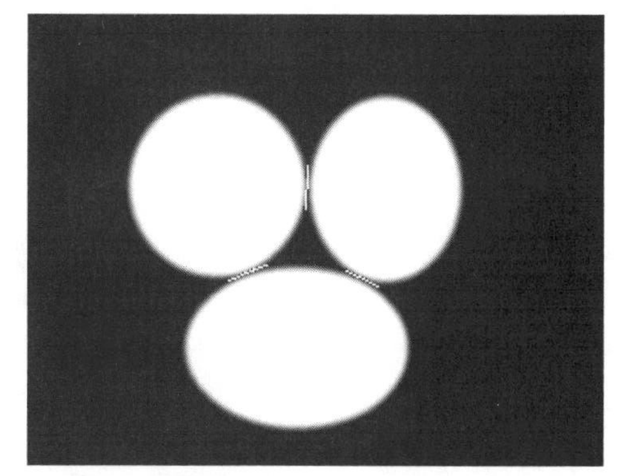

Figure 5.29b *Watershed edge detector image showing object-dividing boundaries.*

Feature Extraction

Once an image has been cleanly segmented into discrete objects of interest, the next step in the image analysis process is to measure the individual features of each object. Many features can be used to describe an object. Generally, the ones that are the simplest to measure and require the fewest in number are the best to use. With these measurements, we can compare the information with known measures to classify an object into one of many categories.

Figure 5.31 shows the outlines of objects overlaid upon the original image. The following discussions dwell primarily on measuring the attributes of an object contained within the bounds of an outline.

Brightness and Color Features

Brightness and color features of an object can be extracted by examining every pixel within the object's boundaries. The histogram of these pixels is a good place to start. The histogram shows the brightness distribution found in the object. For color objects, the red, green, and blue pixel component values of the object can be converted to the hue, saturation, and brightness color space. Then, looking at the histogram of the hue-component image will instantly show the predominant hue of the object. For a color-sorting application, this feature alone can be all that's necessary to classify the difference between two objects, such as red Red Delicious and green Granny Smith apples traveling on a conveyer line.

Statistics of the brightnesses in an object can also be useful measures. The mean brightness, for instance, represents the average brightness of an object. The standard deviation brightness is a measure of how much the object's brightnesses vary from the mean value. The mode brightness is the most common brightness found in the

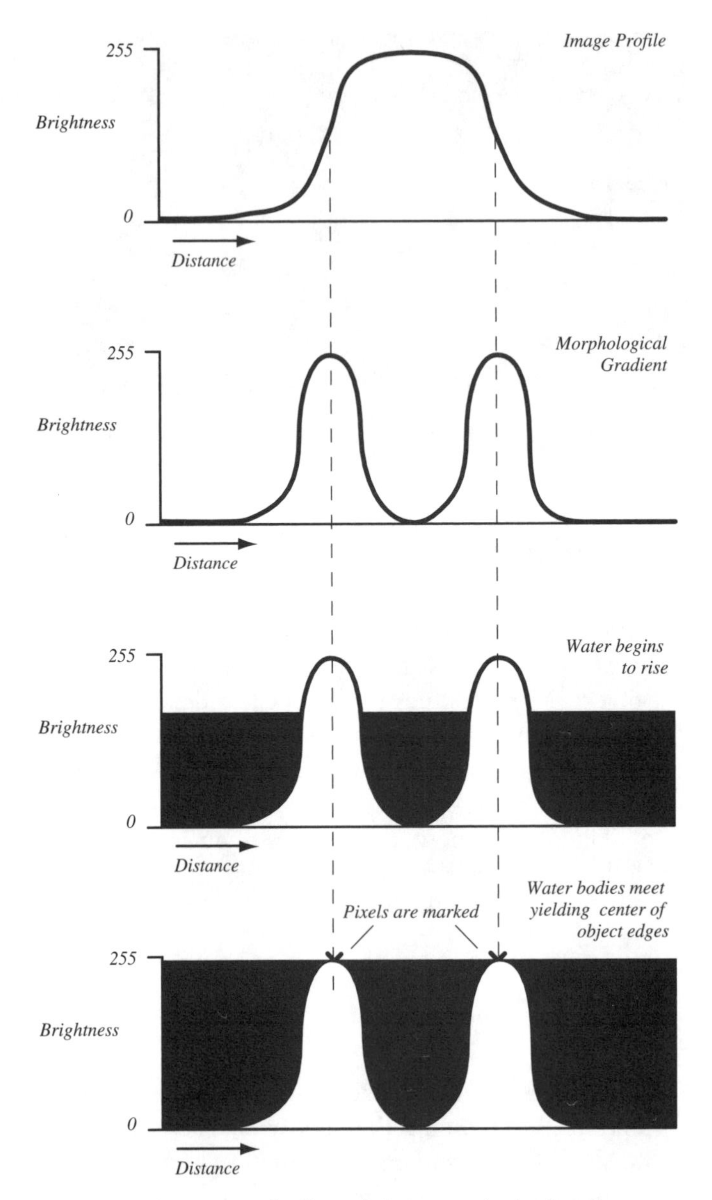

Figure 5.30 *One-dimensional effect of the watershed edge detector on a morphological gradient image.*

object. The sum of all pixel brightnesses in an object relates to the energy, or aggregate brightness, of an object. This measure is called an object's zero-order spatial moment, and is discussed later in this chapter. Although a little obscure, the application of these statistical measures to brightness histograms, or color-component histograms, can help in classifying the brightness or color characteristics of an object.

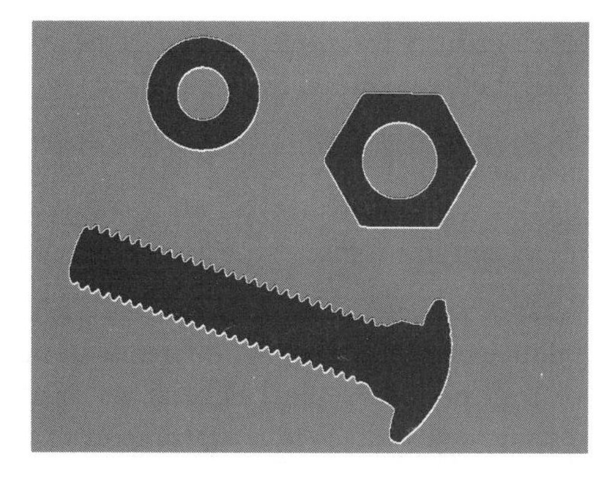

Figure 5.31 *Binary image of washer, nut, and bolt objects with overlaid outlines.*

Texture Features

The texture measure of an object can often be used to discriminate between the surface finish of a smooth or coarsely textured object. A convenient way to determine the texture of an object is to examine its spatial frequency content. A smooth object has only small variances in its brightnesses, which is equivalent to having primarily low spatial frequencies. A coarsely textured object has lots of minute variations, which means that it will contain high spatial frequencies.

Three methods can show frequency content, and hence the texture of an object. A high-pass filter will yield an image of the object's high-frequency components. If the result has a lot of high frequencies, then it has a course texture. If the result does not contain high frequencies, then it has a smooth texture.

Similarly, a Fourier-transform operation can be used. Again, the presence of high frequencies in the frequency image represents coarse texture, and the absence of high frequencies represents a smooth texture.

Sometimes, just looking at the standard deviation of the brightnesses in an object is sufficient to determine its texture. Because the standard deviation is a measure of how much the brightnesses vary from the mean brightness, a high value often indicates a lot of variance, and hence a coarse texture. Similarly, a low standard-deviation value often indicates a smooth texture.

Shape Measures

The most common object measurements made are those that describe shape. Shape measurements are physical dimensional measures that characterize the appearance of an object. The list of different measures can get very long and somewhat abstract. A particular image analysis operation will generally make use of a subset of possible shape measures. In any image analysis application, the goal is to use the fewest necessary measures to characterize an object adequately so that it may be unambiguously classified as required by the application.

Objects can have regular shapes, such as square, rectangular, circular, elliptical, and so on. Sometimes, the shape will be arbitrary, twisting and turning in apparently random ways. Even the weirdest shapes can be characteristic of an object of interest, and hence be important to us. Figure 5.32 examines some common shape measures.

Perimeter—The pixel distance around the circumference of the object. To accurately compute this, where a boundary pixel contacts its neighbor vertically or horizontally, the pixel distance is 1 unit. Where a pixel contacts a neighbor diago-

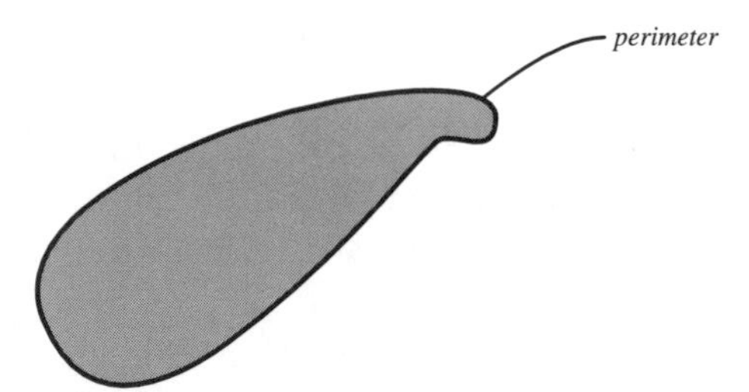

Figure 5.32a *Several common shape measures—perimeter of object.*

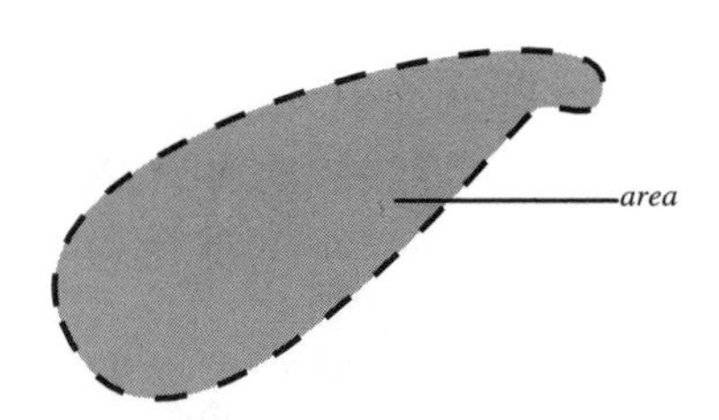

Figure 5.32b *Object area.*

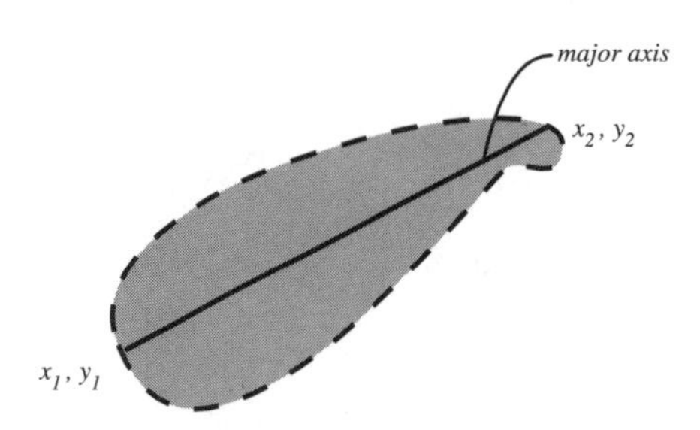

Figure 5.32c *Major axis of object.*

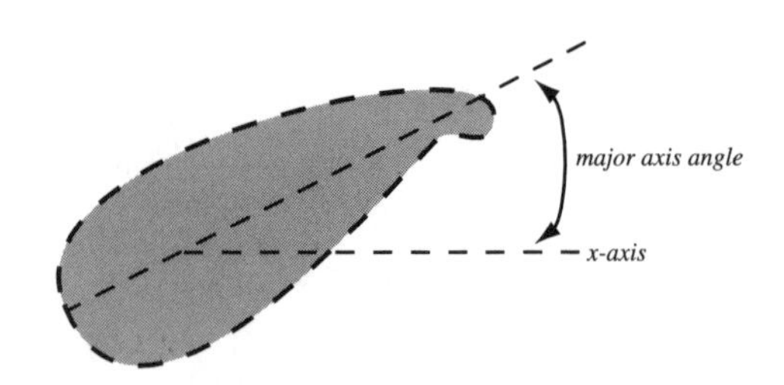

Figure 5.32d *Major axis angle.*

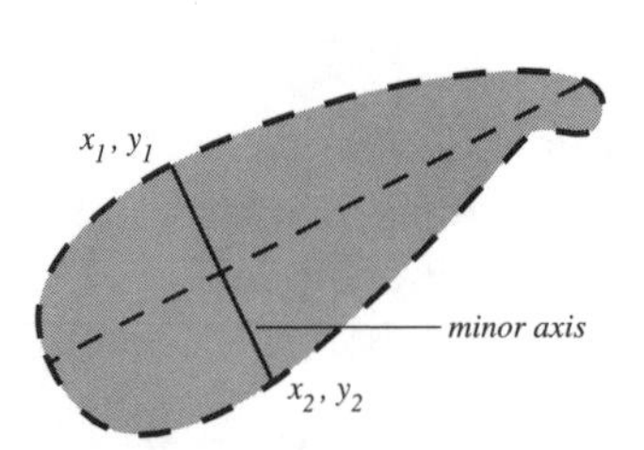

Figure 5.32e *Minor axis of object.*

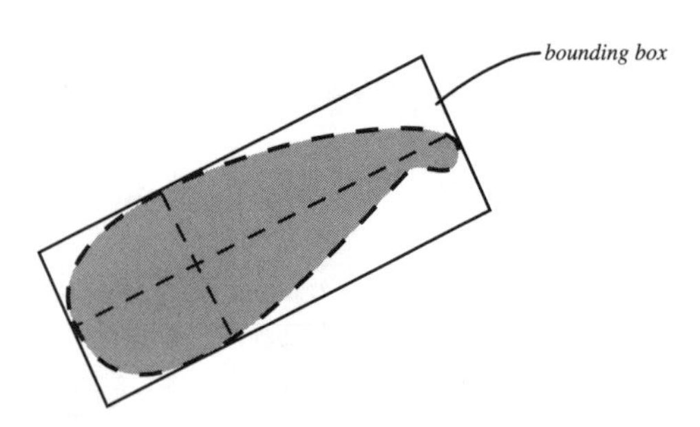

Figure 5.32f *Bounding box of object.*

nally, the pixel distance is the square root of 2, or 1.414 units. The result is a measure of object boundary length.

Area—The pixel area of the interior of the object. The area is computed as the total number of pixels inside, and including, the object boundary. The result is a measure of object size. The area measure usually does not include object hole areas.

Area to Perimeter Ratio—A ratio based on the area and perimeter measures of an object. The result is a measure of object roundness, or compactness, given as a value between 0 and 1. The greater the ratio, the rounder the object. If the ratio is equal to 1, the object is a perfect circle. As the ratio decreases from 1, the object departs from a circular form.

$$\text{Roundness} = (4\pi \times \text{Area})/\text{Perimeter}^2$$

Major Axis—The (x,y) endpoints of the longest line that can be drawn through the object. The major axis endpoints (x_1,y_1) and (x_2,y_2) are found by computing the pixel distance between every combination of border pixels in the object boundary and finding the pair with the maximum length.

Major Axis Length—The pixel distance length between the major axis endpoints. The result is a measure of object length.

$$\text{Major Axis Length} = \sqrt{(x_2 - x_1)^2 + (y_2 - y_1)^2}$$

where (x_1,y_1) and (x_2,y_2) are the major axis endpoints.

Major Axis Angle—The angle between the major axis and the x-axis of the image. The angle can range from 0° to 360°. The result is a measure of object orientation.

$$\text{Major Axis Angle} = \tan^{-1}\left(\frac{(y_2 - y_1)}{(x_2 - x_1)}\right)$$

where (x_1,y_1) and (x_2,y_2) are the major axis endpoints.

Minor Axis—The (x,y) endpoints of the longest line that can be drawn through the object while maintaining perpendicularity with the major axis. The minor axis endpoints (x_1,y_1) and (x_2,y_2) are found by computing the pixel distance between the two border pixel endpoints.

Minor Axis Width—The pixel distance length between the minor axis endpoints. The result is a measure of object width.

$$\text{Minor Axis Length} = \sqrt{(x_2 - x_1)^2 + (y_2 - y_1)^2}$$

where (x_1,y_1) and (x_2,y_2) are the minor axis endpoints.

Minor Axis Width to Major Axis Length Ratio—The ratio of the width of the minor axis to the length of the major axis. This ratio is computed as the minor axis width distance divided by the major axis length distance. The result is a measure of object elongation, given as a value between 0 and 1. If the ratio is equal to 1, the object is roughly square or circularly shaped. As the ratio decreases from 1, the object becomes more elongated.

Minor Axis Width to Major Axis Length Ratio
= Minor Axis Width/Major Axis Length

Bounding Box Area—The area of the box that would entirely surround the object. The dimensions of the bounding box are those of the major and minor axes.

Bounding Box Area = Major Axis Length × Minor Axis Length

Number of Holes—A count of how many holes exist within the interior of the object.

Total Hole Area—The total pixel area of the interior holes in the object. The area is computed as the total number of pixels inside each hole. The result is a measure of the size of the object's hole content.

Total Hole Area to Object Area Ratio—The ratio of the area of all the holes to the area of the object. This ratio is computed as the total hole area divided by the object area. The result is a measure of object perforation, given as a value between 0 and 1. If the ratio is equal to 1, the object is entirely a hole. As the ratio decreases from 1, less area of the object is occupied by holes.

Total Hole Area to Object Area Ratio = Total Hole Area/Object Area

These shape measures represent the most commonly used measures, and are generally sufficient for many image analysis applications. Of course, with some thought, other measures may become obvious as appropriate for a particular need. For instance, dividing the total-hole-area measure by the number-of-holes measure would provide an average-hole-area measure that might be useful for an analysis application requiring overall hole statistics.

Spatial Moments

The *spatial moments* of an object are statistical shape measures that do not actually characterize the object specifically, as do the previous shape measurements. Rather, spatial moments give statistical measures related to an object's characterizations.

The *zero-order spatial moment* is computed as the sum of the pixel brightness values in an object. In the case of a binary image, this is simply the number of pixels in the object, because every object pixel is equal to 1 (white). Therefore, the zero-order spatial moment of a binary object is its area. For a gray-scale image, an object's zero-order spatial moment is the sum of its pixel brightnesses. This is a measure that is related to the object's energy.

The *first-order spatial moments* of an object contain two independent components, x and y. They are the x and y sums of the pixel brightnesses in the object, each multiplied by its respective x or y coordinate location in the image. In the case of a binary image, the first-order x spatial moment is just the sum of the x coordinates of the object's pixels, because every object pixel is equal to 1 (white). Likewise, the y spatial moment is the sum of the y coordinates of the object's pixels. For a gray-scale image, an object's first-order spatial moments are as defined above. The first-order spatial moments of an object represent the object's mass and how it is spatially distributed.

The two most common image object measurements that use spatial moments are object *area* and *center of mass*. As stated above, an object's area is computed as its zero-order spatial moment. An object's center of mass can be computed as the first-order spatial moments (x and y) divided by the zero-order moment, or the object area. There are two forms of the center of mass—one that considers pixels to have uniform weight, as in a binary image, and one that weights pixels based on their brightness values. The second form considers pixels that are black to have a weight = 0, those that are white to have a weight = 255, and pixels with brightnesses in between to have a weight corresponding to their respective gray levels. The definitions for the center of mass measures are as follows:

Center of Mass (Centroid)—The balance point (x,y) of the object where there is equal mass above, below, left, and right. If we think of the pixels in an object as having a unit weight, then the center of mass is the point where the object will perfectly balance on the tip of a point, as shown in Figure 5.33. For simple objects like circles and squares, the center of mass is at the center of the object. Symmetrical objects, like rectangles and ellipses, have a center of mass at the centers of the major and minor axes.

$$\text{Center of Mass}_x = \text{Sum of object's } x\text{-pixel coordinates/Number of pixels in object}$$

$$\text{Center of Mass}_y = \text{Sum of object's } y\text{-pixel coordinates/Number of pixels in object}$$

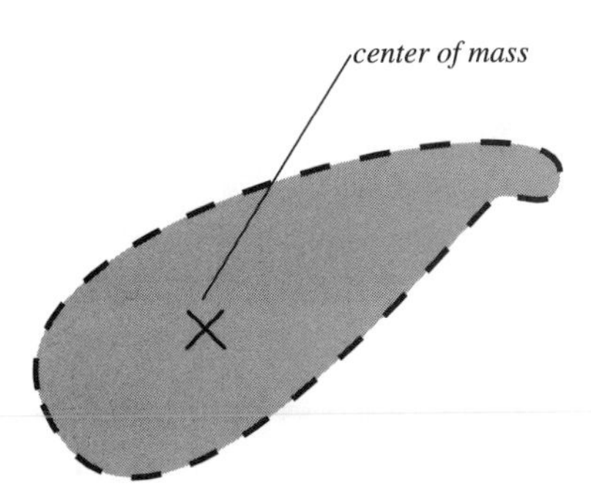

Figure 5.33 *Object center of mass.*

Brightness-Weighted Center of Mass (Gray Centroid)—The balance point (x,y) of the object where there is equal brightness mass above, below, left, and right. If we think of the pixels in an object as having a weight dependent upon their brightness, then the brightness weighted center of mass is the point where the object will balance perfectly on the tip of a point, as shown in Figure 5.34.

Brightness-Weighted Center of Mass_x = Sum of object's (x-pixel coordinates × pixel brightness)/Sum of pixel brightnesses in object

Brightness-Weighted Center of Mass_y = Sum of object's (y-pixel coordinates × pixel brightness)/Sum of pixel brightnesses in object

We can also compute higher-order spatial moments. For instance, the second-order moments produce object orientation information. Spatial moments of an order that is greater than two produce abstract information that is difficult to tie specifically to physical object characteristics. Generally, the zero- and first-order moments are used in practice to provide good distinction between differently shaped objects.

Boundary Descriptions

Object shape measures come in a variety of forms, as we have seen, and can be extended to include others, as desired. However, the most precise way to define the outline of an object is to examine it and record specifically how it behaves. There are numerous techniques for doing this, from the most fundamental method of explicitly listing boundary pixel locations to various methods of more concise boundary description. Generally, each method describes an object's shape for use in subsequent object classification tasks. In this section, we will examine several object boundary description techniques.

Explicit Description

Fundamentally, an outline is composed of a single-pixel-wide sequence of pixels that follow the perimeter of an object. Each pixel adjoins a neighboring pixel either to its left, right, top, or bottom, or on one of its four diagonals. In this way,

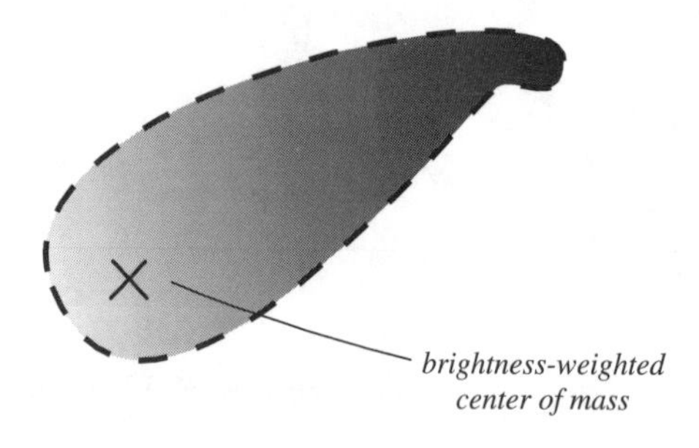

Figure 5.34 *Object brightness-weighted center of mass.*

we can say that the outline is fully connected into an unbroken progression of boundary pixels. As in any image, each pixel has an (x,y) location that defines its position within the image.

By creating a sequential list of the (x,y) boundary pixel locations, we can create a description of an object boundary, as shown in Figure 5.35. This boundary description can be stored, and later used to recreate the outline by simply going through the list and plotting the points. In this way, the list of (x,y) boundary locations can be said to uniquely characterize the shape of an object's perimeter. This is the most fundamental and precise representation of an object's shape.

The accuracy of the (x,y) list of boundary pixels is, of course, related to the spatial resolution of the object's image. The spatial resolution of the imaging system inherently limits this accuracy. For instance, let's suppose that an image has the spatial dimensions of 640 pixels × 480 lines and covers a distance of 10 centimeters across its horizontal field of view. We can calculate the dimensions of each pixel as

Pixel width = 10 centimeters / 640 pixels = 0.0156 centimeters/pixel

and the density of pixels per centimeter as

Horizontal pixel density = 640 pixels / 10 centimeters = 64 pixels/centimeter

Assuming that the imaging system has square pixels, as do most, the pixel height is equal to the pixel width, and the vertical pixel density is equal to the horizontal pixel density. In this case, the inherent accuracy of the boundary depicted by a single-pixel-width outline, compared with the actual object, is ±0.0156 centimeters.

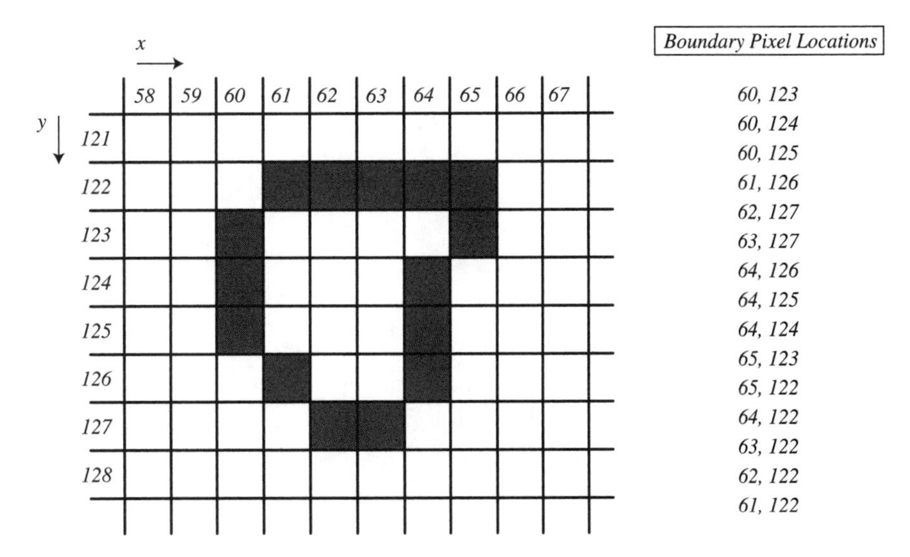

Figure 5.35 *The explicit description of an object boundary is listed as the sequential (x,y) boundary pixel locations.*

The explicit description list of boundary pixels is not well suited for use in subsequent object classification operations. This is because its description is very sensitive to how an object appears within an image. Object location, orientation, and size can change the boundary pixel locations, making comparison of one list to another practically impossible. While the explicit description is still useful to store and recreate the outline of an object's boundary, other boundary representation techniques are better suited for classification operations.

Chain Codes

Object boundaries can be described in more concise terms than an explicit (x,y) pixel location list. Instead, the boundary can be followed and recorded using a *chain code*. First, we pick a starting pixel location anywhere on the object boundary. The goal is to find the next pixel in the boundary. From the earlier definition of an outline, we know that there must be an adjoining boundary pixel at one of the eight locations surrounding the current boundary pixel. By looking at each of the eight adjoining pixels, we will find at least one that is also a boundary pixel. Depending on which one it is, we assign a numeric code of between 0 and 7. This code is assigned based on the illustration shown in Figure 5.36.

For instance, if the pixel found is directly to the right of the current location, a code of "0" is recorded. If the adjoining pixel is to the upper right, a "1" is recorded, and so on. Once the code is recorded, we repeat the process by looking for the next boundary pixel, as shown in Figure 5.37. The result is a list of codes showing which direction was taken in going from each boundary pixel to the next. The resulting chain code list describes the object's boundary; no information is lost between the chain code list and the (x,y) pixel location list. This is because each and every boundary pixel is still fully represented. Only the way each boundary pixel is described is different.

In following the object outline, it is important to remember two pieces of information. First, the operation must remember the direction of the previous

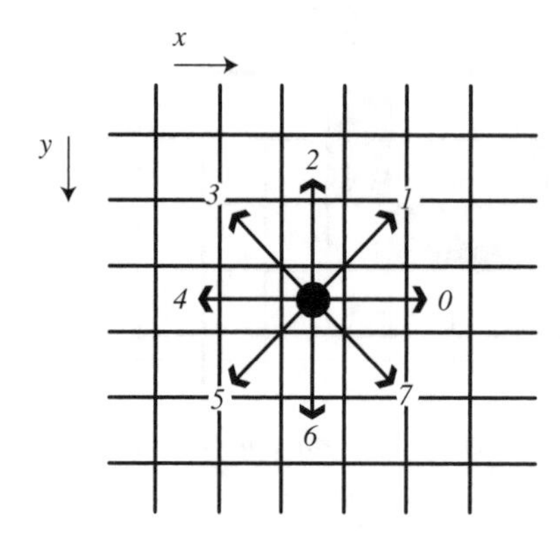

Figure 5.36 *The chain code direction assignments.*

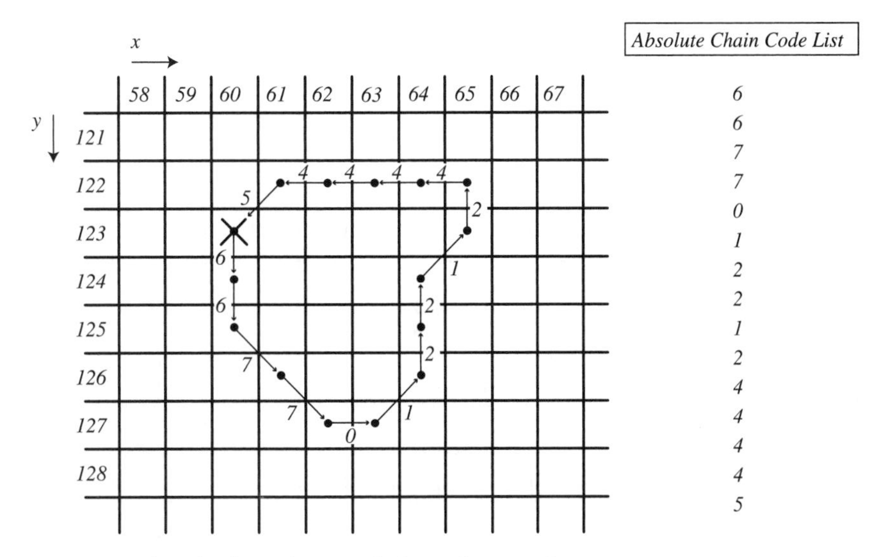

Figure 5.37 *The absolute chain code boundary tracking process.*

location from the current location. Using this information, the operation must make sure that the next boundary pixel found is not actually the last boundary pixel, or the operation may begin to backtrack over the boundary already traversed. Also, the operation must remember the absolute (x,y) location of the starting pixel. This way, the operation can check to see whether the next boundary pixel found is the original starting point; if so, the entire outline has been followed and the operation is complete.

The outline of the object can be recreated by simply going through the chain code list and repeating the original movements found when the boundary was followed. In this way, the chain code represents the object boundary and can be said to characterize exactly the perimeter of the object.

The chain code described above is referred to as an *absolute chain code*. Each code represents the absolute direction, relative to the "0" direction, taken from one boundary pixel to the next boundary pixel. A variation of the chain code can be used to record relative movements. This is referred to as a *relative chain code*. This version of the chain code records the direction from the current boundary pixel to the next, based on its relative direction rather than its absolute direction. Figure 5.38 illustrates the relative chain code method of object boundary description.

The absolute chain code boundary representation is like the explicit boundary representation—it is sensitive to the way an object appears in an image. The relative chain code, however, is not sensitive to object location and orientation, due to its relative nature. Relative chain codes can be very appropriate for use in subsequent classification operations.

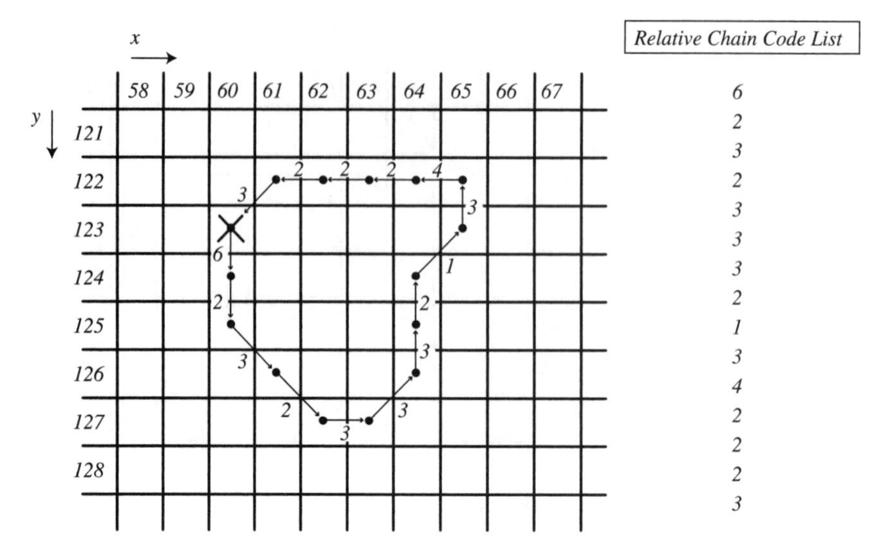

Figure 5.38 *The relative chain code boundary tracking process.*

Line Segment Representation

Many times it is not necessary to record the entire, full-resolution boundary of an object, as the (x,y) pixel location and chain code lists do. Rather, a reduced form of the outline may be sufficient. Generally, an image analysis application will only require the essential characteristics of an object's outline to later classify it as belonging to one of several different classes. For instance, one approach to classifying an object as either a bolt or nut may only require a very primitive outline description, portraying the object as either long and narrow or short and roughly circular. A chain code boundary description that codes every boundary pixel location can be excessive for an application like this. In fact, using a chain code boundary description can actually hinder the object classification task because of its exacting detail.

A standard way of reducing the description of an object boundary is to replace individual boundary (x,y) locations with line segments. By using shorter or longer line segment lengths, the resulting boundary description can be generated with an arbitrary range of accuracy from coarse to fine. The required accuracy is usually determined by the application's requirements.

The *line segment boundary representation* technique can be described as follows. Instead of recording every (x,y) location in the outline, only a limited number of (x,y) locations are recorded, each separated by a number of boundary points other than one. We place a line segment to fill the distance between the chosen (x,y) locations, as shown in Figure 5.39. In this way, the boundary is approximated by the line segments that replace the original (x,y) boundary pixel locations. The number of boundary pixels skipped can be selected based on the accuracy desired

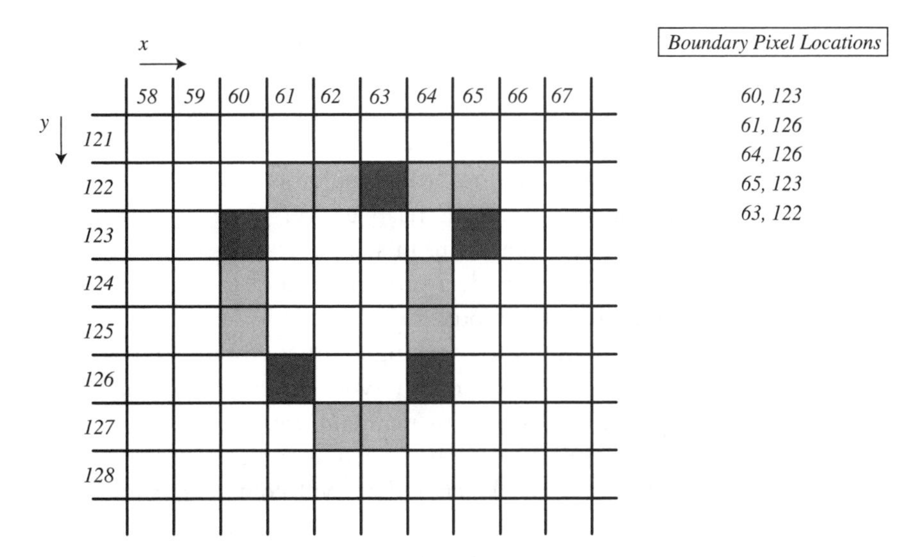

Figure 5.39 *The line segment boundary description replaces multiple (x,y) boundary pixels with line segments, producing a reduced list of (x,y) boundary pixel loca-*

in the new boundary representation. The result of the line segment boundary representation is, again, a list of (x,y) pixel boundary locations that has been reduced in size.

The resulting list can be further processed to an even simpler form. Line segments can be described by a length and an angle, as shown in Figure 5.40. This is

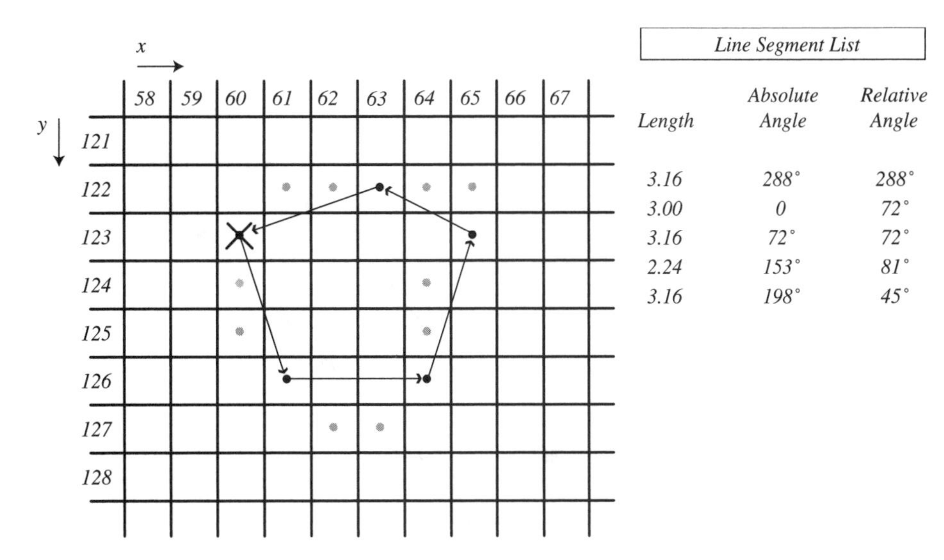

Figure 5.40 *Line segments can be represented more generally as a list of line segment lengths and angles.*

similar to a chain code representation, except that the line segment length can be any value (it's not limited to a length of one), and the angle can be any angle (it's not limited to one of eight directions). The angle can be described in absolute terms relative to a reference angle, such as the horizontal axis. Or, the angle may be a relative measure of the angle between the last line segment and the current line segment. As in the relative chain code representation, the relative angle measure can be more useful in subsequent classification operations. The line segment approach can provide a very simple and concise, yet sufficiently descriptive, representation of object outlines.

In the interest of creating a concise boundary description, we usually want to represent the object outline with as few line segments as possible. It is often still important, however, to maintain enough detail in the boundary description to portray important features such as minute wavers and concavities. In most real-world cases, an object outline will contain portions with little direction change (like straight regions) and portions with a lot of direction change (like curved regions), as shown in Figure 5.41. A variation on the basic line segment approach, called *variable line segment boundary representation*, can be used to better represent small boundary changes without using small line segments throughout the description.

This technique combines multiple short line segments and substitutes them with a single longer line segment over portions of the boundary that have small direction changes. A variable number of boundary points are covered by a single line segment, depending on how much boundary direction change occurs. This

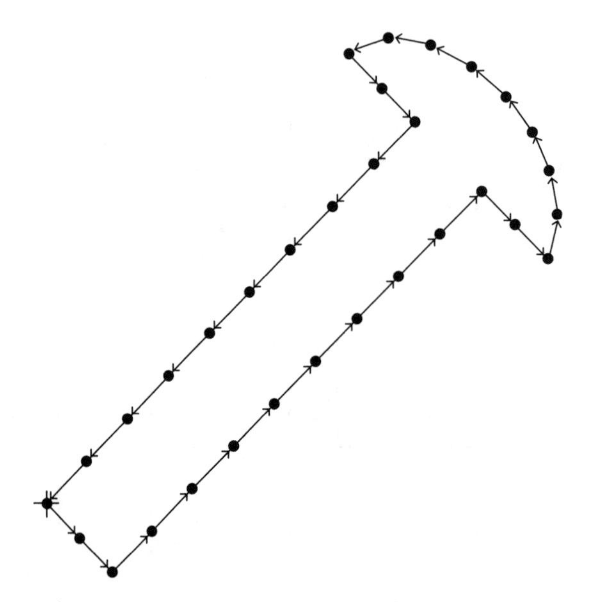

Figure 5.41 *A line segment object boundary often contains regions of little change and regions of great change.*

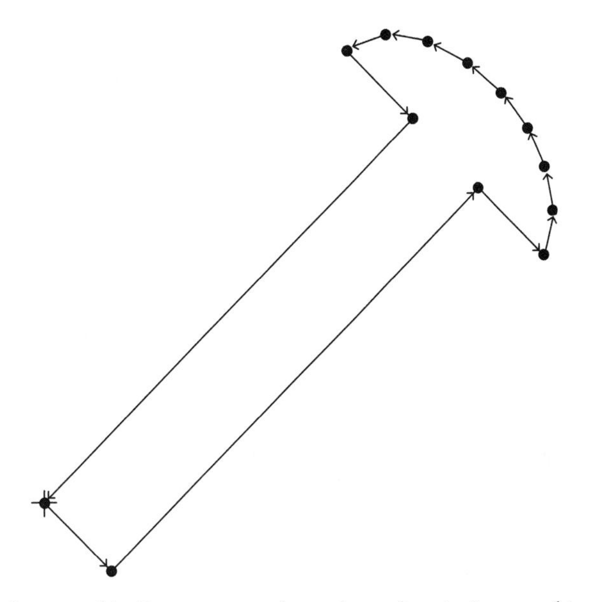

Figure 5.42 *The variable line segment boundary description combines adjoining line segments with similar direction angles.*

way, the line segment boundary description will not unduly record multiple short line segments where the boundary direction changes only a little.

We implement the variable line segment approach by first performing a line segment boundary representation operation, using line segment lengths that are short enough to catch all outline details required by the particular image analysis application. Then, we compare the angles of each pair of adjoining line segments. Where the angles are similar or identical, the two line segments are combined to create a single line segment representing their combined length and overall angle, as shown in Figure 5.42. The end result is a line segment boundary representation that can be significantly more concise, while maintaining a high level of accuracy.

The relative line segment boundary representations are excellent for use in subsequent classification operations. Like the relative chain code, they are insensitive to object location and orientation within an image. Additionally, their concise descriptions streamline the object classification process.

Fourier Descriptors

A more complex boundary description technique, known as Fourier descriptors, creates concise object boundary representations using the techniques of the Fourier transform. As with line segment representations, the resulting boundary description can be generated with an arbitrary range of accuracy from coarse to fine.

The Fourier descriptors method treats the (x,y) points of an object outline as two functions and decomposes them into values representing their frequency components. These values are called the *Fourier descriptors* of the boundary. This process operates mathematically on an image very much like the Fourier transform discussed in Chapter 4. However, the frequency-transformed values (the Fourier descriptors) don't really relate to the physical nature of the object boundary; rather, they are an abstract description of the boundary. Nonetheless, the resulting descriptors can be inverse-transformed back to the original (x,y) boundary points, thus recreating the boundary.

Likewise, the Fourier descriptors can be compared with other known sets of descriptors to classify an object. Many times it is sufficient to make this comparison with a reduced set of descriptors, eliminating the smaller values from the comparison. In this way, the boundary representation by Fourier descriptors can be a compact means for describing object boundary shapes.

Object Classification

At this point in the image analysis process, we have segmented our original image into discrete objects and measured them using a variety of shape measurement schemes, including specific shape dimensions, shape statistics, and shape boundary descriptions. To complete the image analysis process, we must also be able to classify the measurements. This means that we must be able to compare the measurements of a new object with those of a known object or other known criteria. In this manner, we can determine whether an object belongs to a particular category of objects.

As an example, let's suppose that an image analysis operation is required to watch the flow of ring and spade terminals on a conveyer, as shown in Figure 5.43. Let's further say that the analysis application's goal is to determine which of two groups an object belongs in, so that the object may be sorted into bins of like

Figure 5.43 *Binary image of ring and spade terminals.*

objects. One approach is to use the number-of-holes shape measure, which can immediately identify the object as a spade terminal (no holes) or a ring terminal (one hole). The classification is simple: if the object has one hole, it's a ring terminal; if it has no holes, it's a spade terminal. Of course, the problem can get much more involved if we require more specific object classification, such as precise dimensional measurements or specific shape features.

Comparing Measures with Known Features

Before we can embark on the classification of objects within an image, we must have some knowledge of what we are looking for. Sometimes, our knowledge of the objects to be analyzed will be sparse. This might be true in the case of an early analysis of a new chemical substance or biological sample, where the contents are relatively unknown. Hence, we may be interested only in rough categorization of objects based on some ad hoc characteristics, like rough shape and ranges of length or size.

Usually, though, we will know the exact characteristics of the objects that we are interested in classifying. These cases include applications such as machine vision and image interpretation, where we know the precise feature details of the objects of interest. In these cases, we categorize the objects based on known characteristics, such as their outline traits or their precise length or size.

In either case, the object classification process involves comparing a set of measured features of an object with some established criteria. How close the comparisons are determines whether an object is considered a part of a class. This classification process is done in three steps.

First, we determine the object features that we wish to use to classify the object. Second, we set tolerances, establishing how close the feature measurements must be to the established criteria for a match. And third, we create classification groups, or categories, to which an object will be assigned depending on how its feature measurements compare with the established criteria.

The process can best be illustrated by looking at a few examples:

1. Determine whether a cell in a particular biological sample is normally shaped. For it to be normal, it must be nearly circular—within 5 percent.

   Feature Measures—Roundness = $(4\pi \times \text{Area})/\text{Perimeter}$

   Tolerances—Because the cell must be within 5 percent of circular to qualify as normal, the roundness measure must be greater than or equal to 0.95.

   Classes—Normal, if the roundness measure is within the tolerance limits
   Abnormal, otherwise

2. Determine whether a machine-stamped sheet metal part is good or bad. For it to be good, it must be rectangular with dimensions of 4 centimeters high × 3 centimeters wide, to within ±200 micrometers. Further, the part must have a single hole of diameter 0.5 centimeter, to within ±100 micrometers, at the center of the part.

   Feature Measures—Major Axis Length
Minor Axis Length
Number of Holes
Diameter of Hole

   Tolerances—Major Axis Length = 4 cm ± 200 μm
Minor Axis Length = 3 cm ± 200 μm
Number of Holes = 1 ± 0
Diameter of Hole = 0.5 cm ± 100 μm

   Classes—Good, if all measures are within the tolerance limits
Bad, otherwise

3. Determine whether the specific shape and size of a molded part match an original sample part to within 10 percent of the original part, measured using a 10-segment line segment boundary description.

   Feature Measures—Line segment boundary description

   Tolerances—Line segments must all be within ± 10 percent in the length and the direction angle of the original sample part

   Classes—Good, if the shape is within tolerance to the original part
Bad, otherwise

Often, we have the option of determining what measures are needed to classify an object adequately. Generally, an attempt is made to select the measures that are computationally the easiest to derive, and later to compare with an object being analyzed. For instance, it is preferable to measure, and later classify, simple dimensions of an object, rather than its entire shape boundary. Of course, the selected simple dimensions must be sufficient to classify the object into the categories required by the application. In the first example above, it may very well be excessive to use a boundary description to determine whether the cell is round. Rather, the area and perimeter length can be computed, yielding a roundness measure that is sufficient to classify the cell as normal or abnormal.

Perhaps brightness information alone is adequate to classify image objects. As discussed in Chapter 4, brightness slicing techniques can be used to classify several ranges of object brightness. A technique called *gray-level classification* does this by mapping each brightness range to a distinct gray level in the resulting image.

Advanced classification techniques using *fuzzy logic* and *neural network* concepts have been evolving. These techniques vary in the way matches between object measures and known criteria are determined. They still define the object-feature measures of interest and resulting classification categories, as described earlier. But, instead of simple tolerance bounds checking and logical combination of the results, these techniques use more advanced methods to determine matches. Fuzzy logic techniques provide "fuzziness" to the logical reasoning, providing improved classification results when many features are involved with complex interrelationships that are difficult to describe using traditional logic techniques. Neural networks have the unique ability to be "trained." Instead of programming the criteria for classification matches, neural network-based systems are trained by literally showing them many images of known object characteristics and telling them the correct classifications. In this way, they "learn" the feature measures and tolerances for each classification category.

Measure Invariance

In machine vision and most image interpretation applications, objects are measured against specific dimensional requirements. If the object meets the requirements, it is a member of a class (for instance, the "good" class); otherwise, it is not (the "bad" class or another class). We classify objects based on various object dimensional measures, statistical measures, and boundary descriptions.

As we discussed earlier, when we classify an object, we compare measures taken from the image of an object against known measures that place it into a particular class. Each comparison is made with tolerances set by the requirements of the application. These tolerances provide a window of acceptance that allows for the acceptable dimensional accuracy of the object being classified and the accuracy of the imaging system. When the measures compare within tolerance, a match is determined and the object is classified accordingly.

In the real imaging world, a variety of things can cause an object to appear distorted from its actual physical appearance. In particular, the object may take on geometric distortions such as rotation to a random angle, translation to a random location in the image, and scaling to appear smaller or larger. Brightness and color distortions are also common. All of these distortions can disrupt measures like object area, perimeter, and color, and will certainly make a shape description different. The troubling part is that the object may still be exactly what we are looking for to fit a particular class. The worst thing that can happen in an image analysis operation is for an object that physically meets a class's criteria to be missed. This can lead to a "good" object's getting improperly classified as a "bad" object or vice versa.

One way to avoid some erroneous classifications is by opening up the tolerances on a measure. For instance, if a classification is being done using an object's major axis length, the tolerance on the length measure can be extended to a greater number, say from ± 10 percent to ± 20 percent. Hence, if the object being measured appears a little shorter or longer, its major axis length measure will still be within tolerance and the object will be correctly classified. The downside to this approach is that objects whose physical dimensional measure is really not within an acceptable tolerance can end up improperly classified. In most image analysis applications, this error can be just as bad as the alternative error.

The way around these types of classification errors is to select the measures carefully and use them in a way that reduces or eliminates their sensitivity to certain imaging variances. In this way, we can classify objects more accurately.

Measure invariance refers to the trait of a particular measure to be insensitive to a particular variance. When a measure is insensitive to a variance, the measure's value will not change in response to the variance; the measure is invariant. Let's look at an example. Let's say we have decided that an object of interest can be sufficiently discriminated as "passing" or "failing" based solely on its length and width features. We further decide that the (x,y) endpoints of its length and width will be used to classify these features. By examining an image of the object, we record the (x,y) endpoints of both features. We propose that every time we find an object that has the same length and width endpoints, it is a member of the "pass" class. Otherwise, it falls into the "fail" class. Some significant problems will arise with this approach, as shown in Figure 5.44. What if an identical "passing" object is rotated in the image? What if it is translated to another part of the image? What if it is slightly smaller or larger? Even though the object may be a "passing" object, these variations in the way the object appears within the image will cause the length and width endpoints to vary, and the object will be misclassified as a "fail."

Instead of using the length and width endpoint measures alone, more appropriate measures can be derived that are invariant to object rotation, translation, and scaling variations. First, length and width distances can be used instead of the

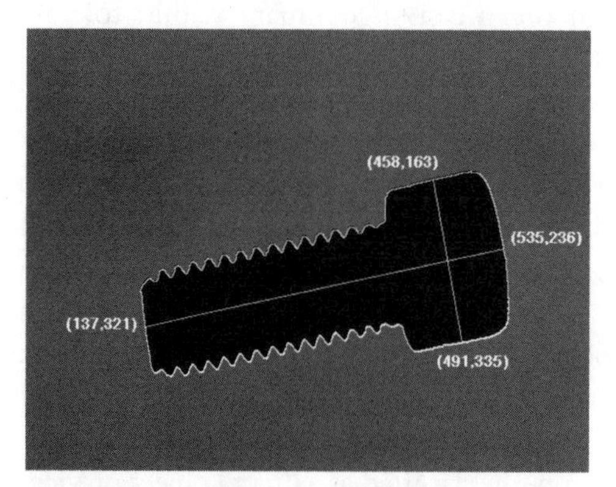

Figure 5.44a *Original binary bolt image with major and minor axes overlaid. The major axis endpoints are (137,321) and (535,236). The minor axis endpoints are (458,163) and (491,335).*

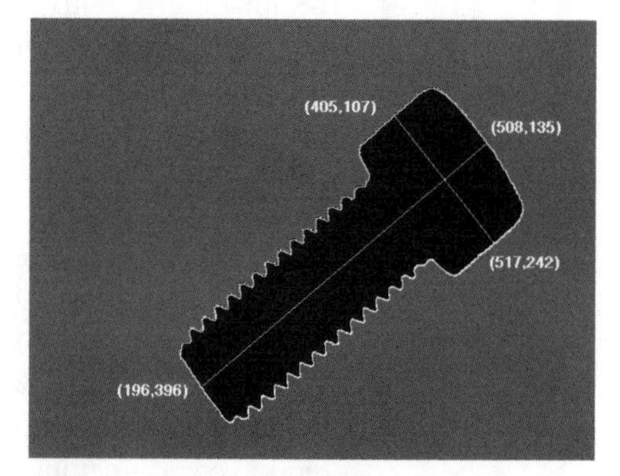

Figure 5.44b *Image of the same object but with a rotated position from the original image. The major axis endpoints are (196,396) and (508,135). The minor axis endpoints are (405,107) and (517,242).*

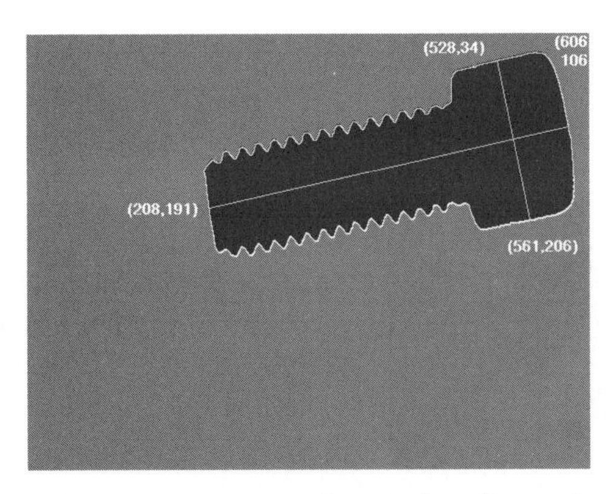

Figure 5.44c *Image of same object but with a translated position from the original image. The major axis endpoints are (208,191) and (606,106). The minor axis endpoints are (528,34) and (561,206).*

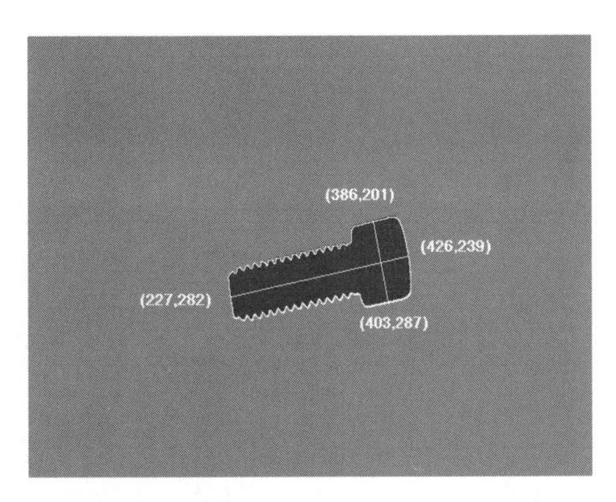

Figure 5.44d *Image of same object but with a scaled size relative to the original image. The major axis endpoints are (227,282) and (426,239). The minor axis endpoints are (386,201) and (403,287).*

absolute length and width (x,y) endpoint locations. This immediately eliminates any sensitivity to object rotation or translation, as shown in Figure 5.45. No matter where the object appears in the image and no matter what orientation it has, the length and width distance measures will be the same.

Second, we can add scaling invariance by normalizing the length and width distance measures by a known distance measure, called a *unit distance*. This is a dynamic calibration procedure, meaning that every time the measures are made, the scaling distortion is automatically removed. Distance reference marks in the image are used to represent a known distance. The measured length and width distances are then divided by this known distance to provide the two measures, length/unit

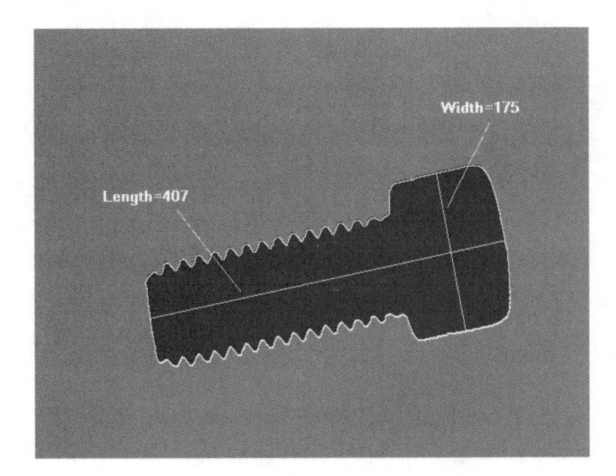

Figure 5.45a *Original binary bolt image with major and minor axes overlaid. The major axis length is 407. The minor axis width is 175.*

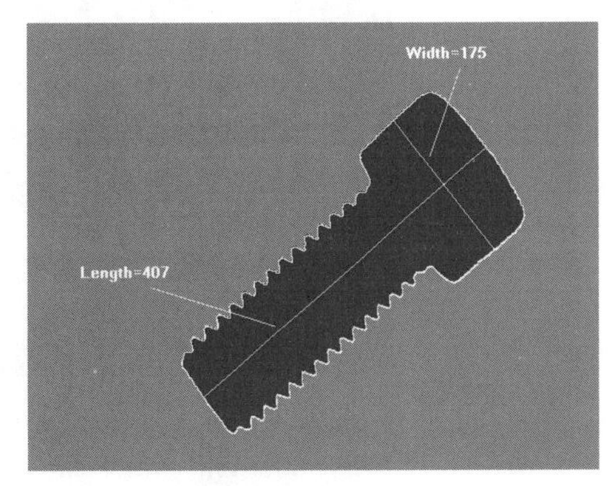

Figure 5.45b *Image of same object but with a rotated position from the original image. The major axis length is still 407. The minor axis width is still 175.*

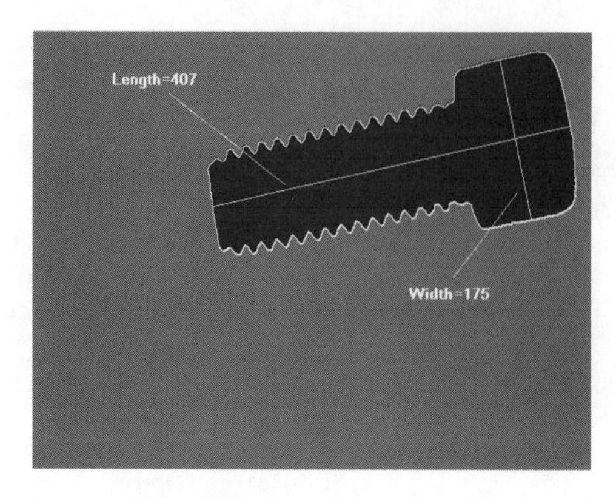

Figure 5.45c *Image of same object but with a translated position from the original image. The major axis length is still 407. The minor axis width is still 175.*

distance and width/unit distance. If the imaging system changes—like if the camera moves or the lens changes magnification—the unit distance will change along with the object's measures. The normalizing process removes the scaling distortion introduced into the raw length and width measures. This process is illustrated in Figure 5.46. Generally, most image analysis applications involve some sort of dynamic calibration processes to remove the effects of imaging system variations.

Another way to remove scaling distortions is to simply use a length-to-width distance ratio. This provides a single measure that guarantees that the length and width distances are correct relative to one another. This technique can be good at providing object scaling invariance if the length and width measures do not have to meet absolute dimensions.

Some boundary shape descriptions are very susceptible to rotation, translation, and scaling variations. The explicit boundary description and absolute chain code

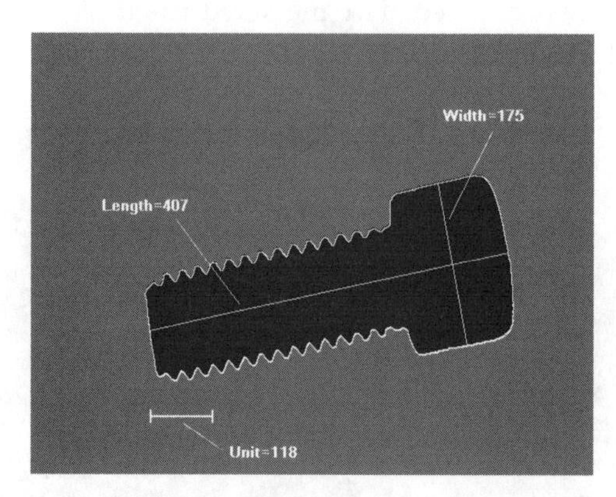

Figure 5.46a *Original binary bolt image with major and minor axes overlaid and a unit distance marking of length 118 pixels. The major axis length/unit distance is 407/118 = 3.45. The minor axis width/unit distance is 175/118 = 1.48.*

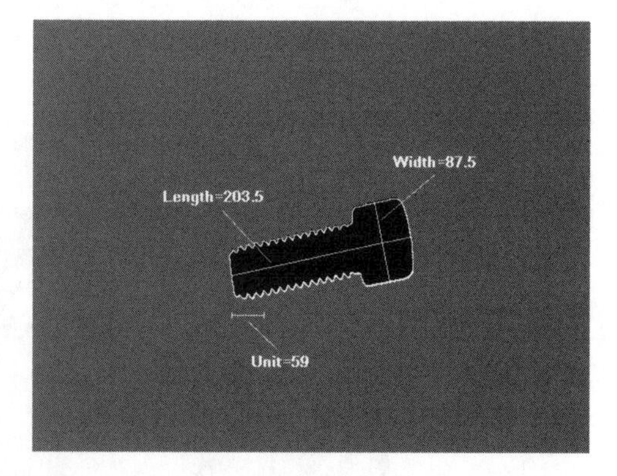

Figure 5.46b *Image of the same object but with a scaled size relative to the original image. The major axis length/unit distance is 203.5/59 = 3.45 (remains the same). The minor axis width/unit distance is 87.5/59 = 1.48 (remains the same).*

techniques fall in this category. They should be avoided for shape classification operations. On the other hand, the relative chain code and relative line segment boundary descriptions are invariant to translation and rotation variations. They can provide good shape descriptions that are not sensitive to these conditions.

Looking at the line segment technique, the invariance occurs because the angle description is made relative to the preceding line segment. If the object is rotated or translated, the relative angles will not change because all line segments are equally rotated or translated. Further, by scaling the line segment distances to a relative distance, scaling invariance can also be introduced to these shape descriptions. Figure 5.47 illustrates making the line segment boundary description invariant to object rotation, translation, and scaling.

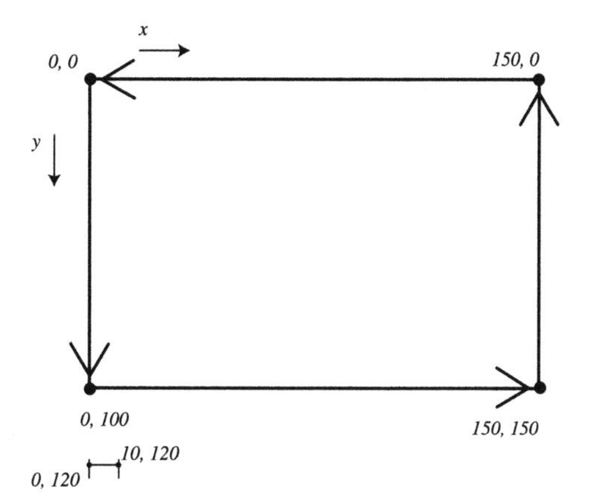

Boundary Pixel List	Line Segment List	
	Length / Unit Distance	Relative Angle
0, 0	100 / 10 = 10	270°
0, 100	150 / 10 = 15	90°
100, 0	100 / 10 = 10	90°
100, 100	150 / 10 = 15	90°

Figure 5.47a *Boundary line segments are listed as having a direction angle that is relative to the preceding line segment and a length that is their measured length/unit distance.*

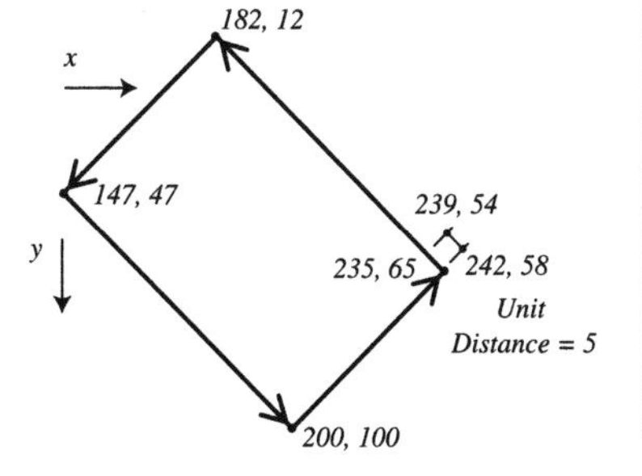

Boundary Pixel List	Line Segment List	
	Length / Unit Distance	Relative Angle
200, 100	50 / 5 = 10	270°
235, 65	75 / 5 = 15	90°
182, 12	50 / 5 = 10	90°
147, 47	75 / 5 = 15	90°

Figure 5.47b *The same object with rotation, translation, and scaling has the identical line segment list. Hence, the boundary description is invariant to these geometric variations.*

Figure 5.48a *Top-view image of an F-15 aircraft.*

Figure 5.48b *Image of an identical aircraft appearing with a different rotation, translation, and scale. In order to find this identical object, the measures used must be insensitive to these geometric variations.*

One example, where shape description invariance is important, is automated military-image interpretation. Reconnaissance imagery can have odd look-angles, magnifications, and distances relative to the objects being imaged. As a result, objects appear with random geometric variations. Let's say that an application has the goal of finding the shape of a particular object—like an aircraft or ship—anywhere in an image, regardless of its location, size, or orientation. An application like this might be used to automatically find objects of military threat. It is essential that the object shape description be entirely insensitive to these geometric variations, as illustrated in Figure 5.48, or else a match may be missed. Of course, once an object of interest is found, a following application could then determine the object's exact location, size, and orientation within the image so that appropriate action could be taken.

Image Matching Techniques

Instead of measuring the dimensions or shapes of objects, another technique is sometimes used to classify objects. *Image matching* techniques compare portions of images against one another. This is done by a process known as *spatial* (or *two-dimensional*) *cross-correlation.*

Referred to as *matched filtering*, the technique involves the pixel-by-pixel comparison of a small reference image containing an object of interest with an image under analysis. The result is an image showing bright spots where the image matches the reference image. The brighter the spot, the better the match. Hence, when the brightness is great enough, a match in objects is determined.

The matched filter is implemented in a similar manner to pixel group processes, discussed in Chapter 4. Instead of having a mask of weight values, matched filters use an *image mask* composed of a reference image. The image mask appears as a small image depicting the object that we wish to find in an image. The mask dimensions are generally much greater than the 3 × 3 and 5 × 5 pixel sizes used in spatial convolution. Rather, they can be any size; the only limitation is the computational effort necessary to compute each output pixel value.

The mechanics of the matched filter are similar to spatial convolution in that the image mask is moved over the input image, pixel by pixel, placing resulting pixels in the output image. The process is illustrated in Figure 5.49. At each pixel location, the input pixels are compared with the pixels of the image mask. A resulting output pixel is created, where a dark value represents a poor match between the two and a bright output value indicates a good match.

If we assume an image mask size of 50 × 50 pixels, the image mask can be represented as follows:

$$\begin{matrix} M_{0,0} & M_{1,0} & M_{2,0} & M_{3,0} & \cdots & M_{49,0} \\ M_{0,1} & M_{1,1} & M_{2,1} & M_{3,1} & \cdots & M_{49,1} \\ M_{0,2} & M_{1,2} & M_{2,2} & M_{3,2} & \cdots & M_{49,2} \\ \vdots & & & & & \\ M_{0,49} & M_{1,49} & M_{2,49} & M_{3,49} & \cdots & M_{49,49} \end{matrix}$$

The resulting output pixel value is computed as the sum of the input pixels in the group, each multiplied by its respective image mask pixel value. This is the process of spatial cross-correlation. The resulting output values can become enormous when all brightness values tend to be equal. On the other hand, the output values will be relatively small when there is not good correlation between the two. In the case of a 50 × 50 pixel image mask, the equation for the matched filter is as follows:

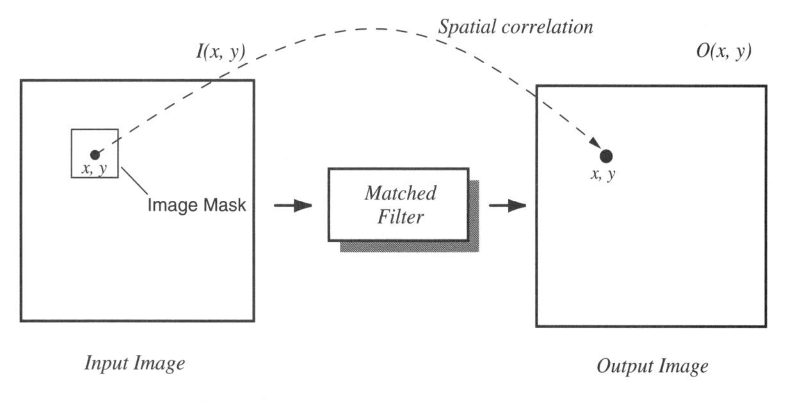

Figure 5.49 *The matched filtering process for the pixel at location* I(x,y)*, creating the output pixel at location* O(x,y)*.*

$$
\begin{aligned}
O(x,y) = {} & (M_{0,0} \times I(x-24,y-24)) + (M_{1,0} \times I(x-23,y-24)) + \\
& \qquad \ldots + (M_{49,0} \times I(x+25,y-24) + \\
& (M_{0,1} \times I(x-24,y-23)) + (M_{1,1} \times I(x-23,y-23)) + \\
& \qquad \ldots + (M_{49,1} \times I(x+25,y-23) + \\
& (M_{0,2} \times I(x-24,y-22)) + (M_{1,2} \times I(x-23,y-22)) + \\
& \qquad \ldots + (M_{49,2} \times I(x+25,y-22) + \\
& \quad \vdots \\
& (M_{0,49} \times I(x-24,y+25)) + (M_{1,49} \times I(x-23,y+25)) + \\
& \qquad \ldots + (M_{49,49} \times I(x+25,y+25)
\end{aligned}
$$

where it is implied that every input pixel group is processed through the equation, creating a corresponding output pixel value.

Using the cross-correlation operation, the matched filter equation will produce output pixel values that range from 0 to enormous numbers, far in excess of 255. As a result, the output values must be reduced to smaller values if we want to display and visually analyze the resulting image.

6 Image Compression

Image compression and *decompression* operations are used to reduce the data-content size of a digital image. The goal of these operations is to represent an image, with some required quality level, in a more compact form. Image compression operations seek to extract essential information from an image so that the image can be accurately reconstructed. Nonessential information is discarded.

We generally do image compression as a prelude to either electronic image storage or transport. This is because both of these operations are sensitive to the amount of data in an image. *Image storage* refers to the electronic storage of an image's data, typically on magnetic or other permanent media. *Image transport* refers to the electronic transfer of an image's data over a data link.

If the amount of data necessary to represent an image can be reduced, then the amount of time to transport it is also reduced. Likewise, the amount of storage space required to store the data is reduced. This way, image compression can yield a significant savings. For instance, the compression of image data by a ratio of ten to one will allow the transport of ten compressed images in the same time required for one uncompressed image.

Still-image compression schemes can be divided into two general groups, lossless compression and lossy compression, as shown in Figure 6.1. Lossless image compression preserves the exact data content of the original image. Lossy image compression preserves some specified level of image quality, but will not preserve the absolute data content of the original.

In addition to still-image compression, there are techniques that address the compression of related sequences of images. This type of image compression is known as motion compression. Motion compression schemes take advantage of the fact that, in a sequence of related images, only the parts of the images that are changing need be compressed. The other portions of the image can be referred back to information contained in a previous image in the sequence.

In this chapter, we will first look at some of the fundamentals of image compression. Then, we will examine a variety of common image compression schemes. Like other digital image processing operations, image compression tech-

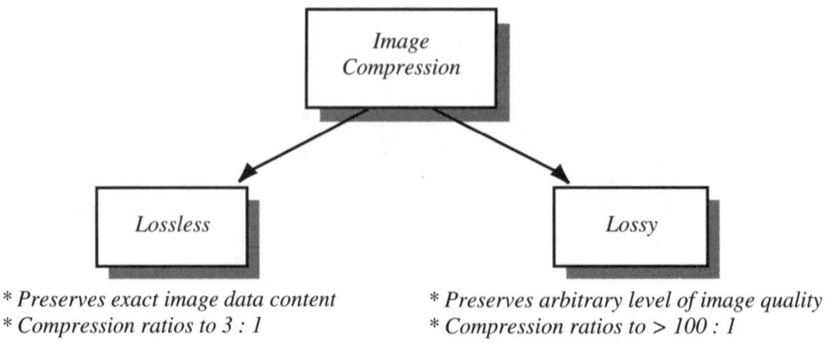

Figure 6.1 *Image compression techniques can be grouped into two major categories: lossless and lossy methods.*

niques are constantly evolving and being refined. As a result, numerous variations and additions to these basic techniques have been developed.

Image Compression Fundamentals

Before embarking on the study of the basic image compression forms, we must first cover some fundamentals. In general, these fundamentals hold true for all image compression schemes.

All image data compression schemes are twofold—they involve both a compression operation and an inverse decompression operation. The compression operation converts the original image data into a compressed image data form. The decompression operation converts the compressed image data back to its original uncompressed form, as shown in Figure 6.2. Image compression and decompression operations are often called *image coding* operations because the processes use data coding methods to represent an image in a new, more concise form.

In Chapter 3, we discussed the quantization of image pixel brightnesses to digital values. This is the first form of coding to which an image is subjected. It is referred to as *pulse code modulation* (*PCM*). The number of bits assigned to each pixel brightness determines the data size and quality of the resulting digital image. The image is coded

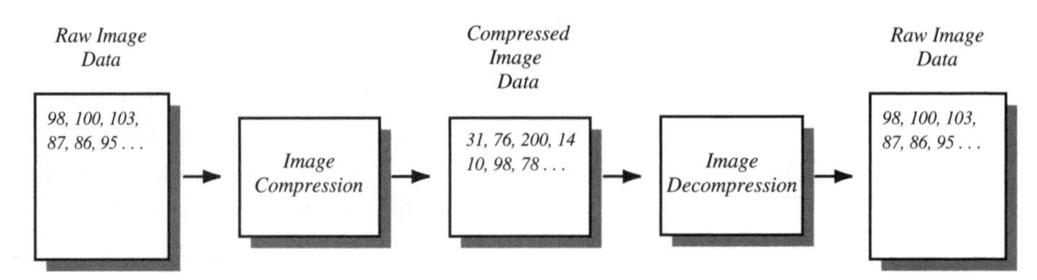

Figure 6.2 *The flow diagram for the image compression and decompression operations.*

to a sequence of pixel brightness values. The coding techniques discussed in this chapter go a step farther, however. They apply more efficient forms of coding to these pixel brightness values, resulting in more efficient digital image representations.

We measure the amount of compression achieved in an image compression operation by dividing the data size of the original image by the data size of the compressed image. The result is called the *compression ratio*. The higher the compression ratio is, the smaller the compressed image has become. In all cases, we want to maximize compression ratios in image compression operations while still meeting application requirements. These requirements include compressed-image quality (for lossy compression schemes), time to compress and decompress the image, and the computational effort.

Image compression operations should not be confused with image file-interchange formats. Many times the two terms are used interchangeably. Generally, compression schemes define an algorithm for the compression and decompression of image data. They rarely define file-interchange formats. File-interchange formats define the data structures for organizing an image file. They call upon the use of compression schemes in their definitions, but usually do not define the associated compression algorithms. Image file-interchange formats are discussed later in Chapter 8.

Image Redundancy

Image compression schemes are based on the fact that any set of data can, and generally does, contain redundancies. This holds true for images as well. Data redundancies can exist as repeated patterns and other forms of common brightness information between multiple pixels of the image. The goal of image compression is to characterize these redundancies and code them to a new form that requires less data than the original.

Let's look at a very rudimentary compression example, as illustrated in Figure 6.3. Say, for instance, that a 640 pixel × 480 line × 256 gray-level image is composed of nothing but vertical lines, each having a brightness value of 255 and a width of 10 pix-

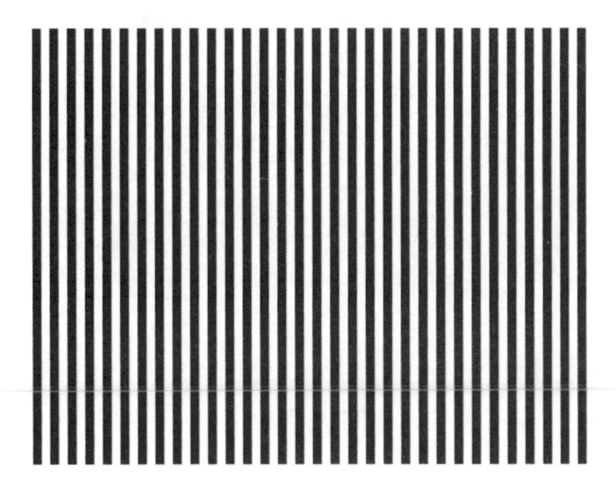

Figure 6.3 *This image of pure white and black vertical lines can be described in very simple terms, thus providing compression of the image data.*

els. Further, each line is separated by a 10-pixel-wide background with a brightness of 0. This image appears as 32 white lines separated by 32 black lines. The image has a data size of $640 \times 480 \times 8$ bits = 307,200 bytes. Obviously, the image contains significant redundancy. In particular, the 32 sets of white and black lines are all equal; only their locations in the image are different. This image can be compressed to a description of "32 vertical white lines interleaved with 32 vertical black lines, each line having a width of 10 pixels." This compressed image representation comprises 104 characters, which equals 104 bytes of data—a compression ratio of nearly 3,000:1.

Although this example is a little far-fetched, it illustrates how a particular image can be represented in two different ways with significantly different data-size requirements. All images have some amount of data redundancy. The trick is to determine the redundancy form and then to compress the image to its smallest form. In practice, this is virtually impossible. This is because the computational effort and time to determine the absolute best way to compress a particular image can become excessive. Instead, many techniques have been developed, each of which is effective for a range or class of image types and quality requirements.

If we apply a compression technique to an image once, it might seem like we could obtain additional compression by repeating the technique several times. For instance, maybe a 2:1 compression scheme could become 4:1, 8:1, or better, depending on how many times we repeated the compression operation. Unfortunately, this does not work because the compression operation removes a particular form of redundancy from the image data. Once it is removed, it cannot be removed again. Although there is some validity to iterative compression approaches, there is no "free lunch," either. Generally, as a compression operation is iteratively applied to an image, the compressed image size reaches a limit where it either doesn't reduce any more or worse, begins to increase in size. If we think about it, it just doesn't make physical sense that we could repetitively compress an image until it was arbitrarily reduced to the size of our choice.

In practice, sometimes a little more compression can be achieved by applying an identical compression operation twice. More commonly, though, further reduction in image size is achieved by successively applying different compression schemes. The schemes used can be mixed and matched based on the best compression efficiency for the types of images involved. This approach often works well because each form of compression removes a different form of redundancy. There is still a limit to how small an image can be compressed. The ultimate size is related to the image's intrinsic redundancy characteristics and the required quality of the decompressed image.

Compression Symmetry

Image compression and decompression operations are not always *symmetrical operations*. This means that, for a particular compression scheme, one operation may

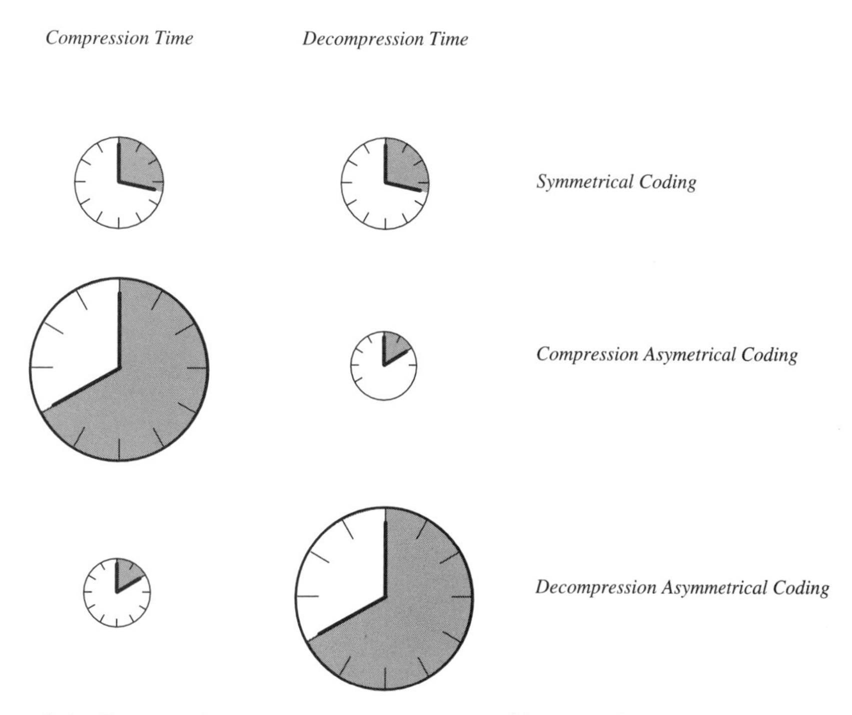

Figure 6.4 *Compression symmetry is a measure of how much time and computational effort is required for a compression operation versus its corresponding decompression operation.*

take longer or require more computational effort than the other, as shown in Figure 6.4. When the two operations require about the same effort, they are called symmetrical. When one operation takes longer than the other, they are called asymmetrical. Depending on the application, we may need very fast compression and may not be concerned with decompression time or computational effort. Alternately, we may require very fast decompression and may not care about compression time or effort. The three forms of image compression symmetry and asymmetry are defined as follows:

Symmetrical Coding (Figure 6.5)

Definition—compression and decompression take similar time and computational effort.

Use—when an image is compressed and decompressed about the same number of times. This is the case when an image is transported from one person to another, as in image interchange applications. Each time an image is compressed, it is transported to the other person and then decompressed. There is no overall advantage

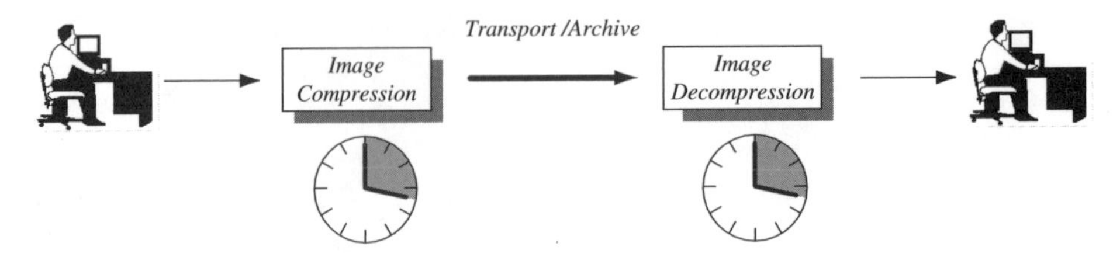

Figure 6.5 Symmetrical coding is used when an image will be exposed to compression and decompression on an equal basis.

for either the compression or decompression operation to be faster and/or more computationally efficient.

Compression Asymmetrical Coding (Figure 6.6)

Definition—compression takes more time and/or computational effort than decompression.

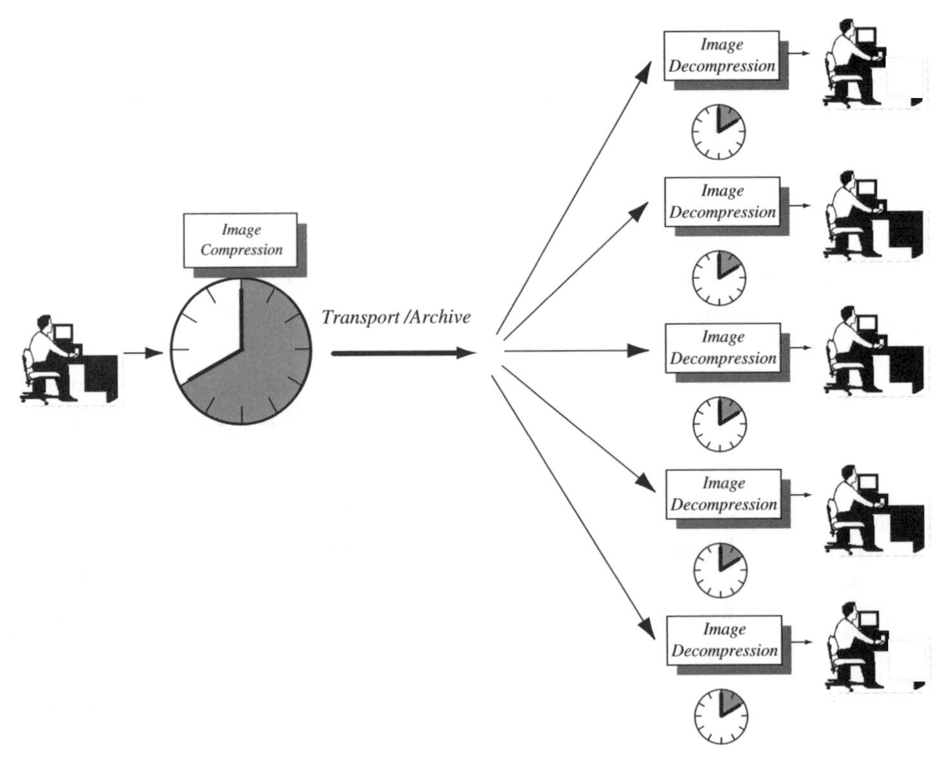

Figure 6.6 Compression asymmetrical coding is used when an image will be exposed to more decompression operations than compression operations.

Use—when an image is compressed once and decompressed many times. This is the case when an image is transported once to many different users, each of whom must decompress it, such as in image-distribution applications. Each time an image is compressed, it is transported to many other people, each of whom must decompress it. Total compression/decompression time can be significantly reduced by making the decompression operation faster and/or more computationally efficient than the compression operation.

Decompression Asymmetrical Coding (Figure 6.7)

Definition—decompression takes more time and/or computational effort than compression.

Use—when an image is decompressed once and compressed many times. This is the case when images are stored many times, but only rarely, if ever, retrieved and decompressed, such as in image archival applications. Each time an image is retrieved and decompressed, it may have been stored many times (like all the bank

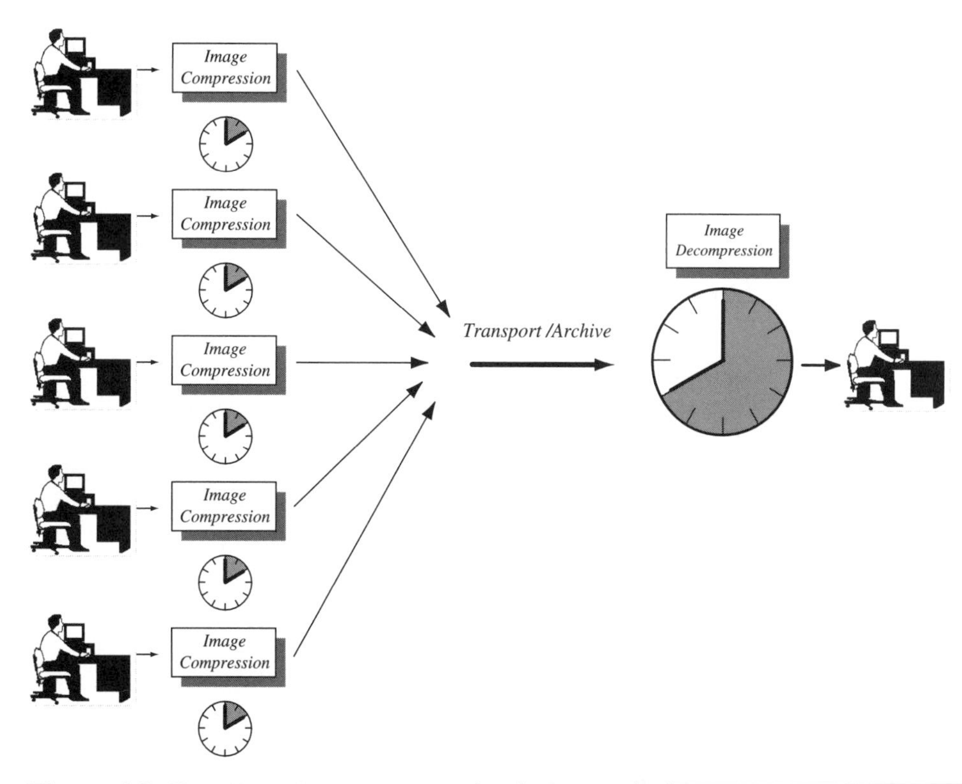

Figure 6.7 *Decompression asymmetrical coding is used when an image will be exposed to more compression operations than decompression operations.*

checks you have ever written), each time requiring compression. Total compression/decompression time can be significantly reduced by making the compression operation faster and/or more computationally efficient than the decompression operation.

Data Explosion

Many different image compression techniques exist. Each has a forte for effectively compressing images of certain types. But each technique also has drawbacks, in the form of occasions when it does not perform well. When a compression scheme is not well suited for use on a particular image type, we might expect that it would provide poor compression characteristics. This can happen. Additionally, though, the phenomenon of *data explosion* can result in a compressed image that is larger in size than the original. Data explosion is obviously an undesirable result when attempting to compress an image to a smaller size.

The data explosion effect generally occurs when a compression scheme attempts to remove a form of image data redundancy that is not present in the original image. This can occur when an image contains patterns with characteristics that are unexpected or unknown to the compression algorithm. Usually, a compression scheme will have an "escape valve," where, if it detects that its results are expanding the size of an image, the scheme will either process the image with another algorithm or pass it in its raw form. We will discuss an example of data explosion later in the section on run-length coding.

Compressed Data Error Effects

Generally, when transporting image data over a data link, we can assume a perfect transmission. In other words, we do not expect data errors or changes of any form. This is usually a good assumption, because generally data-handling protocols watch for *data channel errors* and transparently request retransmissions when necessary. Often these protocols work at low levels, so that the data that an application transmits or receives can be considered error-free.

For the sake of discussion, let's see what happens to image data if an error does occur. For an uncompressed image, single data errors create erroneous pixel brightnesses in a received image. These errors will appear as bright or dark spots, often randomly distributed throughout the image. We visually recognize these errors as impulse noise, similar to that seen in a noisy broadcast television image.

For a compressed image, however, even single data errors can cause disastrous results. Because a compression scheme's job is to reduce the data size of an image, we can envision each piece of data as taking on greater importance in the description of the image. As a result, each piece of data may affect not only the brightness of a single pixel, but also the brightnesses of many pixels. Depending on the robustness of the compression scheme, the entire reconstructed image can show errors throughout it, based on a single data-transmission error to the compressed image data. We will discuss an example of a channel error effect in the upcoming section on run-length coding.

Color Compression Extensions

The compression of color images is usually handled by compressing each color component image individually. In the case of an RGB (red, green, and blue) image, we simply apply the same compression scheme to the three color component images. Likewise, the decompression operation is handled by applying it to the three compressed component images, recreating the original RGB image.

Generally, the RGB color component images carry a lot of redundancy between them. This means that often some of the same image information is contained in each of the color component images. A good compression philosophy can make use of this fact by first converting the image from the RGB color space to a less redundant space.

HSB (hue, saturation, and brightness) space, as discussed in Chapter 3, is a good color space in which to apply image compressions. This is because in RGB space, much of the same image detail information, like edges, is repeated in each RGB component image. So, compressing the RGB component images involves redundantly compressing some of the same information, which is not always efficient. By converting an image to HSB space, most of the image detail information is confined to the brightness component.

The RGB and HSB component images for the same color image are shown in Figure 6.8. The hue and saturation components carry very little detail information. As a result, the hue and saturation component images can be compressed to smaller sizes, yielding an overall compression ratio that is greater for an HSB image than for the same RGB image.

Lossless Image Compression

When a set of arbitrary digital data is compressed, like a text document or numeric accounting data, it is always done so that subsequent decompression produces the exact original data. If the reconstructed data is not exactly like the original, a text document might be missing a few characters here and there, or an accountant's spreadsheet might have a few erroneous entries. Because of the type of data in these examples, close approximation is not good enough. For these

Figure 6.8a *The red component image of a color image.*

Figure 6.8b *The green component image.*

Figure 6.8c *The blue component image.*

Figure 6.8d *The hue component image—the darker gray levels represent the red end of the color spectrum, and the brighter gray levels represent the blue end of the color spectrum.*

Figure 6.8e *The saturation component image—the brighter the gray levels, the more saturated the colors.*

Figure 6.8f *The brightness component image.*

cases, the data must be reconstructed exactly to its original form, or the compression scheme is unusable. The type of compression scheme, where the compressed data is decompressed back to its exact original form, is called *lossless data compression*. It is devoid of losses, or degradations, to the data.

The counterpart to lossless data compression is called *lossy data compression*. Lossy compression schemes introduce degradations to the data they compress; however, they do so in a way that is tolerable for the intended application. Lossy schemes exist because they usually can compress data to much smaller sizes than lossless schemes.

Many imaging applications require the perfection of a lossless scheme when compressing their digital images. These applications are those where very subtle brightness and spatial details are considered relevant to the image quality. Often, an observer cannot even see these subtleties. However, because an image may have subsequent digital image processing operations applied to it, the required image quality may, nonetheless, need to be perfectly maintained.

Lossless image compression is always required when the future analytical use of an image is not known. For instance, space exploration imagery is often studied for years following its origination. Without knowing how the image will be processed in the future, it is impossible to know what lossy degradations will be tolerable. Therefore, it is best not to introduce any degradations in the compression process. In these cases, lossless image compression techniques are used.

A variety of lossless image compression schemes have been developed. Many of these techniques come directly from the digital data-compression world and have been merely adapted for use with digital image data. We will look at several of these methods.

Run-Length Coding

In lossless image compression, there is an intrinsic limitation to how much an image can be compressed. Compression past this point will eliminate some of the information necessary to recreate the original fully in its exact form.

The *entropy* of an image is a measure of this limit. An image's entropy is a measure of its information content. If the entropy is high, an image's information tends to be highly unpredictable. Stated another way, a high-entropy image's information contains a lot of randomness and has little redundancy. If the entropy is low, an image's information is more predictable—it contains little randomness and its redundancy is high.

We can compute an image's entropy as the probability of its occurrence. This is displayed as a number representing the number of bits necessary to represent that probability. For any random image, this would be as follows:

$$\text{Entropy} = \text{Number of Pixels} \times \text{Number of Lines} \times \text{Number of Bits per Pixel}$$

which, for a 640 pixel × 480 line × 8 bit image, would be the following:

$$\text{Entropy} = 640 \times 480 \times 8 = 2{,}457{,}600$$

This is the entropy measure for any random 640 × 480 × 8 bit image. Any one of $2^{2,457,600}$ possible different images can be represented by an image of these dimensions. Conversely, there is a 1 in $2^{2,457,600}$ chance that a particular image of these dimensions will be identical to another.

Because images are rarely made up of totally randomly varying brightnesses, the actual entropy of a normal image will generally be something less than the calculation above. This is because the raw image data quantity will always be higher than the average information data quantity. The actual entropy of an image is the average information quantity of the image. In other words, a 640 pixel × 480 line × 8 bit image can be compressed from a raw form, requiring 2,457,600 bits, to something smaller. The form of compression that does this is called *entropy coding*. Entropy coding techniques reduce image redundancies by using variable-length coding methods. This can be done by coding variable numbers of pixels with fixed-length codes, or by coding fixed numbers of pixels with variable-length codes.

Run-length coding image compression takes advantage of the fact that several nearby pixels in an image will, statistically, tend to have the same brightness value. This form of redundancy can be reduced by grouping pixels of identical brightness into single codes.

The run-length scheme works as follows. The original image is evaluated by starting at the first pixel in the upper left corner, as shown in Figure 6.9. By looking at the first pixel and its following neighbors across the line, the scheme determines how many following pixels have the same brightness value. If the next one or more pixels in the sequence have the same brightness value as the first, they are all repre-

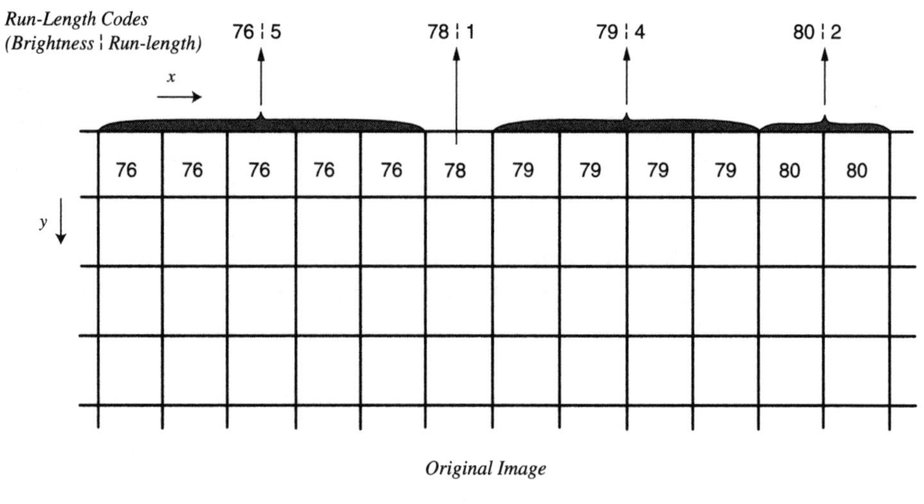

Figure 6.9 *The run-length coding operation.*

sented by a new code. The new code is made up of two values—a brightness value followed by the number of pixels (run-length) that have the same value. The process then moves to the next pixel in the line with a new brightness value and repeats. When the end of the line is reached, the process begins again at the start of the next line. The process continues until the entire image is run-length coded.

If, for instance, 15 sequential pixels had the same brightness value of 234, they would be coded into the two values—brightness | run-length = 234 | 15. The 15 bytes of data needed to represent the original image's 15 pixel brightnesses are compressed to only 2 bytes. The resulting run-length compressed image is made up of 2-byte codes, representing each run's brightness value and length.

A run-length compressed image is decompressed by expanding each code in the compressed image. The brightness of each code is replicated across the decompressed image line for the number of pixels indicated by the adjoining length value.

The run-length technique can be used on binary images with an additional enhancement. Because pixel brightnesses can only be 0 (black) or 1 (white), there is no need to code the brightness value for each run. Instead, we assume that the image begins with a brightness of 0, and simply record the length of the pixel run until a white pixel is encountered. Then, we record the length until a black pixel is encountered, and so on. Because the brightness must be either 0 or 1, there is no need to state the brightness explicitly in the *binary image run-length code.* This modification to the run-length operation for binary images provides even higher compression ratios.

Instead of using the previously discussed run-length technique on gray-scale images, we can use the above binary image run-length technique on each bit plane of a gray-scale image. This is referred to as *bit plane run-length coding.* In this way, we can use the efficiency of the binary image run-length technique while operating on a gray-scale image.

Run-length image compression will provide around 1.5:1 compression ratios on gray-scale images. Ratios from about 4:1 to greater than 10:1 can be expected on binary images, depending on the subject of the image. Applying bit plane run-length coding to gray-scale images will generally yield up to 2:1 compression ratios.

Run-length coding is particularly prone to data explosion problems. Let's say, for instance, that the brightnesses in an image change between every pixel in the image. This is often the case with noisy images. In this example, each pixel of the image will be a distinct run, and will therefore be represented by its own run-length code. Because each run-length code occupies two bytes of data, a brightness value and a run-length, each 1-byte pixel in the original image will be represented by 2-byte run-lengths in the compressed image. The compressed image will be twice as large as the original, as shown in Figure 6.10.

This is an unfortunate artifact of the run-length compression scheme, making it a poor choice for compressing noisy or highly random images. More effective

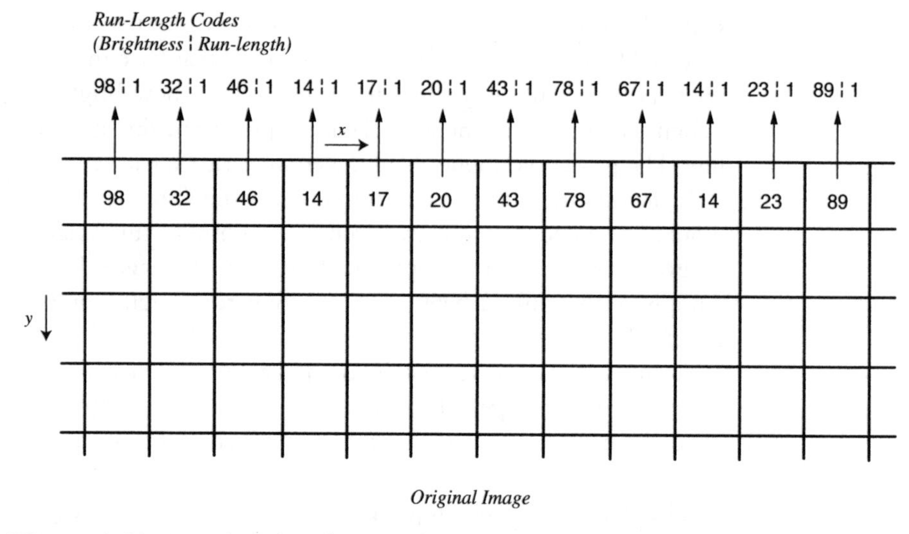

Figure 6.10 *Run-length coding can be prone to data explosion under certain conditions of rapid brightness changes in an image.*

techniques have been developed that modify the basic run-length code's operation for these cases. When high rates of brightness change are detected in a portion of an image, the code can simply revert back to a pass-through mode. In this mode, pixel brightnesses are represented only by their brightness, rather than by a brightness and a run-length value.

Data channel errors can also cause problems for the run-length code scheme. For instance, let's say an image is run-length compressed and then transported over a data link to another location. When received, the compressed image has suffered a data error—one run-length value in the center of the image is missing. When the image is decompressed, the decompression operation begins by expanding each run-length code, as usual. When the run-length code with the missing run-length value is decompressed, however, the brightness value from the next code is erroneously interpreted as the run-length value. As a result, the code is incorrectly expanded to either a longer or shorter length than it should be. Worse, the following codes are all interpreted backwards as well (run-length | brightness value), as shown in Figure 6.11. The remainder of the image will be hopelessly scrambled.

The run-length decompression scheme does not respond well to even a single data error, making it a poor choice for compressing images that are likely to encounter subsequent data errors. Techniques have been developed that modify the basic run-length operation, making it more robust in these cases. By embedding additional codes that indicate the start of a new line, the decompression operation can resynchronize to the correctly oriented data (brightness value | run-length) and resume correct decompression. This way, only a single line of the decompressed image will be corrupted by the data error.

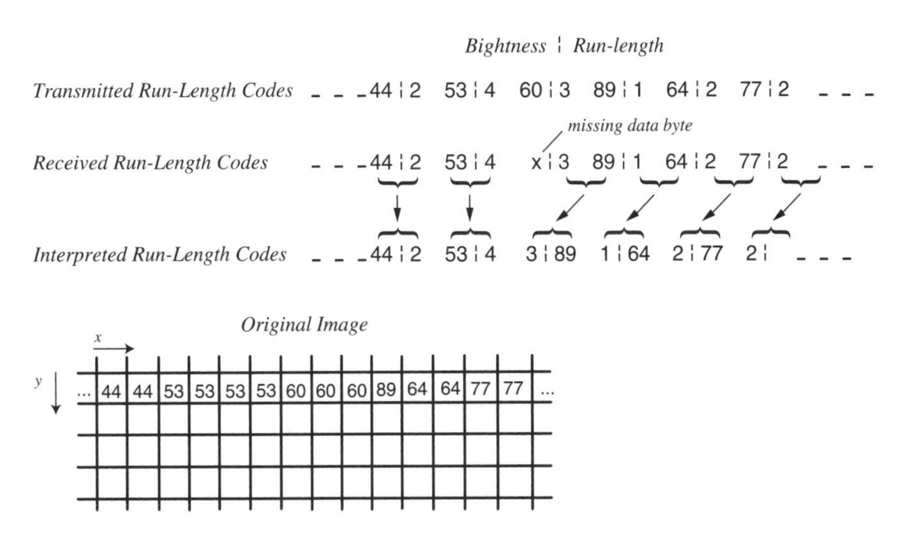

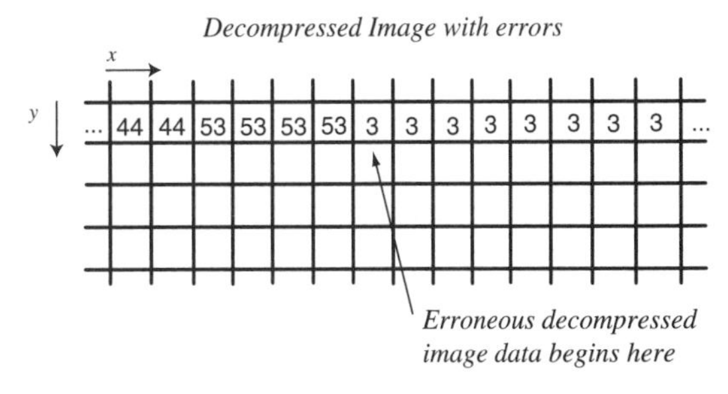

Figure 6.11 *Data-channel errors can cause significant decompression problems of a corrupted run-length coded image.*

Huffman Coding

Another common entropy-coding compression technique is *Huffman coding*. Huffman coding converts the pixel brightness values in the original image to new variable-length codes, based on their frequency of occurrence in the image. In this way, brightnesses that occur frequently are assigned shorter codes, and brightnesses that occur infrequently are assigned longer codes. The result is that the compressed image will require fewer overall bits to describe the original image.

The Huffman compression scheme begins by looking at the brightness histogram of an image. With the histogram, the frequency of occurrence for every brightness in the image is available. By ordering the brightness values by their frequencies of occurrence, we are left with a list where the first value is found most often in the image, and the last value is found least often in the image. With this list, the Huffman coder assigns new codes to each brightness value. The assigned codes are of varying lengths; the shortest codes are assigned to the first (most frequent) values in the list and, eventually, the longest codes are assigned to the last (least frequent) values in the list. Finally, the compressed image is created by simply substituting the new variable-length brightness-value codes for the original 1-byte brightness-value codes. Of course, the Huffman code list that couples original brightness values to their new Huffman variable codes must be appended to the image for use by the Huffman decompression operation, as shown in Figure 6.12.

Huffman codes are assigned by creating a Huffman tree that pairs the brightness values based on their combined frequencies of occurrence. The Huffman tree ensures that the longest codes get assigned to the least frequent brightnesses and vice versa. Using the brightnesses ranked in order of their frequencies of occurrence, the two at the bottom of the list (least frequent) are paired together and labeled 0 and 1. The paired brightnesses are represented by their combined frequencies of occurrence. Then, the next two lowest frequencies of occurrence, made up of brightnesses (or previously paired brightnesses), are determined and paired. Again, the new pair is labeled 0 and 1, and is represented by their combined frequency of occurrence. This continues until all brightnesses and paired brightnesses have been paired. The result is a tree that, when followed, indicates the new Huffman binary code for each brightness in the image.

Figure 6.13 shows a 640 pixel × 480 line image, where each pixel is represented, for simplicity, by a 3-bit brightness value. The image's histogram shows the actual number of pixels in the image with each of the eight brightness values. The brightnesses are ordered based on their frequencies of occurrence and then paired into a Huffman tree, as described above. Although all the pixels in the original image were coded as 3-bit brightness values, the Huffman codes are as small as 1 bit and can be as large as 7 bits. The longest Huffman code can never be greater than the number of different brightnesses in the image (in this case, eight) minus 1. Even though a Huffman-coded image can have some brightnesses with very long codes, their frequencies of occurrence are always statistically low.

The data size of the original image can be computed as 640 × 480 × 3 bits. The Huffman-coded image data size can be computed as the sum of the eight frequencies of occurrence multiplied by the respective number of bits in their code.

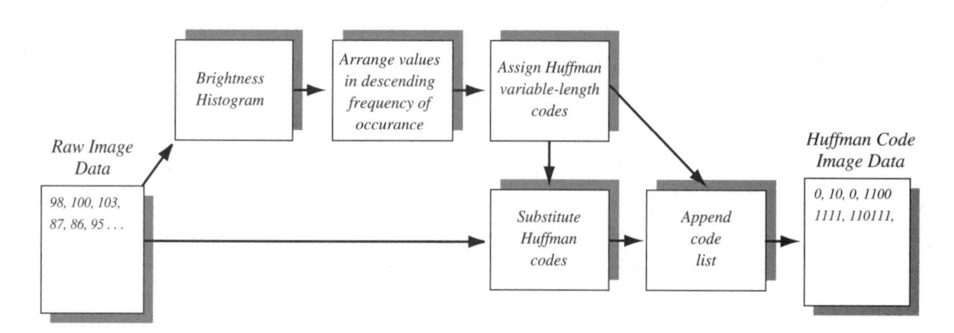

Figure 6.12 *The flow of the Huffman coding operation.*

Figure 6.13a Original 640 pixel × 480 line lighthouse image, with 3-bit brightness values, to be Huffman coded.

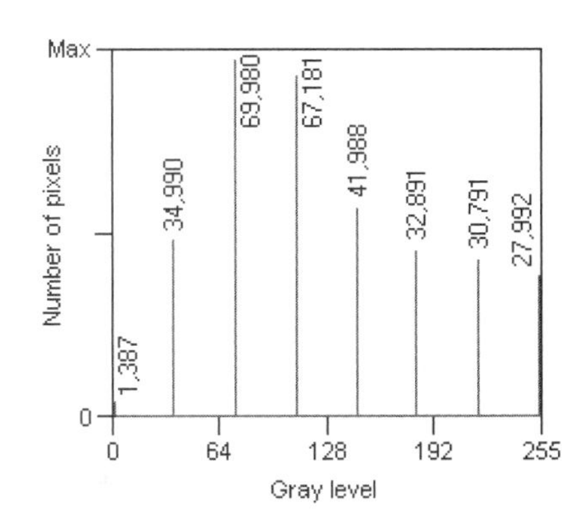

Figure 6.13b Histogram of original image with the number of pixels at each gray level.

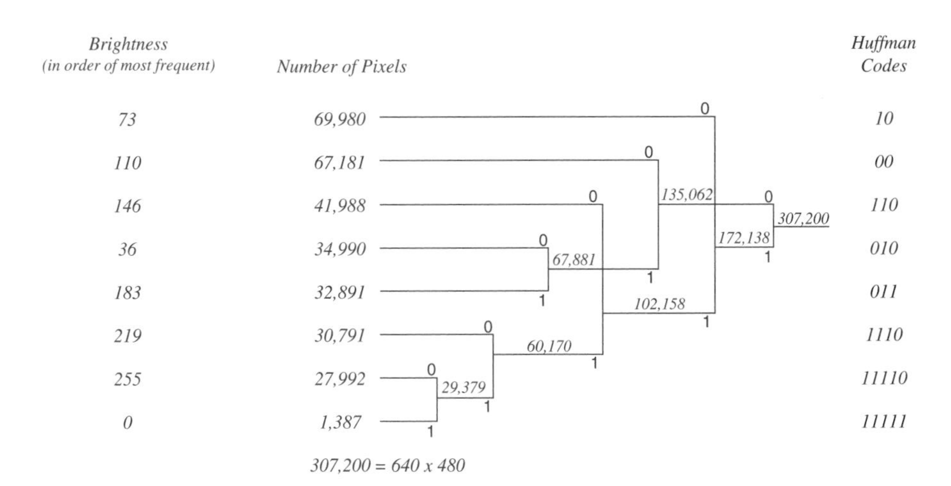

Figure 6.13c The creation of the Huffman tree. The smallest codes are assigned to pixel brightnesses with the highest frequencies of occurrence.

Huffman image decompression reverses the compression process by substituting the original fixed-length 1-byte-long brightness values for the variable-length Huffman-coded values. The original image is exactly recreated. Huffman image compression will generally provide compression ratios around 1.5:1 to 2:1.

Modified versions of the Huffman coding scheme can be used to allow code changes throughout the image. This is done by computing image histograms over regions of the image, rather than over the entire image. Codes are then assigned based on the brightness frequencies of occurrence in each region. This modification of the basic Huffman coding scheme allows the codes to change based on the brightness distributions found in different parts of the image. The codes are therefore more efficiently adapted to the regions of the image, yielding improved compression ratios. Of course, every time the Huffman codes are changed, the changes must be appended to the image for later decompression.

Lossless Predictive Coding

In contrast to entropy coding techniques, *predictive image coding* techniques can also provide excellent compression ratios. Predictive techniques work on the principle that a pixel's brightness can be predicted based on the brightness of the preceding pixel. Using this, a predictive coder codes only the difference information between two pixels.

The fundamental form of predictive image coding is *differential pulse code modulation (DPCM)*. As the name implies, the difference between pixel brightnesses is the quantity that is coded.

The DPCM compression scheme operates on the entire image, pixel by pixel. The first pixel, in the upper left corner, remains unchanged; it is coded exactly as its original brightness. The process moves to the second pixel in the line. The preceding pixel's brightness value is subtracted from the current pixel's brightness. The result of the subtraction is the new coded value for the second pixel in the image. This process repeats across the line. At the start of the next line, the process begins over. The process continues until the entire image is DPCM-coded. The lossless DPCM compression and decompression operations are illustrated in Figure 6.14.

As an example, let's say the first five pixels of an image line have brightness values of 23, 48, 76, 56, and 83. We'll assume that the image originates with 8-bit brightness values. The DPCM-coded values are as shown in Table 6.1.

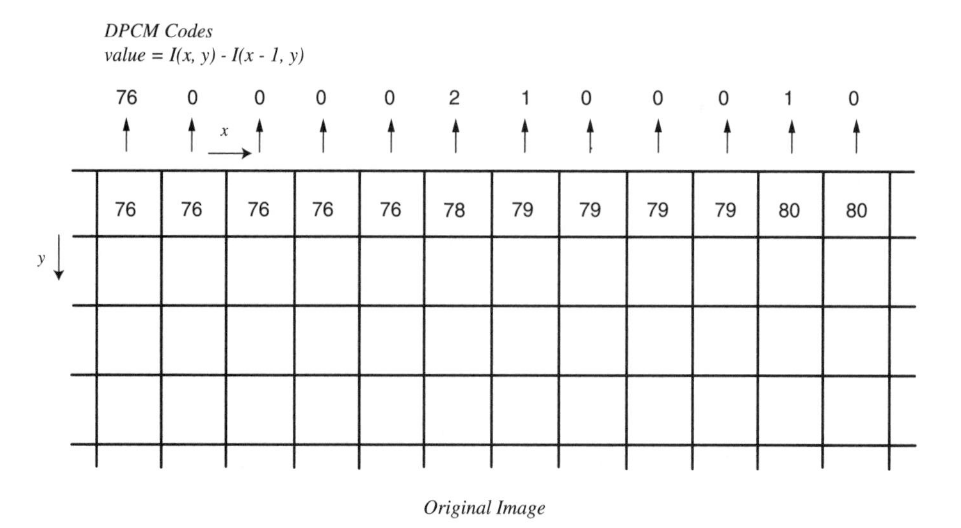

Figure 6.14 *The lossless DPCM coding operation.*

Table 6.1 Lossless DPCM Coding

	ORIGINAL (8-BIT VALUES)	DPCM-CODED (6-BIT VALUES)
Pixel #1	23	23
Pixel #2	48	48 – 23 = 25
Pixel #3	76	76 – 48 = 28
Pixel #4	56	56 – 76 = –20
Pixel #5	83	83 – 56 = 27
	40 bits	30 bits

The five original pixel brightnesses are compressed from 5 × 8-bit values (40 bits) to 5 × 6-bit difference values (30 bits).

The DPCM operation works with the assumption that neighboring pixels will be similar or highly correlated. As a result, their differences will usually be very small values. Looking at the values in the above example, none are greater than 31 or less than –32. These difference values can be coded using 6-bit numbers instead of 8-bit numbers, yielding a compression ratio of 8/6 = 1.333:1. If all difference values were under 16, only 4-bit numbers would be necessary, yielding a 2:1 compression ratio.

In the above example, using a 6-bit code, there could be conditions where the difference between two adjacent pixels is either greater than 31 or less than –32. Although it may be rare, this is likely to occur within an image with normally distributed brightnesses. Therefore, what happens if the difference between two pixels happens to be greater than the limit of available coding bits?

A 6-bit code can only represent difference values between –32 and 31. If the difference between two pixels is greater than these limits, something different must be done to code the difference value. This case is referred to as a *code overload* condition. Commonly, a code—such as –32 in this case—will be reserved to indicate the overload case. When this code is used, it indicates that the absolute, rather than differential, brightness value follows as an 8-bit value. In this way, when a difference of greater than 31 or less than 31 occurs (–32 is now used to signify the overload case), the current pixel's brightness is coded as an 8-bit value instead of a differential 6-bit value. Revisiting the earlier example, but with an overload condition, the coding is as shown in Table 6.2.

The five original pixel brightnesses are compressed from 5 × 8-bit values (40 bits) to 5 × 6-bit difference values plus 1 × 8-bit absolute brightness value (30 + 8 = 38 bits).

Of course, the penalty for this is that not only are 8 bits used for the absolute brightness, but also 6 bits are used to code the condition that an 8-bit absolute value follows. This means that the original 8-bit pixel brightness ends up getting coded as a 6-bit + 8-bit = 14-bit value. If many of these overload conditions exist in a particular image, a data explosion situation can occur, as described earlier.

Table 6.2 Lossless DPCM Coding with a Code Overload Case

	ORIGINAL (8-BIT VALUES)	DPCM-CODED (6-BIT VALUES)
Pixel #1	23	23
Pixel #2	48	48 – 23 = 25
Pixel #3	86	86 – 48 = 38 (overload) → –32 and 86 (8 bits)
Pixel #4	56	56 – 86 = –30
Pixel #5	83	83 – 56 = 27
	40 bits	38 bits

The DPCM compression technique, like other methods, relies on the statistics of the pixel brightnesses in the image to be "normal." For a "normal" image, it is statistically rare that difference values will exceed the selected coding limit (In this case, 6 bits). Hence, the overall average compression ratio will generally create good compression.

A DPCM-compressed image is decompressed by repetitively adding the current pixel difference value to the last pixel's computed brightness. If the current value is a code overload value, then the current pixel's brightness is the following absolute 8-bit brightness value.

The DPCM compression operation works best on images that do not have an inordinate number of large brightness swings between adjoining pixels. When applied to normal images, DPCM coding can provide around 2:1 compression ratios. For images with long runs of constant-valued pixels, compression ratios will increase significantly.

Additional prediction schemes have been developed that use more than just the preceding pixel's brightness to predict the current pixel's brightness. By looking at several preceding pixels, even pixels in adjacent lines, a predictor can adapt to local image brightness trends and make better assumptions of what the current pixel's brightness will most likely be. The better the decompressor can predict the current pixel's brightness value, the less information it needs from the compressor to represent the difference between the predicted and actual values. As a result, the difference between current pixel brightnesses and predicted brightnesses can be coded as smaller values, thus leading to smaller compressed images.

Lossless Block Coding

Pixel block compression techniques code a group of pixels at a time, searching for image redundancy in the form of repeated patterns. Block coding is based on the statistical likelihood that brightness patterns will probably repeat throughout an image. A block of pixels can be defined in the form of a two-dimensional array of pixels or a one-dimensional line of pixels. Generally, the one-dimensional approach is used for its simplicity. Pixel block sizes can range from one pixel to hundreds of pixels.

Block coding is implemented using a table of different pixel brightness patterns, called a *codebook*. The original image is evaluated pixel by pixel across each line, beginning in the upper left of the image. The evaluation searches for brightness patterns that match one of the patterns in the codebook. When a matching pattern is found, the block of pixel brightnesses is represented, in the compressed image, as a code pointing to the entry in the codebook. The codebook size is kept small enough to ensure that the number of code bits necessary to point to a codebook entry is smaller than the average number of bits replaced in the original image's pattern.

The codebook will generally be different for every image, making it possible to take advantage of the particular qualities of the image being coded. Because no information exists about the patterns that will be found in a particular image, the codebook initially starts out with 256 entries (for a 256-gray-level image), each representing a single-pixel pattern of a brightness from 0 to 255. Obviously, this codebook will only provide a 1:1 compression ratio, but it is a good place to start. As the original image is evaluated, new patterns are entered into the codebook. This way, the codebook grows in size, limited by how many code bits we wish to use to point to the patterns in the compressed image. Often, as more patterns are added to the codebook, older, shorter patterns are dropped from the codebook. This leads to combining multiple short patterns into single longer patterns. In this way, the code becomes more efficient as the image is compressed.

The codebook must be attached to the compressed image file for later use by the decompressor. The decompression operation builds the decompressed image by simply replacing the codes in the compressed image with the patterns indicated in the codebook. Lossless pixel block-coding techniques can generally provide between 2:1 and 3:1 compression ratios. The computational requirements for block coding, however, are higher than for other techniques like run-length and DPCM coding. When more time or compute power is a reasonable tradeoff for smaller compressed images, block-coding methods can be superior to other single-pixel approaches.

A common lossless block-coding method for images is the *Lempel-Ziv-Welch* (*LZW*) scheme. Based on the Lempel-Ziv algorithms for lossless data compression, it operates, in principle, as described in the previous block-coding discussion. The codebook is initialized with the 256 single-pixel brightness entries. Each pixel in the original image is examined one by one, stepping across lines of the image. As a new pixel is evaluated, the compressor appends the brightness to the current pixel block, expanding the block by an additional pixel. If the new block is not in the codebook, it is added. If the new block is in the codebook, the process repeats by adding more and more pixels to the current block until the block is not found in the codebook. At that point, the new block is added to the codebook. Ultimately, the codebook will reach its size limit. For instance, if 10-bit codes are being used to point to codebook entries, then only $2^{10} = 1024$ different pixel block patterns can exist in the codebook.

Once the codebook is full, the LZW compressor matches pixel blocks to codebook blocks, looking for the longest matches. The pixel block is then represented in the compressed output image by the appropriate code that points to the codebook entry.

Extensions to the basic block-coding codebook approach provide for scrapping the codebook during an image and restarting a new one. This way, if an image has a dramatic statistical change somewhere within it, the compressor is not crippled with a codebook that is no longer efficient. This approach works well with images that have changes in their spatial characteristics in different regions of the image. An example might be at a breakpoint in an image between a scene's foreground and background, where each portion of the image has different attributes of spatial content, brightness, or color.

Further, a special code can be set aside to provide a code overload-like function, where, if a block cannot be found in the codebook and the codebook is full, the absolute pixel brightness values are used. The overload code is used to signify that the following values are absolute brightness values rather than a pointer to a pixel block in the codebook.

Following the block-coding process, it is common to Huffman-code the coded image. This way, the most common block codes are reassigned short codes, and the least common are assigned long codes. This secondary Huffman coding process can improve the overall compression ratio for the image coding operation.

Lossy Image Compression

Lossy image compression schemes are not inferior to lossless schemes. In fact, for certain applications, they are far superior. Lossy schemes do not compress an image to a form that can be precisely reconstructed back to the original image form. Some loss in the digital data of the original image will occur. The important thing, however, is that the image can still maintain its visual integrity, and hence some arbitrary level of quality.

All forms of lossy image compression involve discarding image data. Except for truncation coding, image data is not discarded directly from the image, however. Instead, the image is first transformed to another form, then portions of it are discarded. The methods of transforming and discarding image data are what distinguishes the different lossy compression schemes.

The greatest advantage to lossy compression schemes is their ability to compress an image to a much smaller data form than lossless schemes. Lossy schemes can push the compression ratios to 10:1 without visually noticeable degradations. They can reach 100:1 and greater compression ratios with some noticeable degradations.

For many applications, the degradations introduced by lossy compression schemes can be easily accepted in exchange for their incredibly high compression ratios. For instance, television broadcasters do not need to maintain absolute image data integrity when transporting video programming to viewers. While the video images meet extremely high digital quality standards at the originating television studio, the viewer is pleased with accepting much lower quality. In other words, certain degradations introduced by lossy compression will not degrade the visual quality beyond its acceptance by the end-user, the viewing consumer. After all, most consumers are already content with VHS-format videotape image quality, which can run down to as low as one-half the intrinsic quality of the original video images.

Many lossless image compression schemes have been developed. Generally, each is tuned to meet the quality requirements of a specific application. We will now look at several of these methods.

Truncation Coding

This form of image coding is the simplest form of lossy image compression. It works by discarding image data using spatial downsampling and brightness resolution reduction. In other words, *truncation coding* techniques directly throw away image data to achieve a smaller image data size.

At first, truncation coding may appear to be trivial and not practically useful. After all, it is just a technique for the brute-force removal of image data. But, it can be a viable technique in some cases. For instance, when an image is over-resolved for a particular application, truncation coding can be done without impacting the application. For instance, let's say we wish to store and later electronically transport an image to a customer. Further, let's say that the image has the dimensions of 640 pixels × 480 lines, and that it is destined for a printing job in which it will appear with the dimensions of 1.5 inches × 1.125 inches. Using a halftone printing process, with a halftone resolution of 133 dots per inch, the image needs only the following rough spatial resolution:

1.5 inches × 133 dots (pixels) per inch = 200 pixels
1.125 inches × 133 dots per inch = 150 lines

Therefore, the image can be reduced in spatial size by about 10 to 1, without any perceptible degradation in the finished product. So, by simply downsampling the image from the dimensions 640 × 480 to 200 × 150, we can yield a compression ratio of about 10:1.

Truncation coding can be done to either the spatial resolution or the brightness resolution of an image. For spatial reduction, we remove a regular pattern of pixels from the image using downsampling techniques, as described in Chapter 4. For instance, if we remove every other pixel and line from an image, its data size will reduce by a factor of 4. We can decompress such an image by one of two methods. In the first, we simply reconstruct the image at a reduced size—one-fourth the size of the original, for the above example. The other method interpolates the missing pixels, creating an approximation to the original image, at its original size, as shown in Figure 6.15.

Figure 6.15a *Original lighthouse image.*

Figure 6.15b *Spatial-truncation coded image with a compression ratio of 4:1. The image is decompressed to one-quarter size.*

Figure 6.15c *Same spatial-truncation coded image, but decompressed to full size with nearest neighbor pixel interpolation.*

Figure 6.15d *This error image shows the differences between the original image and the spatial-truncation coded image with nearest neighbor interpolation. The error image is created by subtracting the decompressed image from the original.*

Figure 6.15e *Same spatial-truncation coded image, but decompressed to full size with bilinear pixel interpolation.*

Figure 6.15f *The error image for the spatial-truncation coded image with bilinear interpolation.*

Brightness resolution truncation coding is done by truncating all image pixel brightness values to a representation using fewer data bits. The low-order bits are the ones truncated, of course, because they have the least significance to the brightness value. If, for instance, 5 bits are truncated from each pixel's brightness, an image shrinks by a factor of 8 bits/3 bits = 2.667. We can decompress a brightness-truncated image in a couple of different ways, depending on the application, as shown in Figure 6.16. In the first method, we simply reconstruct the image with the reduced brightness resolution. Sometimes, this method may show brightness posterizing effects, though. Using another method, we can add a 5-bit noise pattern, called *dither noise*, to the pixel brightness values. This creates 8-bit pixel brightnesses with a random noise pattern that diffuses the posterizing effect, resulting in a decompressed image that is generally more pleasing to view.

Figure 6.16a *Original lighthouse image.*

Figure 6.16b *3-bit brightness-truncation coded image with a compression ratio of 2.667:1.*

Figure 6.16c *The error image for the brightness-truncation coded image.*

Figure 6.16d *Same brightness-truncation coded image, but decompressed with an added 5-bit dither noise to diffuse the posterizing effect.*

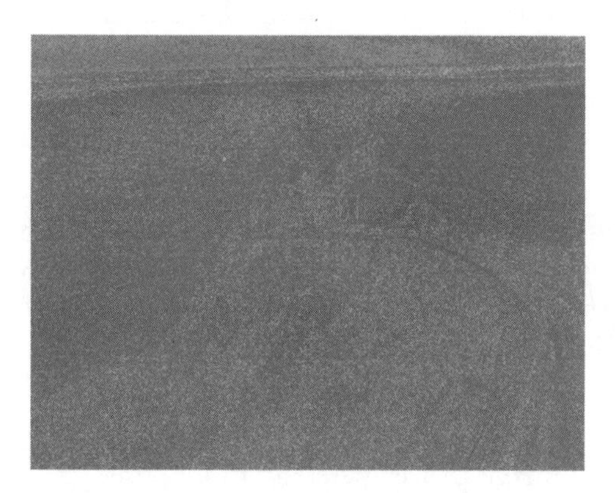

Figure 6.16e *The error image for the brightness-truncation coded image with added dither noise.*

A motion sequence of multiple images can be truncation-coded as well. Instead of discarding spatial or brightness data as described before, whole image frames are discarded on a regular basis. If every other frame is discarded from a image sequence, the sequence reduces in data size by a factor of two. One way to decompress the sequence is to simply replicate image frames to fill in for the discarded frames. This technique can cause the sequence to appear jerky, though. Alternately, the decompression process can use interpolation techniques to recreate the discarded image frames, using a composite mix or average of the images before and after each discarded frame. This method can cause motion blurring in some cases. Optimally, a blend of the two decompression techniques is used to minimize the jerking and blurring artifacts.

Several times throughout this book, we have discussed an operation that is related to truncation coding—binary contrast enhancement. As we have seen, sometimes it is desirable to highlight aspects of an image using the binary contrast enhancement operation. This is especially true when processing images of binary subjects, like text documents or line drawings. We can view this operation as a "smart" brightness truncation compression scheme. It sets the pixel brightnesses in an image to one of two levels—black or white—depending on whether the brightness is below or above a threshold brightness level. In this way, the binary contrast enhancement operation intelligently reduces the number of bits representing pixel brightnesses from eight to one. This, of course, represents a compression ratio of 8:1.

Truncation coding provides precisely predictable compression ratios. They are based directly on how much image data is discarded, as illustrated in the previous examples.

Lossy Predictive Coding

Lossy DPCM compression and decompression schemes are identical to the lossless form of DPCM, discussed earlier, except that the code overload condition is not handled as a special case. Instead, the maximum difference code is used for successive pixels until the coder catches up with the actual brightness value of the original image. This is illustrated on a one-dimensional image profile in Figure 6.17. As a result, the decompressed image brightness will appear somewhat smeared wherever sharp brightness transitions are encountered in the original image.

Let's go back to the example used in the lossless DPCM discussion. With the code overload condition, the coding looked as shown in Table 6.3.

The five original pixel brightnesses were compressed from 5 × 8-bit values (40 bits) to 5 × 6-bit difference values plus 1 × 8-bit absolute brightness value (38 bits).

The lossy DPCM coding of the same set of pixel brightnesses is shown in Table 6.4.

The five original pixel brightnesses are compressed from 5 × 8-bit values (40 bits) to 5 × 6-bit difference values (30 bits). However, when the image is decom-

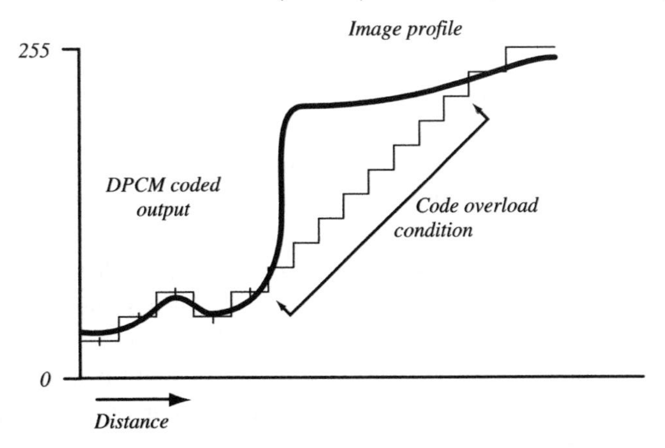

Figure 6.17 *When a rapid brightness change is encountered, the lossy DPCM compressor must use its maximum code values until it catches up with the actual signal. This causes brightness smearing, because the reconstructed brightnesses lag their position in the original image.*

Table 6.3 Lossless DPCM Coding with Code Overload Handling

	ORIGINAL (8-BIT VALUES)	DPCM-CODED (6-BIT VALUES)
Pixel #1	23	23
Pixel #2	48	48 – 23 = 25
Pixel #3	86	86 – 48 = 38 (overload) → –32 and 86 (8 bits)
Pixel #4	56	56 – 86 = –30
Pixel #5	83	83 – 56 = 27
	40 bits	38 bits

Table 6.4 Lossy DPCM Coding without Code Overload Handling

	ORIGINAL (8-BIT VALUES)	DPCM-CODED (6-BIT VALUES)
Pixel #1	23	23
Pixel #2	48	48 – 23 = 25
Pixel #3	86	86 – 48 = 31 (overloaded)
Pixel #4	56	56 – 79 = –23
Pixel #5	83	83 – 56 = 27
	40 bits	30 bits

pressed, the third pixel has the erroneous brightness value of 23 + 25 + 31 = 79, off by 7 gray levels. The fourth pixel corrects the erroneous condition because the brightness value is once again within range. If the brightness step had been greater, it may have taken several pixels before the coder would have caught up. Erroneously decompressed brightness values (overload conditions) result in image smearing of sharp brightness transitions in the decompressed image.

Lossy DPCM almost always adds some degradation to a compressed image, but can provide increased compression ratios of about 3:1 or more over its lossless counterpart, with only minor distortion effects. Usually, we can minimize the visual effect of its distortions through proper code length selection, yielding results acceptable for many applications. Figure 6.18 shows a lossy DPCM-compressed image.

A special case of lossy DPCM, called *Delta Modulation* (*DM*), uses only a single bit to code the difference in brightness between adjacent pixels. Each pixel is coded as having a brightness of less than the previous pixel (0) or greater than the previous pixel (1). DM images can show major effects of brightness errors whenever significant brightness transitions are encountered. Generally, for a 640 pixel × 480 line image, whenever transitions in excess of 32 gray levels are present, significant smearing will be noticeable. The advantages of the DM coding technique are an extremely simple coder and decoder implementation and a fixed 8:1 compression ratio (for 256-gray-level images).

Figure 6.18a *Original lighthouse image.*

Figure 6.18b *Lossy DPCM-coded image using a 5-bit code length.*

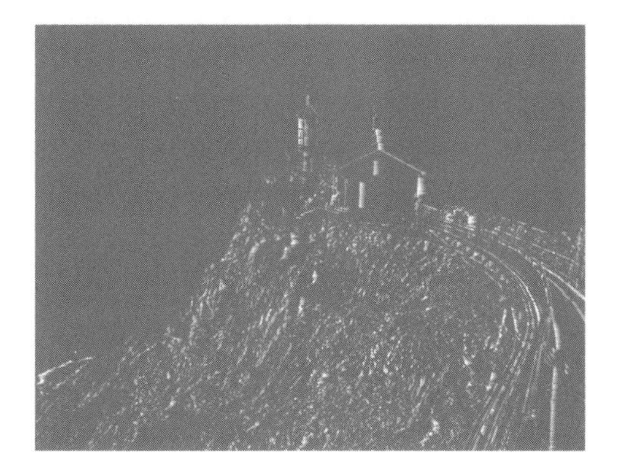

Figure 6.18c *The error image for the lossy 5-bit DPCM-coded image.*

Lossy Block Coding

Lossy pixel block-coding schemes are similar to the lossless forms discussed earlier, except that pixel blocks are not perfectly represented by the codebook blocks. When a compressed image is decompressed, it will not represent the exact data of the original image. Instead, some errors will occur, although in most cases they can be kept to a visual minimum.

Lossy block-coding techniques use a codebook approach, just like their lossless counterparts. They do not, however, attempt to find identical pixel block matches in the codebook. Rather, they look for good representative matches, where the error between the block being coded and the codebook block is minimized. These techniques measure the error between pixel blocks and codebook blocks using a measure that is based on the intended use of the decompressed image. For instance, if the decompressed image will be used by a human observer, then an error measure that minimizes image distortion, as perceived by the human, would be best to use. This way, image qualities that are important to the human observer, such as edges and good contrast, can be maximized. If, however, the decompressed image is to be used for some other application, like one involving image analysis, the error measure may completely different. In this case, the error measure would be selected to maintain other image qualities that are important to the application.

A common form of lossy block-compression is a technique known as *vector quantization* (*VQ*). Vector quantization methods generally operate on relatively small two-dimensional pixel block sizes, like 4 × 4 pixels. Each pixel block of the image is processed one by one. First, the average value of the pixel block's brightness is computed and subtracted from each of the pixel brightness values. This leaves a pixel block of residual brightness values, representing the difference between each pixel's brightness and the average brightness of the block. The block of residual pixel values is then compared with those in the codebook. The code that points to the closest match in the codebook is used. The block of pixels in the input image is replaced in the output image by the block's average brightness value and the code pointing to the correct residual pixel block in the codebook. This process continues over the entire image, yielding a compressed output image.

The VQ codebook is not created during the compression process, as in the lossless form of block coding. Rather, it is created first, using the statistics of a group of training images that are similar in content to the image being compressed. This way, there is a strong likelihood that a pixel residual block pattern stored in the codebook will be close, although rarely identical, to block patterns in the image being compressed.

Another form of lossy block coding is *fractal compression*. Fractal compression methods also divide the image into individual pixel blocks for evaluation. The codebook is created by selecting additional prototype blocks that represent the characteristics of the image. To compress an image, each pixel block is compared against the codebook blocks for the closest match. What is unique in fractal compression is that instead of just selecting the closest match between pixel blocks and codebook blocks, the matching process can apply a geometric transformation to the codebook blocks to make them better match the pixel block being evaluated. This allows each codebook block to be stretched, shrunk, and rotated, creating more flexibility and accuracy in the matching process. Fractal compression schemes provide good compression because natural images tend to have many repeating patterns that vary in size, rotation, and position.

As in lossless block coding, the lossy block-coding codebook must be attached to the compressed image file for later use by the decompressor. The decompression operation is similar to the lossless technique. It builds the decompressed image by replacing the block average brightness values and codes in the compressed image with the residual block patterns from the codebook, summed with their block average brightness values. The lossy form of pixel block-coding techniques can generally provide 10:1 and greater compression ratios. The computational requirements for block coding, however, are higher than for other techniques. As is typical with image compression schemes, a compromise must be made between compression ratio and required compute power and time. Huffman coding of the pixel block codes can, again, provide additional compression following the lossy block-coding process.

Transform Coding

Transform coding is really a form of lossy block coding, but it is unique in the way pixel blocks are coded. Transform compressors do not use codebooks to store block patterns. Instead, the blocks are transformed from the spatial domain to the frequency domain. The fundamental frequency component values of the frequency domain image are the codes stored as the compressed image.

As we discussed in Chapter 4, frequency transforms provide the ability to transform, and thereby represent, an image in the frequency domain as its fundamental frequency components. We also discussed how the frequency image could be transformed back to the spatial domain, recreating the original image. This principle is the foundation to transform compression techniques.

In the frequency domain, the fundamental frequency components, represented by pixel brightnesses, tend to clump in regions, especially around the low frequency zones. As a result, there are generally large areas of the frequency image where the frequency components have a very small or 0 value. This occurs because the frequency transform process removes a lot of redundancy from the image. The frequency domain version of the image is generally a very efficient representation of the original image.

Transform image compression techniques exploit this efficient nature of the frequency image by simply eliminating components from the frequency image that have very small values. Because the small-valued components' contribution in the inverse frequency transform is very small, when the image is inverse-transformed back to the spatial domain, the removal of the small-valued components causes little distortion. Additionally, other components can be reduced in their resolution without significant inverse transform effects.

In transform coding, a pixel block is usually a two-dimensional block of pixels with relatively small dimensions, such as 4×4 or 8×8 pixels. The block of pixel brightnesses is treated as a miniature image and frequency-transformed to a frequency image of equal size. The fundamental frequency components are then evaluated, discarding the ones with small values and reducing the resolution of all. The remaining frequency components become the codes that are stored as the compressed image.

The decompression operation is merely the inverse frequency transform of the remaining stored frequency components. The quality of the decompressed image is related to which, and how many, of the components are removed from the frequency image. Also, the resolution reduction of the saved components will affect the decompression quality. Generally, significant compression ratios, starting on the order of 10:1, can be achieved with good resulting image quality. The quality of the decompressed image will vary from minor, imperceptible distortions to conspicuously obvious levels of distortion. The quality is inversely proportional to the compression ratio. A transform-coded image is shown in Figure 6.19.

Many frequency transforms exist and can be used in transform image compression. In selecting a good transform for image compression purposes, it is important to choose one that is efficient in removing image redundancies. This means that we would like to use a transform that produces the fewest number of frequency components, and thus creates the smallest compressed image size. For image compression operations, it turns out that the *discrete cosine transform* (*DCT*), works very well for a wide range of arbitrary image types. The DCT is similar to the Fourier transform discussed in Chapter 4. Its primary difference is that it is better at compactly representing images of very small size. Because small pixel blocks are generally used in block coding, this becomes a very important characteristic of the DCT, making it well suited for image compression operations. The

Figure 6.19a *Original lighthouse image.*

Figure 6.19b *Transform-coded image with a compression ratio of 20:1.*

Figure 6.19c *The error image for the transform-coded image.*

DCT does, however, require more computational power than the Fourier transform.

In our discussion of the Fourier transform in Chapter 4, we noted that the Fourier transform models the spatial image as though it were periodic. This means that the Fourier transform sees the image as if it wrapped around from its right side back around to the left side and from its bottom back up to the top. As a result, the fundamental frequency components created in the Fourier frequency transform process can include additional ones that model the edge discontinuities in the image—discontinuities that don't really exist in the image. These extra components are undesirable for a compression operation because more of them must be saved to reconstruct the original image accurately from the compressed image.

One way to make the Fourier transform more appropriate for image compression operations is to window the pixel block first, before applying the transform, as discussed in Chapter 4. But because typical pixel block sizes used in compression are small—like 4×4 or 8×8 pixels—the windowing process ends up throwing away significant image information. This can lead to poor decompressed image quality.

For transform-coding schemes created specifically for a particular image type, different frequency transforms can be evaluated for their compactness. For use on a wide range of arbitrary image types, however, the DCT's compact frequency representation of small image blocks has made it a common transform-coding choice.

Motion Compression

Motion compression is a particular case of image compression where image redundancy between sequential image frames is removed. It is applicable only for a sequence of related images, like those encountered in motion picture and video production, medical image sequences, and teleconferencing applications. This form of image compression is referred to as *interframe coding*.

In natural image time-sequences, such as those created by the applications noted above, the differences between succeeding image frames are small and can be attributed mostly to individual object and camera motions. In these cases, bulk portions of an image are identical between two succeeding frames. This, of course, implies that a lot of redundancy exists between the frames. Interframe coding techniques remove this redundancy by referring to a previous frame for information, rather than recoding the same information multiple times. Whenever an entire scene changes, however, motion compression must treat the new frame as unrelated to a previous frame and rely on the standard single-image coding techniques described earlier in this chapter.

Several interframe coding techniques have evolved. Each attempts to code, in a more concise form, image information that is duplicated between two image frames. In this way, the resulting individual image sizes are reduced. The determination of duplicated image information between two frames is made by measuring the motion between the frames. This process is called *motion estimation*.

Motion-estimation techniques generally involve the block-by-block comparison of two images. The current image frame is compared against a *reference frame*, which is either the preceding frame or an earlier frame. When no difference—or a difference below some set limit—is detected, the block is determined to have no motion. When a difference above the limit is detected, the block is determined to have motion. When a block has motion, comparisons are made to determine how it moved between the two frames. Sometimes, however, the block will be entirely new to the frame. The result of the motion estimation process tells us which blocks in the current image frame have motion—are different from the corresponding blocks in the reference frame—and how they moved between the frames. When a block is entirely new to the current image frame, we must recode its image information. All other blocks can be coded by simply referencing the corresponding block in the reference frame. This process is shown in Figure 6.20.

Motion compression schemes generally sit on top of still-image block-coding compression operations. The first image in a sequence is coded as though it is a stand-alone image. The following images are coded the same way, but with the addition of special block codes. These motion-related codes are used to indicate that either a block is identical to the corresponding block in the reference image, or

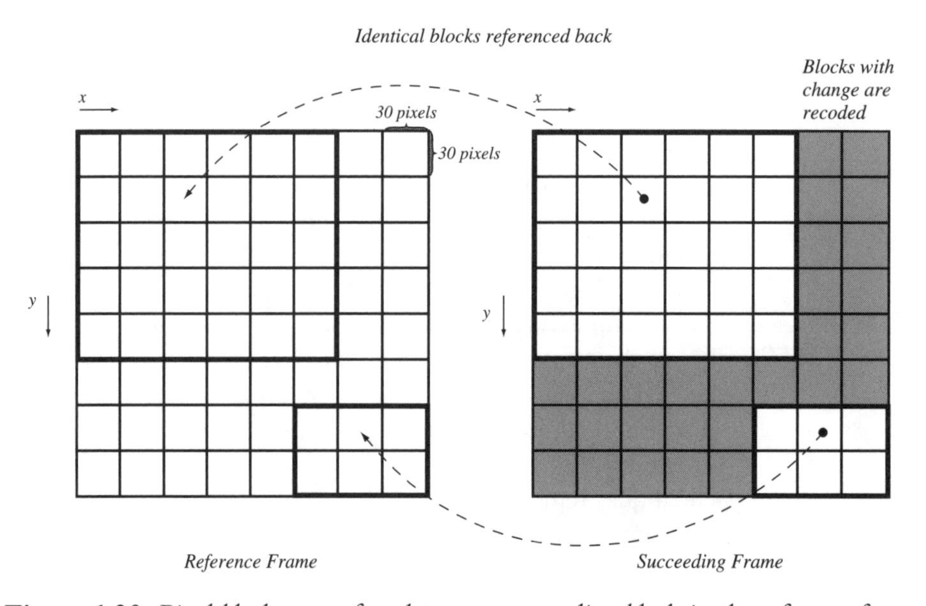

Figure 6.20 *Pixel blocks are referred to a corresponding block in the reference frame when possible; otherwise, they must be recoded.*

that it is a moved version of a block in the reference image. The special block codes are used whenever a block can refer to a correspondingly similar block in the reference frame, rather than be recoded. The block motion conditions are as follows:

1. Block has no motion—Identical to same block in reference frame. The "no change" block code is used.
2. Block has motion—Identical to another block in reference frame. The corresponding reference-frame block location is used as the block code.
3. Block has new motion—New block that moved into frame and was not in reference frame at all. The block is recoded.

Although lossless motion compression schemes can be implemented, lossy forms are far more prevalent. This is because most motion applications require very high compression ratios and can accept certain image degradations. Lossy motion compression schemes, when applied to time-related motion sequences, can typically achieve compression ratios in excess of 100:1 with very good quality.

Standard Image Compression Schemes

We have discussed several techniques of lossless and lossy image data compression. Each method eliminates a unique form of data redundancy from an image's data, hence reducing its data size. Many of the methods discussed can be set up and adjusted in a number of ways. For instance, the number of bits used in a lossy DPCM coder or the number of frequency-component values retained in a transform coder can be arbitrarily selected to achieve the desired decompressed image quality. As a result, there is a need to standardize the implementation of an image data compression scheme so that equipment or software produced by multiple vendors will work together.

A number of standardized image compression techniques have evolved to support the requirements of different industries. A good example includes the compression standards created for facsimile (fax) machines to transmit a document's image. Many standards have also been created to compress common still images as well as motion image sequences. Let's look at some of the most common nonproprietary image data compression schemes used today.

CCITT Group 3/4 and Joint Bi-level Image Experts Group (JBIG)

The most widely used standardized image compression schemes are the *CCITT Group 3* and *Group 4* schemes. These schemes were established by the *International*

Telegraph and Telephone Consultative Committee (*CCITT*) for the transmission of binary images. These compression standards are lossless and are used by all common fax machines to transmit text and line-drawing document images.

The Group 3 and 4 compression standards are based on run-length and Huffman coding techniques. First, the runs of black and white pixels are compressed to run-lengths. The run-lengths are then Huffman coded. The Huffman coder assigns shorter codes to more frequently occurring run-length values and longer codes to less frequent values. Additionally, comparisons for common image data are made between lines, further eliminating line-to-line redundancy. An end-of-line signal is embedded into the compressed image data to allow resynchronization in the presence of data errors. The Group 4 standard is a simplified version of the Group 3 scheme that provides improved compression ratios.

Unfortunately, the Group 3 and 4 schemes can cause data explosion when applied to highly detailed images such as those containing halftones. A newer standard, called the *Joint Bi-level Image Experts Group* (*JBIG*), improves the compression ratios of these types of binary images. The JBIG standard was established by a joint effort of the *International Organization for Standardization* (*ISO*), *International Electrotechnical Commission* (*IEC*), and CCITT. JBIG-compressed images progressively decompress so that intermediate, lower-resolution images can be displayed without decompressing the entire compressed image data file. The JBIG standard augments, and may eventually replace, the Group 3 and 4 standards.

Joint Photographic Experts Group (JPEG)

The *Joint Photographic Experts Group* (*JPEG*) standard, established jointly by the ISO/IEC and CCITT organizations, is one of the most important image data compression standards of the 1990s. The JPEG image data compression standard handles gray-scale and color images of varying resolution and size. It is intended to support many industries that need to transport and archive images. The JPEG standard is used in graphics arts, desktop publishing, medical imaging, color fax, and countless other applications. This standard is commonly used in a lossy mode. However, there is also a lossless mode with reduced compression performance.

The JPEG image compression scheme uses several cascaded compression modes. First, an image is transformed to the frequency domain using the discrete cosine transform. Then, resulting smaller-valued frequency components are discarded, leaving only the larger-valued components. The remaining frequency components are DPCM coded and then Huffman coded. The lossless version of JPEG uses only the DPCM and Huffman coding portions of the standard.

The JPEG compression scheme is adjustable. For instance, the number of retained frequency components can be changed, producing variable compression ratios and inversely proportional decompressed image quality. The JPEG algorithm can be fine-tuned to meet an application's requirements of compressed image data size and decompressed image quality.

CCITT Recommendation H.261 and Moving Picture Experts Group (MPEG)

Motion image sequences have their own image compression standards. The *CCITT Recommendation H.261* standard, also known as *p×64* (pronounced "p times 64"), defines a compression scheme intended specifically for motion sequences of related images. The primary use for the H.261 scheme is in real-time video teleconferencing applications.

The H.261 scheme is based on transform coding—using the discrete cosine transform—and Huffman coding. H.261 can provide low-quality, video-rate compression that requires only a 64 Kbits/second transport link. High quality video-rate images can be produced using a 2 Mbits/second link.

A follow-on standard to the H.261 scheme is the *Moving Picture Experts Group* (*MPEG*) image compression standard, established jointly by the ISO/IEC and CCITT organizations. MPEG is intended for the mass distribution of motion video sequences, such as motion pictures and television programming. The MPEG scheme is compression-asymmetric, meaning that the compression algorithm requires more processing time and computing power than the decompression algorithm. This way, MPEG minimizes decompression hardware by placing a heavier burden on the compression side.

MPEG is divided into two distinct efforts—*MPEG-1* and *MPEG-2*. MPEG-1 is intended for low-resolution image sequences like those composed of 320 pixel × 240 line images. It performs well using 1.5 Mbits/second transport links. MPEG-2 is intended for higher resolution applications, such as full-resolution, 640 pixel × 480 line video image sequences. The MPEG-2 standard requires transport links between 4 and 10 Mbits/second, depending on the required decompressed image quality.

Like JPEG and H.261, the MPEG motion image compression standard also uses discrete cosine transform and Huffman coding techniques. Additionally, interframe coding methods are used to make MPEG's compression performance better than schemes intended solely for still-image compression.

7 Image Synthesis

Image synthesis operations create images from other images or non-image data. The goal of these operations is to synthesize digitally, through computed methods, a resulting image that did not exist before. Image synthesis operations generally create images that are either physically impossible or impractical to acquire—or, in some cases, that do not even exist in a physical form.

Image synthesis operations see daily use in both the medical diagnostic imaging and computer-aided design fields. There have been many advancements in medical diagnostic imaging since the creation of computed tomography. These techniques can create physically unobtainable cross-sectional images of the human body. Other industrial computed tomography uses have since evolved.

The computer-aided design field has exploited image synthesis operations by creating realistic images of mechanical assemblies and architectural designs. The use of these images, before an assembly or structure is ever built, allows a deeper level of visualized understanding to play an active role in the design process.

Computer visualization of abstract data sets, such as scientific and financial data, is unique, because it creates images of things that are not real in the physical sense. However nonreal, the resulting images can powerfully convey to the human observer data trends that might otherwise go unnoticed.

The image compositing operation, discussed in Chapter 4, is a form of image synthesis. This technique creates a resulting image that did not physically exist prior to the operation. The graphic arts and the film and video production fields use these operations to create special effects that are difficult or costly to realize physically.

Image synthesis techniques and uses are emerging at a fast pace, finding numerous commercial interests. The computer graphics field is pursuing the development of many of the techniques and principles of three-dimensional image rendering and visualization. In this chapter, we will discuss the basics of several image synthesis operations.

Tomographic Imaging

Most of us have had medical conditions requiring an image of internal body parts for diagnosis. Conventionally, an X-ray image can provide a view of these objects. The X-ray image is an aggregate image where all the objects between the X-radiation source and the film appear overlaid upon one another. As the X-rays pass through the body, they are absorbed in differing amounts, depending on the density of the objects encountered. Soft tissues absorb small amounts of the radiation, while bones absorb more. As a result, in the final film image, soft-tissue objects appear with darker brightnesses and hard bony objects appear with lighter brightnesses, as illustrated in Figure 7.1.

Because the X-ray image is an aggregate image, it is often difficult to distinguish the appearance of one object from another. This is especially true when looking at multiple soft-tissue objects having low absorption characteristics that overlay one another.

Also, an X-ray can sometimes only be taken from a single angle, as with the human torso. As a result, it can be difficult to envision the three-dimensional shape of an object such as an internal organ. If, for instance, a single X-ray image shows an elliptical object, it is impossible to determine whether it is a flattened, two-dimensional ellipse or an inflated, three-dimensional ellipsoid. Of course, knowledge of the imaged object can help in the interpretation, but only a second view from another angle can absolutely provide depth information.

Tomography is a form of X-ray imaging that attenuates the effects of objects that lie in front of or behind a perpendicular plane of interest. It creates a cross-sectional image of a solid object. The technique is implemented by moving an X-radiation source and a piece of film in parallel and opposing directions, pivoting about the plane of interest. Image information along the plane is constructively reinforced on the film, while information outside the plane is not. The resulting image is a cross-section of the object about the pivot plane.

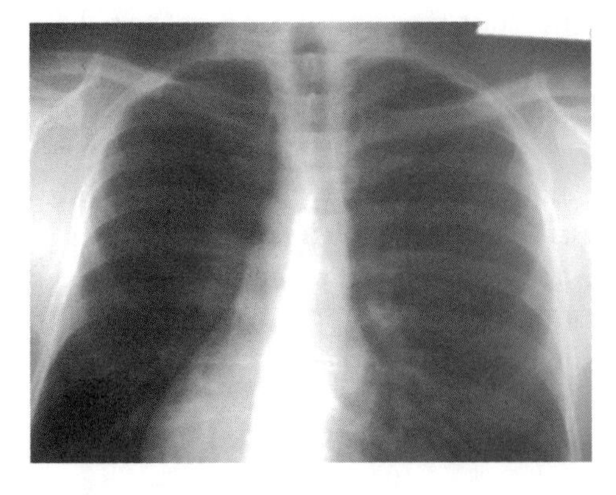

Figure 7.1 *A human chest X-ray image showing dense objects with lighter brightnesses, and objects that are less dense with darker brightnesses.*

Computed tomography uses digital computation techniques to create synthetic cross-sectional images of solid objects. It, too, uses a radiation source and is based on the constructive and destructive reinforcement of many discrete image projections.

Computed tomography is used in three primary modes. The original technique is the *transmissive mode* and uses an X-radiation source. X-rays are transmitted through the imaged object and received at X-ray detection devices. The received signal at the detector is proportional to the density of the elements of the imaged object. This classical form of computed tomography is referred to simply as *CT*. Ultrasonic pressure sources and associated detectors can also be used to create transmissive computed tomography images in much the same way.

The *emissive mode* of computed tomography relies on the emission of a detectable signal from the imaged object. The object can be directly excited or a substance can be introduced that is excited. In either case, detectors receive the emitted signals. Two emissive-mode systems are common, *Magnetic Resonance Imaging* (*MRI*) and *Positron Emission Tomography* (*PET*). MRI tomography excites certain molecules of the imaged object by placing it within a large, changing magnetic field. The molecular reactions are measured by detecting the radio-frequency emissions of the molecules in response to the changing magnetic field. PET imaging introduces a positron-emitting substance into the imaged object. The substance emits a constant flow of positrons. As the emitted positrons come to rest, they interact with an electron and create two gamma-ray photons, each traveling away from the other. Two detectors track the coincidental photons, and thereby determine the location of the positron emission.

The third mode of computed tomography is the *reflective mode*. As in the transmissive mode, a source transmits a signal at the imaged object. Instead of passing through the object, the signal enters the object and is reflected by the internal elements of the object back out to a detector device. The received signal at the detector is proportional to the density of the elements of the imaged object. Reflective computed tomography has the advantage of not requiring the imaged object to be surrounded with sources and detectors. Ultrasonic pressure and radar sources and detectors are used to implement reflective computed tomography.

All forms of computed tomography have found important uses in medical diagnostic imaging applications. The transmissive-mode techniques are also used for nondestructive evaluation of mechanical structures. In the past, mechanical stress testing of critical parts, like turbine blades, could only be evaluated by destructively cutting and grinding parts to uncover stress fractures and material fatigue. Computed tomography can create cross-sectional images of the part under test without destroying it. This allows the part to undergo further tests and analysis, if desired.

We can create the synthetic computed tomography image by backprojecting individual image projections. Because the resulting image geometry is created from the information in the projections, computed tomography can actually be considered a form of image restoration, in its broadest definition. In the following sections, we will discuss the workings of the computed tomography process for the CT X-ray transmissive mode. Emissive- and reflective-mode processing techniques are similar.

Acquiring Projection Images

Computed tomography techniques create a cross-sectional image of an object by gathering numerous projection images around the object at the plane of interest. A projection image is a collapsed one-dimensional image, where each pixel's brightness is equal to the X-ray absorption across the section of the object. By combining the multiple projection views, the cross-sectional image is synthesized.

Each projection is created by pointing a line of X-ray sources at the object and receiving their energy by a corresponding line of detectors on the other side of the object, as shown in Figure 7.2. Each detector creates a brightness that is proportional to the absorption of the material in the object along the line between the detector and its source. In this way, the line of detectors form the pixels of a one-dimensional X-ray image, called a *projection image*.

The X-ray sources and detectors are a part of a circular structure that encircles the object in the plane of the desired cross-section, as shown in Figure 7.3. Following the first projection-image acquisition, the structure is rotated slightly around the object and another projection image is acquired. This process continues 180° around the object. In all, up to 180 or more projections are acquired. While each projection image has little value on its own, when they are combined, the cross-sectional image emerges.

The process rotates only 180° around the object, and not 360°, because of the symmetry properties of the system. The projection acquired at 0° is identical to

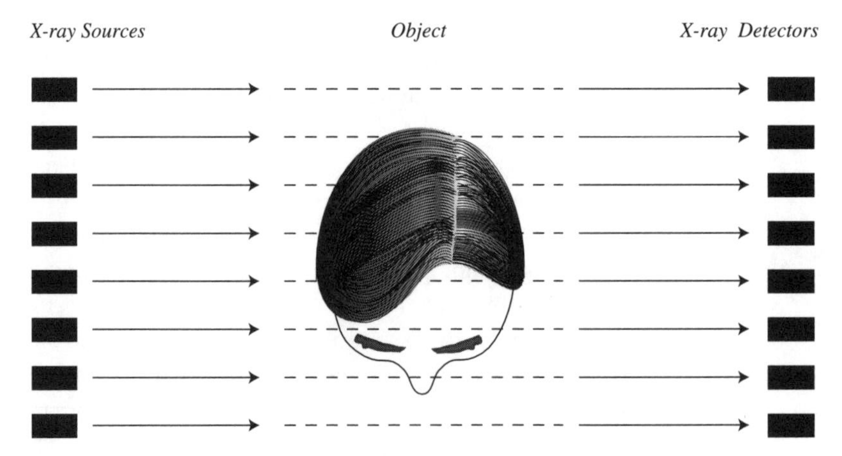

Figure 7.2 *A line of X-ray sources and corresponding detectors create the projection image of the imaged object. The projection image is a one-dimensional line of pixels.*

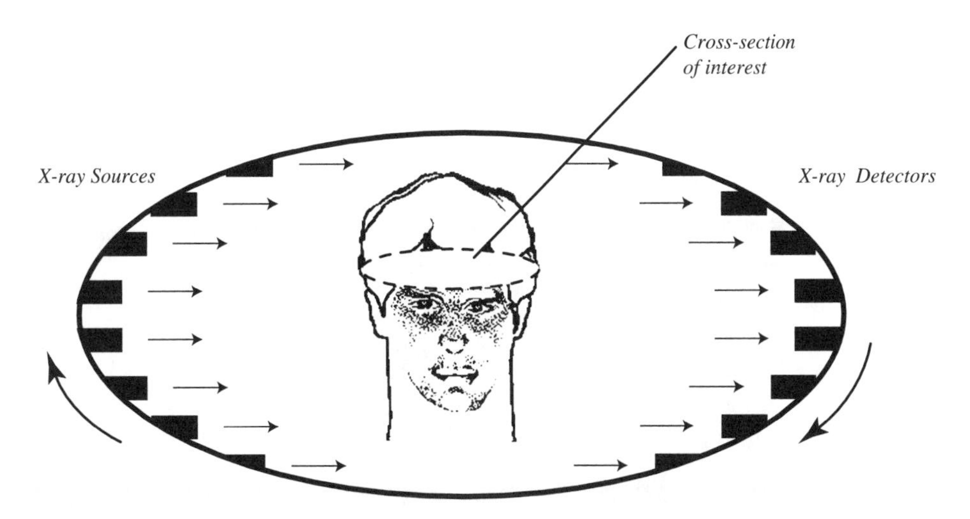

Figure 7.3 *A circular structure that houses the X-ray sources and detectors rotates about the plane of interest, acquiring a projection image at each stop.*

that at 180°, just reversed in order. The acquisitions made between 180° and 360° will simply mimic those acquired between 0° and 180°, providing no new information.

As an example of the projection-acquisition process, let's suppose the goal is to create a cross-sectional image of a human head that cuts horizontally through the head at the eye level. The projection images are taken around the head encircling the horizontal plane of the eyes—the cross-section of interest. The source and detector arrays are incrementally sequenced around the head. At each position, a projection image is acquired, resulting in a set of projection images suitable for reconstructing the desired cross-sectional image.

Most of the time, for diagnostic applications, it is desirable to create a series composed of several different cross-sectional images of an object. The series can then be viewed as a sequence of slices, each a little deeper than the previous slice. To do this, the source and detector structure is set at a starting slice-plane and the projection images are acquired. Then, the structure is moved down until it aligns with the new cross-section of interest, and the projection acquisition process is repeated. This continues until all the projections for all the desired slices are acquired. Figure 7.4 illustrates a typical MRI transaxial head scan image series.

Cross-Sectional Image Reconstruction from Projections

Once we have acquired the image projections for a particular cross-section, the reconstruction process can begin. The cross-sectional image is created by backprojecting the individual one-dimensional projection images. We can think of this process in very simple terms. If we envision the resulting image as a canvas, then each projection's brightnesses are smeared across the canvas at the angle at which they were originally acquired. Each time a new projection is backprojected, the

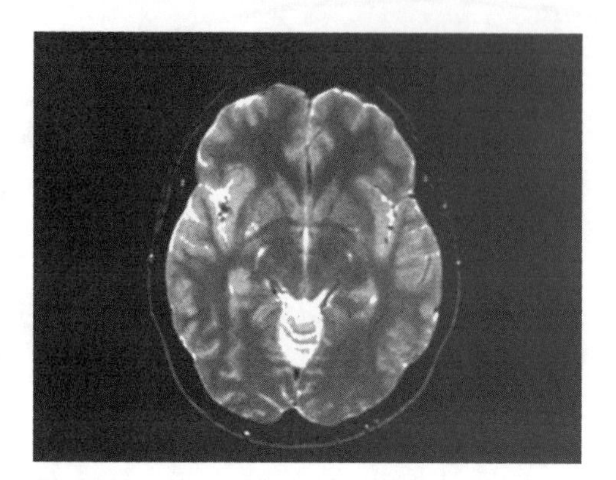

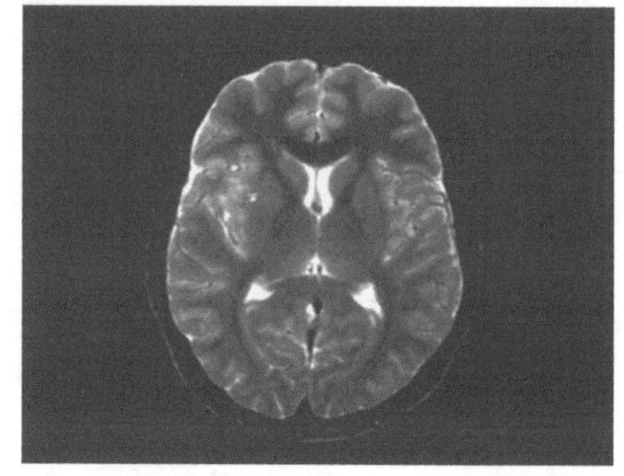

Figure 7.4a *MRI transaxial head scan image series, top slice. The transaxial view is in the horizontal plane relative to a standing subject.* ***Figure 7.4b*** *Second slice.*

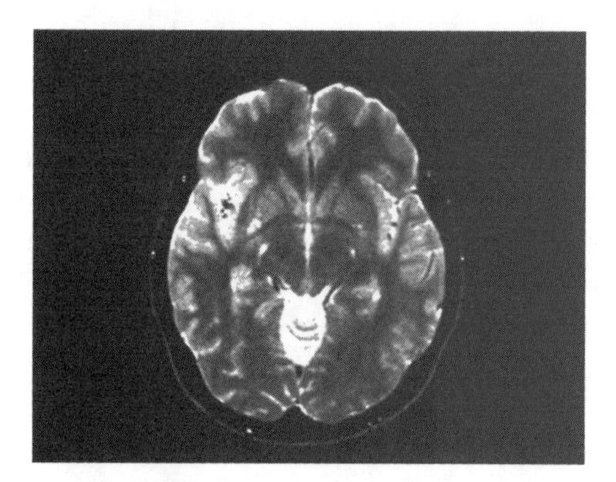

Figure 7.4c *Third slice.*

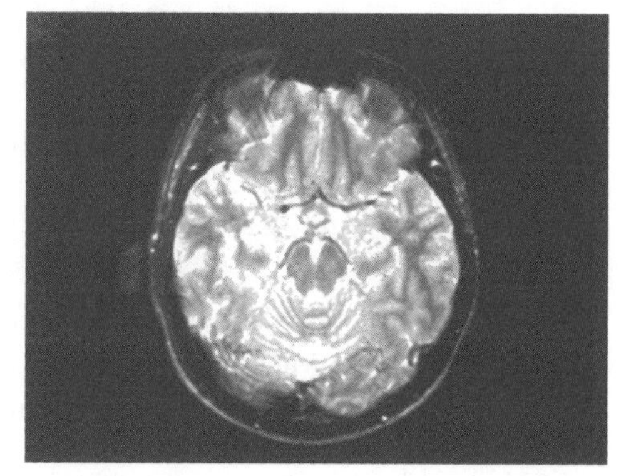

Figure 7.4d *Fourth slice.*

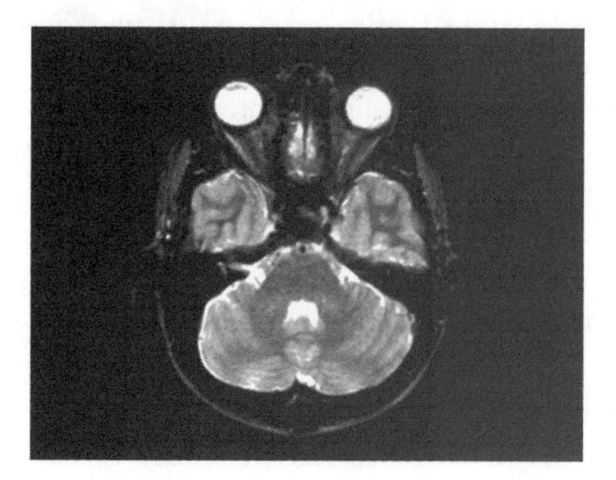

Figure 7.4e *Fifth slice.*

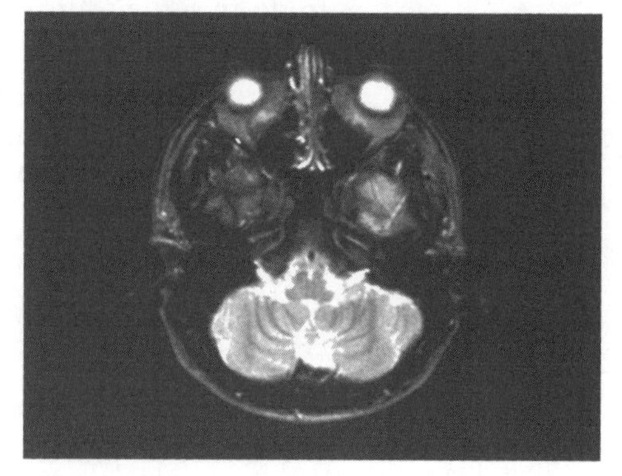

Figure 7.4f *Sixth slice.*

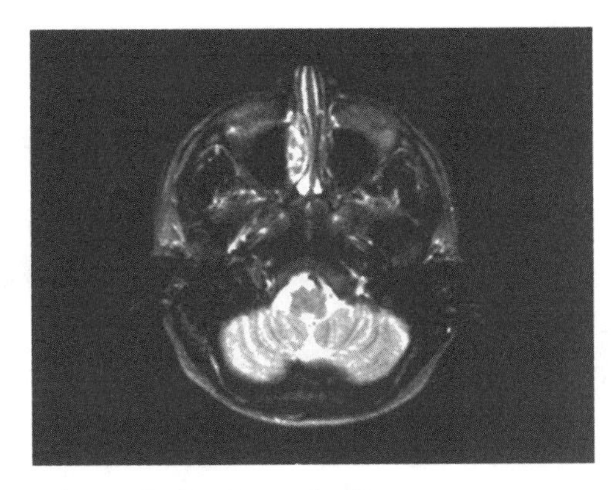

Figure 7.4g *Seventh slice.*

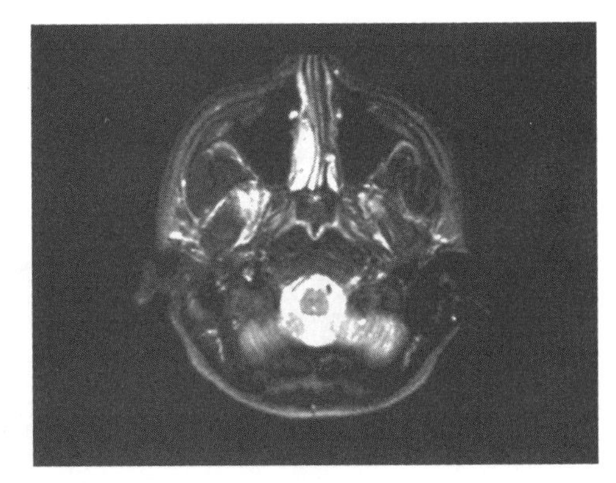

Figure 7.4h *Bottom slice.*

resulting image takes on a little more form. Ultimately, the final image is achieved. Figure 7.5 shows several individual backprojection images of a simple circular

Figure 7.5a *The individual backprojection images of several projections of a circular object—the 0° backprojection.*

Figure 7.5b *The 22.5° backprojection.*

Figure 7.5c *The 45° backprojection.*

Figure 7.5d *The 67.5° backprojection.*

Figure 7.5e *The 90° backprojection.*

Figure 7.5f *The 112.5° backprojection.*

Figure 7.5g *The 135° backprojection.*

Figure 7.5h *The 157.5° backprojection.*

object. Figure 7.6 shows the successive combination of the backprojection images to form the cross-sectional image.

Figure 7.6a *The reconstruction of backprojection images of the circular object—one backprojection image at 0°.*

Figure 7.6b *The sum of the two backprojection images at 0° and 90°.*

Figure 7.6c *The sum of the four backprojection images at 0°, 45°, 90°, and 135°*

Figure 7.6d *The sum of the eight backprojection images at 0°, 22.5°, 45°, 67.5°, 90°, 112.5°, 135°, and 157.5°.*

Figure 7.6e *The entirely reconstructed cross-sectional slice-image created using more than 100 backprojections.*

Looking at the reconstruction of the circular object cross-section, we can see that each projection adds more detail to the resulting image. The first projection (at 0°) appears as a wide streak, and tells us only the width of the object. The second projection (at 90°) confines the object to the bounds of a square. The projections at 45° and 135° add tapers to the corners of the object. As more and more projections are added, the circular form becomes evident. This process can be thought of as adding resolution to the final image.

For a very complex object, like a human head, the backprojection process works the same way. Each backprojection improves the resolution of the final image. The ultimate resolution achieved in the image is related to the number of projections involved. Therefore, the complexity of an imaged object's details dictates the number of projections necessary for a particular reconstruction quality.

We implement the backprojection process by horizontally replicating each one-dimensional projection image into a two-dimensional image. Each two-dimensional image is then rotated to its original angle when acquired. These operations require pixel interpolation resampling to estimate intermediate pixel brightnesses, as described in Chapter 4. Poor interpolation can cause highly corrupted image reconstructions that yield erroneous image artifacts.

All of the rotated two-dimensional projection images are summed together, pixel by pixel, using dual-image point processes. Because each projection has pixels with brightness values between 0 and 255 (8-bit brightness resolution), the overall summation can create pixels with brightnesses in excess of 255. Therefore, the pixels of the resulting summation are each divided by a constant value, placing them in the brightness range of 0 to 255 for display. The backprojection process yields the reconstructed cross-sectional image.

The process of acquiring and backprojecting projection images, as described above, involves *parallel beam projection.* This approach uses parallel sources and detectors to create the projection images. In practice, contemporary CT systems use a *fan-beam projection* approach. Unlike parallel beam projection, fan-beam projection uses a single X-ray source to illuminate a line of detectors, as illustrated in Figure 7.7. This reduces hardware and alignment requirements considerably.

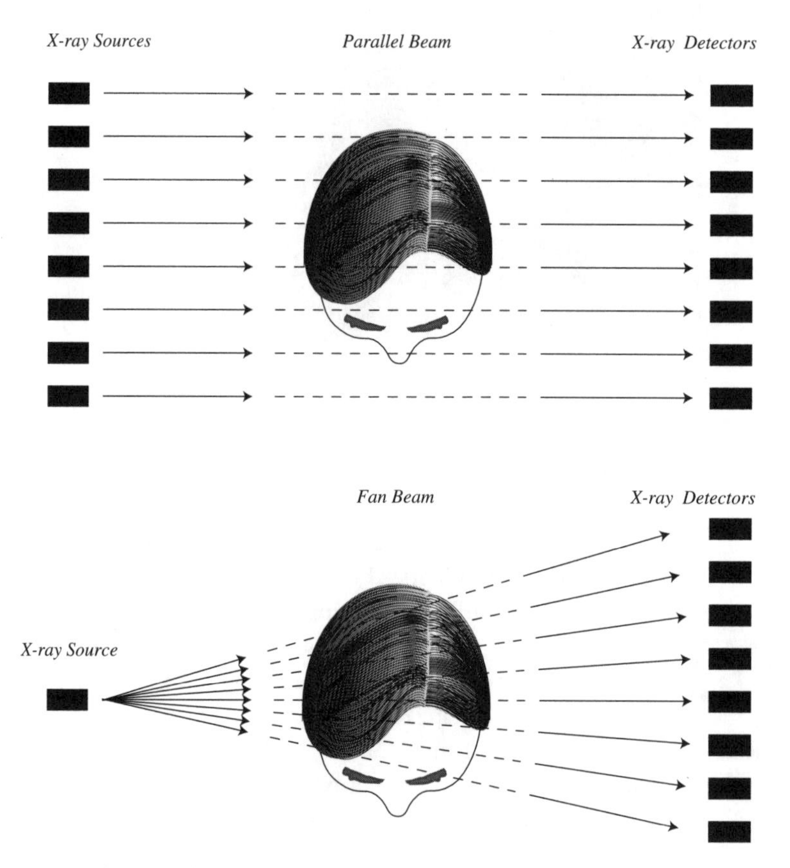

Figure 7.7 *The parallel-beam and fan-beam projections. Contemporary CT systems use the fan-beam method.*

Because the pixels are not parallel, the fan-beam projection approach requires a more complex reconstruction algorithm to create the final cross-sectional image.

Three-Dimensional Scene Construction from Images

We can create images of three-dimensional objects from multiple two-dimensional images of the same scene. As long as the geometric relationship between the images is controlled, or at least understood, we can recreate the depth dimension. This type of operation is used to synthesize images showing attributes that are not readily apparent in the original individual images. In particular, three-dimensional attributes such as depth features can be made more visible.

We will discuss two primary methods of three-dimensional scene construction from images. The first uses stereo image pairs. This technique combines the information from two closely spaced images of the same scene to create the depth-dimension information. The second technique uses object cross-sectional images to form depth information. Both of these techniques can be used to create images of a scene from a three-dimensional viewpoint, or to create a three-dimensional model from which an image of the scene can be created.

Stereo Image Pairing

Human vision relies on a variety of cues to determine depth in an image. With depth information, we are able to view objects as three-dimensional entities. In humans and other two-eyed animals, depth perception comes from several cues, the most prevalent being *stereoscopy*, or *stereo vision*. Other cues include relative object sizes, object shading, object obscuring, object shadowing, and even atmospheric haze.

Stereo vision allows us to judge the relative distance of objects. It is based on the principle of *object parallax*, which is a measure of how two objects relate to one another spatially when viewed from two different viewpoints. In the case of the human, our two eyes provide the two different views, and hence the parallax information necessary for us to perceive depth through stereoscopy.

Figure 7.8 illustrates two objects at different depths, relative to the observer. Each eye sees the objects with a different separation distance. These different separation distances are the observed parallax between the two objects. We can calculate the depth using the following equation:

$$\text{Depth Difference} = \frac{\text{Distance}}{\text{Eye Separation}} \times (\text{Right Separation} - \text{Left Separation})$$

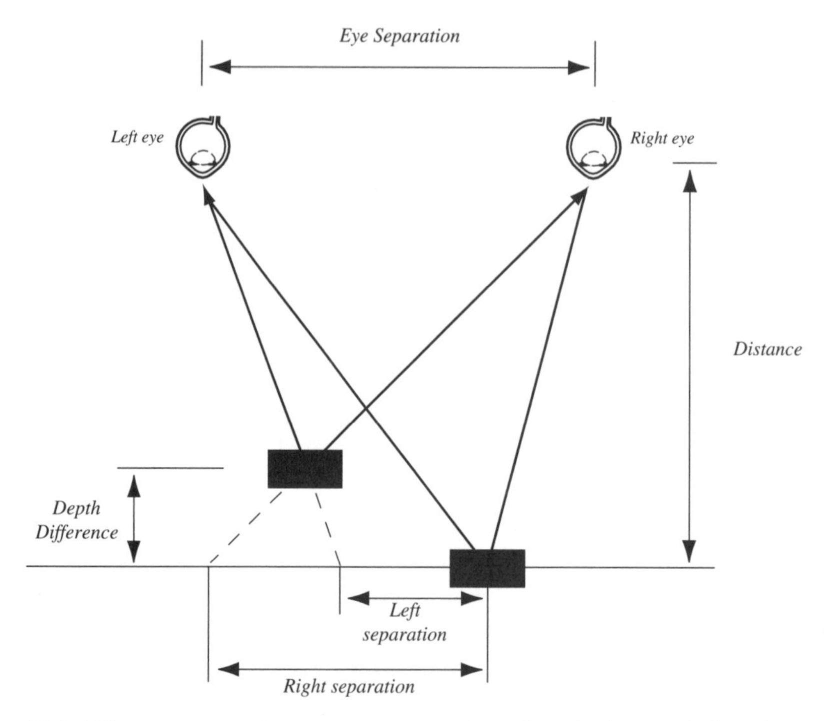

Figure 7.8 *The geometry of stereoscopy, providing the calculation of the relative depth of two objects.*

Using this equation, the eye–brain combination creates accurate depth perception in everything we view. We can create the three-dimensional sensation by viewing two images of the same scene, provided they represent the scene from two slightly different view angles. An image stereo-pair is shown in Figure 7.9. Careful examination shows slight differences in the way features are displaced by

Figure 7.9a *Left side image of stereo image pair.*

Figure 7.9b *Right side image.*

differing amounts in the image-pairs. These minute differences provide the observed parallax that is perceived as depth.

The previous equation can be used to compute depth digitally from a stereo image pair, as well. This technique is used to automatically create elevation maps of Earth using satellite imagery. The operation begins by acquiring two satellite images of the same area on Earth, each separated by some distance. This is done by pointing the satellite's camera at the same location from two different viewpoints on its orbital path. The images can be thought of as a left image and a right image. The operation must also know the distance between the satellite's positions and the altitude of the satellite above Earth when each image was acquired. Then, the separation distances between objects in the two images are measured. The objects measured can include geological features such as mountain peaks, ridges, and rivers, as well as man-made features like roadways and buildings.

We can automatically measure the distance between two imaged objects by searching for object matches in the two images. Techniques like matched filtering, as discussed in Chapter 5, can be used for this. As two objects are located in each image, their distances from one another are measured. These distances—left image separation and right image separation—complete the equation, giving the relative elevation distance between the objects. We continue this process, matching all the scene features between the two images. The result is an elevation image where pixel values represent elevations rather than image brightnesses. The elevation image is more appropriately referred to as an *elevation model*, because it models the three-dimensional attributes of the scene. Later in this chapter, we will use an elevation model to render arbitrary three-dimensional images of terrain.

Image Cross-Section Stacking

We can create a two-dimensional image with an arbitrary three-dimensional viewpoint using multiple-image cross-sections, or slices. First, we must create a series of cross-sectional object images. Then, by visually stacking the slice-images, we can achieve a three-dimensional object appearance. Image cross-section stacking methods use depth cues other than stereo vision to create a three-dimensional sensation.

Acquiring Depth-Slice Images

Cross-sectional images of an object can be created with the techniques of computed tomography, as discussed earlier. Computed tomography is an excellent way to create multiple depth-slices of a three-dimensional object nondestructively. As seen earlier in Figure 7.4, the MRI cranial image series represents image slices taken at sequential depths through the head. Many times, though, the cost, physical size, location, complexity of use, and image quality of computed tomography techniques will make them inappropriate for use in a particular application. In these cases, more direct techniques can be used to create cross-sectional images of objects. The methods of sequential material removal and optical sectioning are often convenient and effective.

Sequential material removal is the classical technique used to create image slices of material samples. This approach is often used to analyze geological, biological, and metallurgical samples. A small portion of the object's material is ground or sliced off, creating a flat face. An image of the face is acquired and stored as the first slice. An additional amount of material is then ground or sliced off. Again, the new face is imaged. This process continues through the entire object, and an image of the newly created slice is acquired each time. The result is a series of images very much like those created by the computed tomography process.

The amount of material removed at each step determines the depth of each slice. Generally, that decision is based on the depth resolution required by the application. Each depth slice is usually cut to the same dimension.

The sequential material removal technique for creating object image slices is common and straightforward. The process produces a convenient image history of each slice's appearance. The downside to the technique is that it is a destructive process. At the end, the object has often been destroyed, or at least reduced to a number of slices.

Another technique for creating object image slices is *optical sectioning*. Although a common microscope can be used with some success, a more practical approach uses a *confocal laser scanning microscope*, as shown in Figure 7.10. This device uses a scanned laser light source to illuminate the object at discrete individual points on a single focal plane. As a result, only the illuminated point of the object at the plane of interest reflects light. By scanning the laser light across the entire plane of interest, the accumulated reflected light forms the slice image. By sequentially focusing the laser light to deeper focal planes, multiple depth-slice images are created. The confocal laser scanning microscope's laser light source and focused beam provide image quality and depth resolution that are superior to those of a common microscope.

For an optical sectioning technique, the imaged object must be of a non-opaque material, such as biological tissue or fluid, because the illumination source is light. The optical sectioning method works well for many applications and doesn't destroy the object sample being imaged.

Stacking Slice Images

With a series of object depth-slice images, we have all the information necessary to create arbitrary three-dimensional views of the object and its interior parts. One way we can view the slices is by simply replaying them as a motion sequence. This gives the viewer a sense of diving into the object by seeing deeper layers in

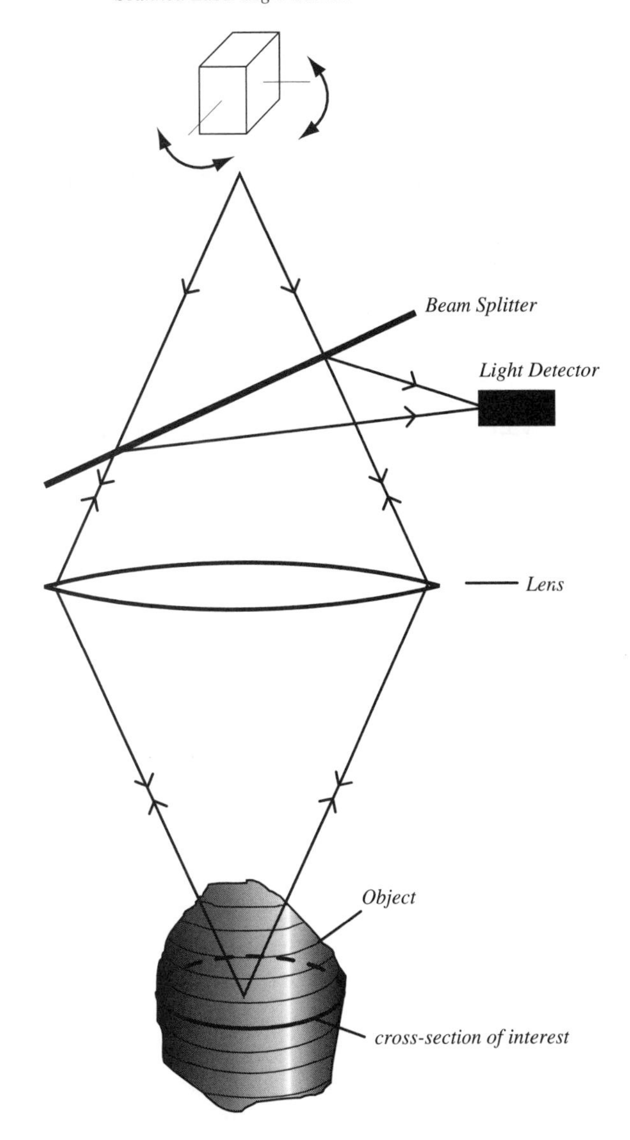

Figure 7.10 *The confocal laser scanning microscope. The illuminating laser light is focused to a single plane of the imaged object at a time.*

rapid succession. The brain then takes over by integrating the information into a three-dimensional perception of the object's composition and structure.

It is often preferable to see the entire three-dimensional object as a single static image. We can do this by stacking tilted perspective views of each slice image into the original geometry of the object. First, we use the geometric transformation operations discussed in Chapter 4 to geometrically transform each slice to a

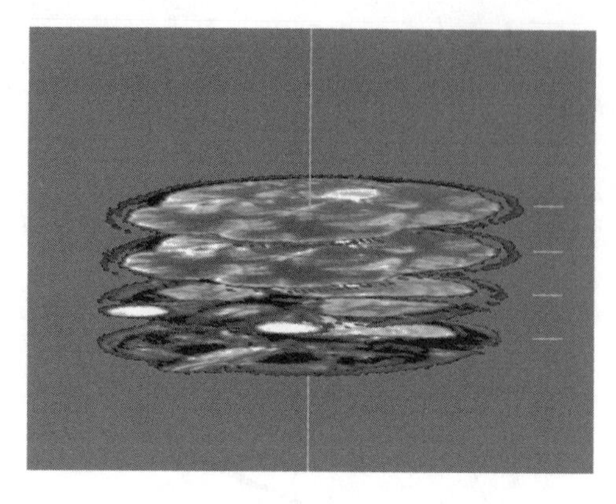

Figure 7.11 *A stack of four slice-images from the MRI transaxial head-scan image series of Figure 7.4.*

desired viewpoint. Then, we translate and composite each slice on top of the previous slice. The stack is built up, bottom to top, from the deepest depth slice to the shallowest. The result is a stacked formation that helps simulate the geometry of the original object, as shown in Figure 7.11.

We can view the exterior of the stacked image object from any arbitrary orientation. This is done by changing the viewpoint of the individual stacked images using geometric transformation operations. Further, the object's interior can be viewed by creating images of the stack that are cut back to an internal level. This is done by masking a portion of several stack images, making them appear transparent over a region of the volume. This allows us to see the image of an internal slice because it is no longer obscured by a higher-level slice.

Volume Rendering Slice-Images

There is another way of looking at the three-dimensional nature of an object using multiple slice-images. Like the other images discussed in this book, each slice-image is composed of a two-dimensional array of pixels, each with an (x,y) location. If we stack the images on top of one another and view them directly from the top, we see the top slice-image. The next slice-image is underneath the top slice-image, and so on. The pixels of each slice-image line up with their corresponding pixels of the same (x,y) location, above and below it.

With this geometry in mind, we can alternatively label the pixels with three-dimensional coordinate locations. They keep their original (x,y) coordinates and take on a new one, z. The z coordinate is simply the number of the slice with which the pixels are associated. Each pixel of the entire slice-image stack now has (x,y,z) coordinates. This way, we eliminate the concept of slice-images, each with (x,y) pixels. Instead, we represent all of the pixels in all of the object slice-images as belonging to a single, three-dimensional object image, as shown in Figure 7.12.

The definition of a pixel, given in Chapter 3, defines it as having an area. It represents the two-dimensional brightness of an original scene over an area with

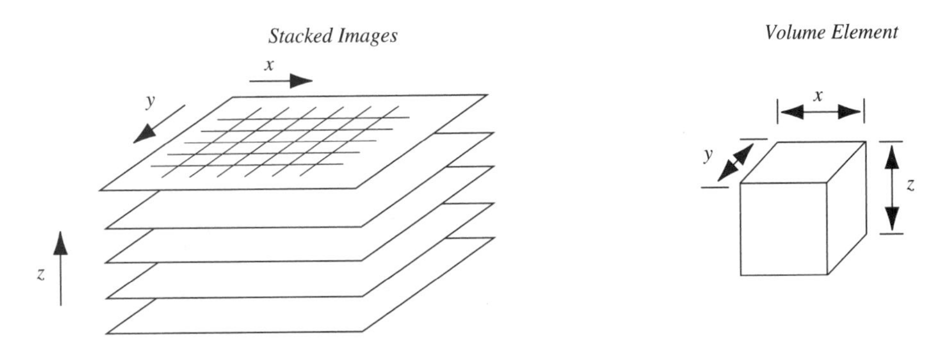

Figure 7.12 *All the pixels of individual slice-images can be combined to create a single three-dimensional image that has pixels with* x, y, *and* z *coordinates.*

the dimensions of 1 pixel wide × 1 pixel high. With the addition of the depth coordinate, pixels with (x,y,z) coordinates represent the brightness of a volume. The volume has the dimensions of 1 pixel wide by 1 pixel high by 1 pixel deep. Pixels that represent volume brightnesses are referred to as *volume elements*, or *voxels*. Our flat, square pixel is now represented as a cube with equal dimensions on each side.

With a single array of voxels defining the three-dimensional object, we can now apply the techniques of *volume rendering* to create images of arbitrary three-dimensional views. In the three-dimensional views of image slices described earlier, there were visually disturbing gaps between the slice-images, making the overall image somewhat difficult to interpret. These gaps were present because the pixel slices were flat; they had no thickness associated with them. Volume rendered images don't show gaps between the slices. This is because the pixels now have depth—a depth of a 1-pixel distance. As a result, volume-rendered images appear solid from top to bottom, as well as from side to side.

The depth resolution (z-axis) of object slice-images is usually coarser than the horizontal (x-axis) and vertical (y-axis) resolutions. This is because it is often physically impossible to acquire images that have such a fine depth. When slices are separated in depth by more than the equivalent of 1 pixel of horizontal or vertical resolution, intermediate slices are often created using pixel interpolation methods. Just like interpolating intermediate pixels in the x and y directions, as discussed in Chapter 4, the intermediate z-direction pixels can also be estimated.

Volume rendering techniques are an emerging area of the computer graphics field that extend the concepts of classical surface rendering (discussed in the following section). With volume rendering techniques, voxels can take on characteristics in addition to brightness and color. For instance, by creating an opacity characteristic, a see-through image can be rendered. Outer layers of an object can be made translucent, exposing the image data beneath them. This makes it possible to see an object's internal structures. Figure 7.13 shows two volume renderings generated from the same set of CT data.

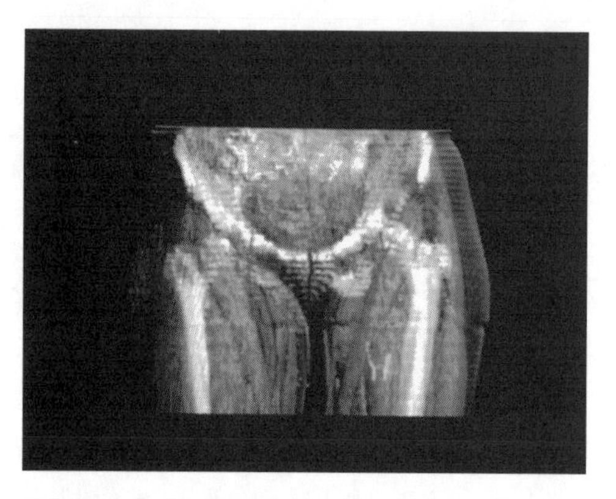

Figure 7.13a *Volume rendering of a human pelvic region created from a series of CT cross-sectional images. Transparency allows the internal tissue and bone to be viewed simultaneously.*

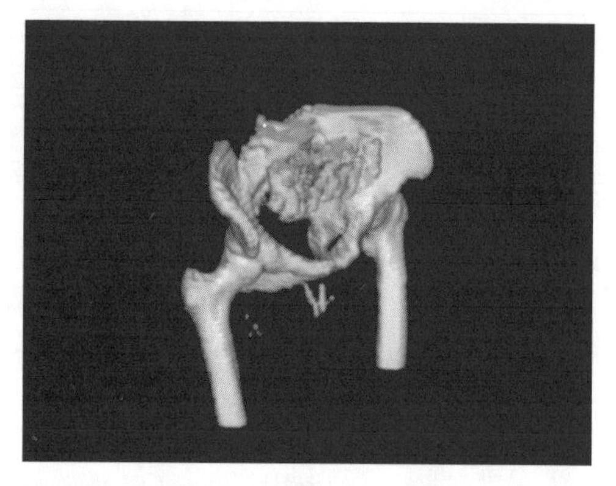

Figure 7.13b *The same CT image data is used, but from a different viewpoint and without transparency, to reveal only the bone structure. (Images courtesy of Vital Images, Inc. and Dr. Dana Mears, Shadyside Hospital.)*

Like the slice-stacking method, voxels can be removed from the object, allowing us to view the internal parts of an object. Further, the volume-rendering viewpoint can be placed inside an object. This allows us to explore details deep inside the object, such as a tumor growth, stress fracture, or material impurity, from the inside out.

Volume rendering is a computationally intensive operation. With enough computing power, however, it can be done interactively, giving us the ability to fly inside an object, in real-time, looking from side to side and all around. Volume rendering techniques have become, and continue to emerge as, one of the most powerful visualization tools for exploring three-dimensional image data sets.

Visualization of Non-image Data

A significant area of image synthesis is the creation of images that do not originate from the data of other images. This realm belongs almost exclusively to the computer graphics field.

Many techniques create three-dimensional data sets that represent an object's model. For instance, the designs of mechanical parts and assemblies are often modeled as three-dimensional objects using a *computer-aided design* (*CAD*) system. Often it is desirable to "see" the model before it is actually fabricated to make decisions about it. The "seeing" process is called *visualization*. A three-dimensional model is visualized by rendering it to a form that simulates its appearance in the real world.

Three-dimensional representations of measured phenomena, such as physical objects and events, or even nonphysical objects and events, can be represented as an image. This form of visualization can bring to light subtle features that may never be seen in the raw data itself. Here, we will discuss some examples of image synthesis operations for non-image data sets.

Computer-Aided Design Model Rendering

Three-dimensional object models are often created in the areas of mechanical, architectural, and biological analysis and design, to evaluate the form, fit, and function of a new part. These models are created using CAD systems and exist only as mathematical descriptions. A model can be visualized in a number of ways using image synthesis operations called *surface model rendering*. The goal of rendering operations is to provide the designer with visual feedback of how the model will appear if and when it is built.

A CAD system models an object as *surfaces*, or *polygons*, as shown in Figure 7.14. Each surface is defined by (x,y,z) control points at important locations such as corners and sharp features. Mathematical equations define how the surface behaves between the control points. Surfaces can take on the characteristics of flat plates, cylinders, spheres, cones, and a wide variety of other geometrical shapes. Additionally, a designer can define a custom surface formula, if desired.

The three-dimensional object model can be rendered to a two-dimensional image from any desired three-dimensional *viewpoint*. This is done by projecting the three-dimensional control points to a two-dimensional *viewplane*. The viewplane is a plane perpendicular to a line between the object and the viewpoint. This is illustrated in Figure 7.15. The viewpoint can be a three-dimensional location either outside or inside the object. The result is a synthetic image that views the object from the desired viewpoint. Four basic methods of three-dimensional model rendering are commonly used to visualize a three-dimensional model's appearance. Examples are shown in Figure 7.16.

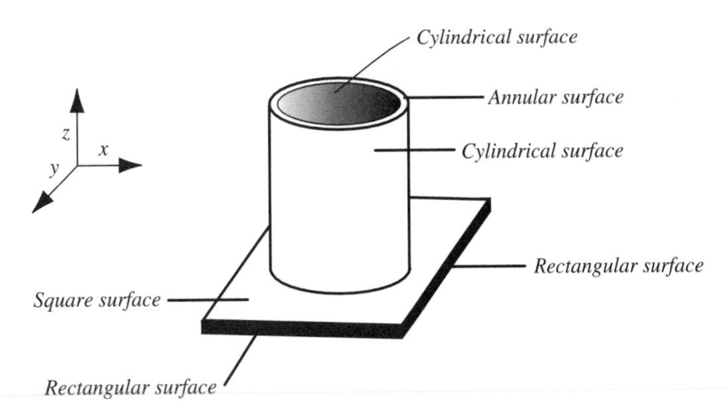

Figure 7.14 *CAD-defined objects are composed of multiple surfaces.*

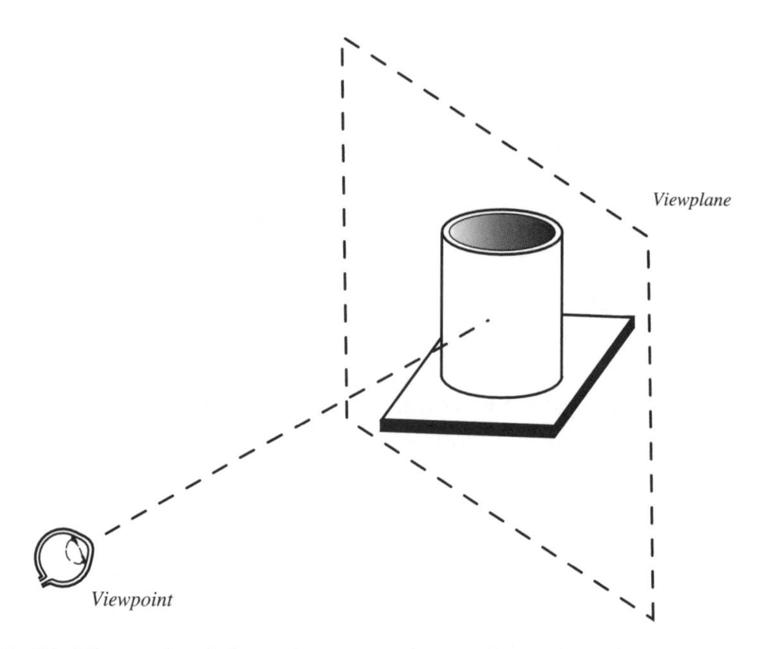

Figure 7.15 *The rendered three-dimensional viewplane of an object is perpendicular to the observer's viewpoint and the object.*

The most primitive method for rendering an object model is a *wireframe*. A wireframe simply shows the edges of all the surfaces in the model. The next step toward realism is a wireframe using *hidden-line removal*. This form makes the wireframe look more realistic by removing the surface edges that would not be visible from the chosen viewpoint. Either way, the wireframe-rendered model gives the observer a rudimentary feeling for the object's dimensions, interaction with neighboring parts, and overall appearance.

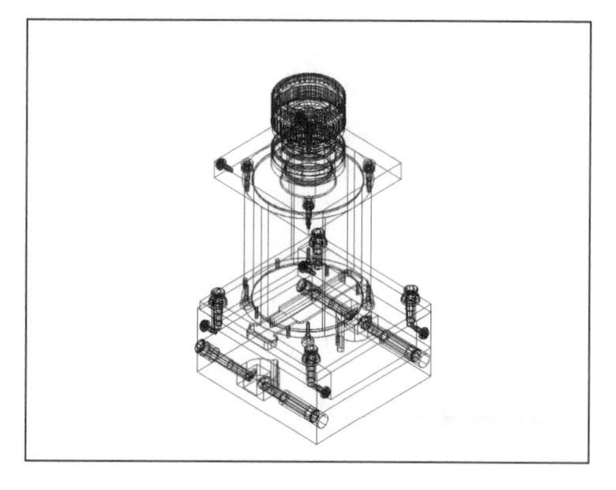

Figure 7.16a *Wireframe rendering of a mechanical camera assembly.*

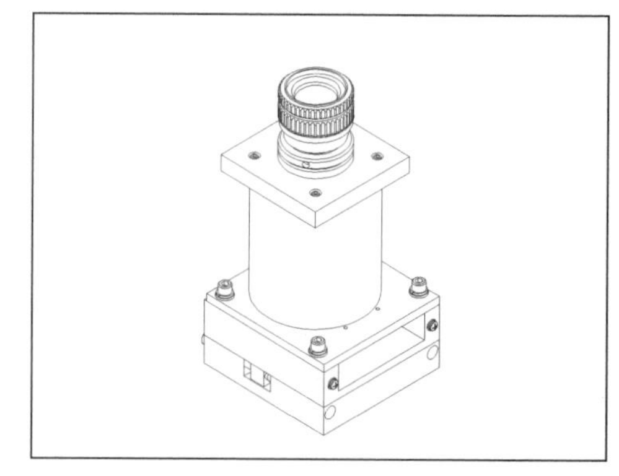

Figure 7.16b *Hidden-line wireframe rendering.*

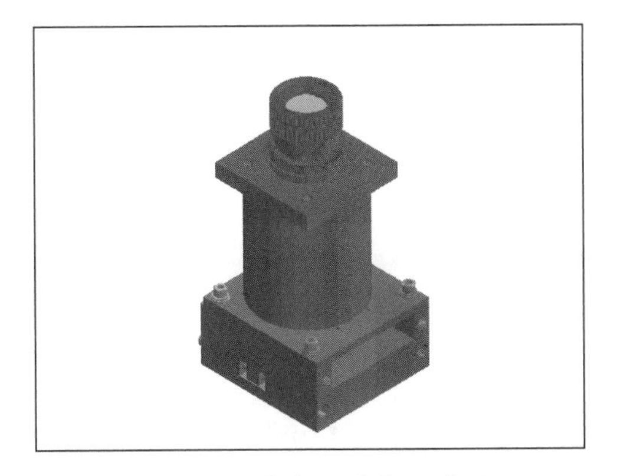

Figure 7.16c *Solid model rendering.*

Figure 7.16d *Shaded solid model rendering.*

The third rendering method is the *solid model* of the object. A solid model fills each surface of the object model with a constant brightness or color, making it appear solid. Surfaces that appear behind other surfaces are removed in the rendering process. The fourth rendering method is a *shaded solid model.* A shaded solid model rendering adds lighting effects, making the rendering look the way the object would look in the real world. If a light source illuminates the object from one side, that side will appear brighter than the other side, which will be seen in a shadow.

IOS 38 3D Model Shading

There are several surface shading methods that simulate the reflective properties of light, and hence the appearance of the shaded model. Common shading techniques include *flat* (or *Lambert*), *Gouraud*, and *Phong* shading. While each of these techniques requires additional computational effort, each lends an increasing level of realism to the rendered model, as shown in Figure 7.17.

We can enhance the shaded solid model by adding object texturing, surface finishes, and reflection characteristics. Further, we can add features to the environ-

Figure 7.17a *A sphere is rendered with a single light source using various shading techniques—a flat- or Lambert-shaded sphere.*

Figure 7.17b *Gouraud-shaded sphere.*

Figure 7.17c *Phong-shaded sphere.*

ment around the model like multiple light sources, with varying qualities and colors, and haze. By paying close attention to the model and its surrounding lighting environment, highly realistic visualizations of objects with real-world appearance can be achieved. This level of realism is referred to as *photo-realistic rendering*. Often, photo-realistic renderings are made of entire scenes, composed of numerous individual objects. The resulting image can truly be mistaken for a photograph of an actual setting. Figure 7.18 shows some examples of complex photo-realistic scene renderings.

Like the earlier-mentioned volume-rendering techniques, surface rendering can also require substantial computing power. With the appropriate equipment, however, real-time image rendering can be accomplished, allowing us to interactively peruse a model at will. Animations of model fly-throughs can be created showing object details of interest. For instance, we may walk (or fly) through a fully furnished house that has yet to be constructed. Or, we may visually explore the molecular structure of a newly hypothesized drug and examine its interactions with other molecules.

Figure 7.18a *Photo-realistic renderings of complex scenes—an office building.*

Figure 7.18b *An outside golf course landscape. (Images courtesy of Autodesk, Inc.)*

Figure 7.18c *An interior scene. (Image courtesy of Intergraph Corp.)*

Figure 7.18d *A complex mechanical part. (Image courtesy of EDS/Unigraphics, Inc. and Lightwork Design, Ltd.)*

Elevation Data Rendering

Earlier in this chapter, we discussed the creation of three-dimensional terrain elevation data. The result was an elevation image, or more appropriately, an elevation model. The elevation model, like other object models, is a data set that describes a surface as three-dimensional control points. Instead of just describing corner points of each surface, an elevation model has control points spaced at regular intervals, each having an x, y, and z coordinate location. It can be thought of as an image with (x,y) located pixels, where each pixel also has a z coordinate that is a measure of the pixel's height. So, instead of pixel values representing a pixel's brightness, elevation model pixel values represent height.

Interestingly, the elevation model's surface representation is identical to an analogy used earlier in Chapters 4 and 5. When we discussed the spatial convolution and gray-scale morphology operations, we used the "pixel mound" analogy. Pixels were viewed as having a height, rather than a brightness. The brighter the pixel, the higher it appeared. This allowed us to see an image with a three-dimensional representation. The elevation model is, in fact, used in this way. Pixel values represent pixel heights, and are in no way related to a brightness interpretation.

Elevation models can be rendered from any three-dimensional viewpoint, much the same way as CAD models. The result is a synthesized image that shows the three-dimensional nature of the modeled terrain. A wireframe rendering is a quick, computationally easy way to show the rough outline and characteristics of the surface. Figure 7.19 shows several viewpoints of a wireframe terrain rendering.

A solid model improves the appearance of the wireframe rendering by filling in the holes with solid brightnesses and colors. The brightnesses and colors used in the solid modeling process can be based on relevant data such as the elevation or, perhaps, the known terrain composition.

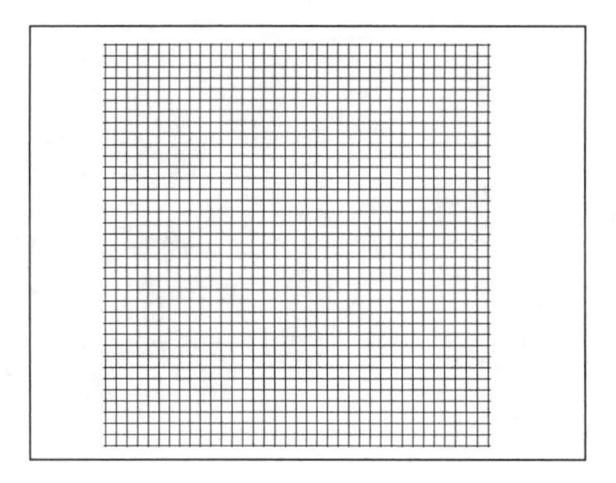

Figure 7.19a *A downward-looking (90° elevation angle) viewpoint showing a wireframe terrain rendering. The rendering appears as a square grid, because the elevation axis is coming directly out at the observer.*

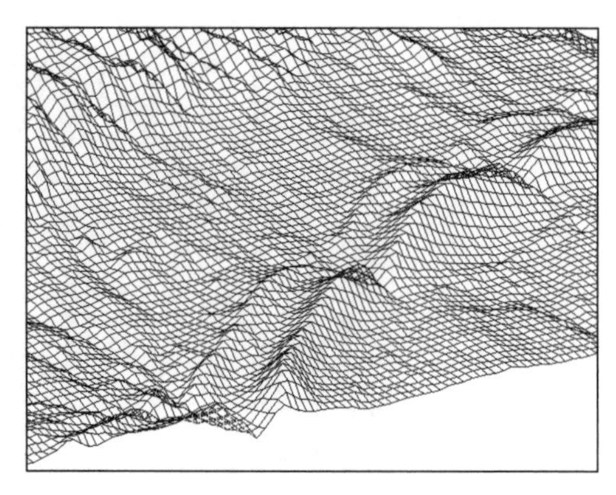

Figure 7.19b *60° elevation angle.*

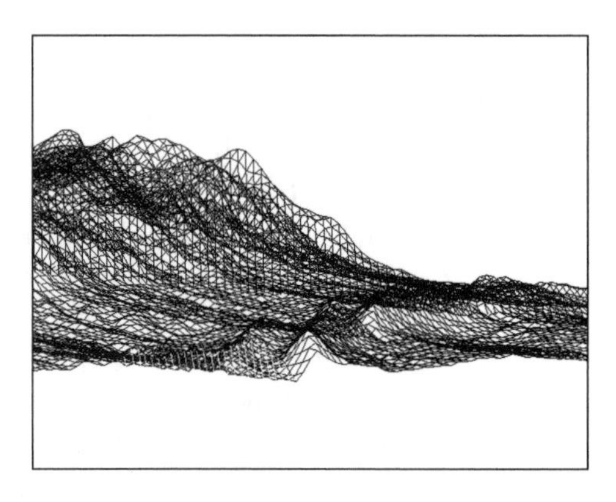

Figure 7.19c *20° elevation angle.*

Shading the solid model adds an additional level of realism by rendering the appearance of the terrain with an imaginary light source, like the sun. But, because each square of the wireframe is surrounded by 4 pixels of varying height, each square may not be planar, as shown in Figure 7.20. This means that the entire square area between any 4 pixels does not point in the same direction. For simple rendering of the model with light sources, it is desirable to have only planar surfaces. We can achieve planar surfaces by dividing each square into two triangles. All of the area within each triangle, by definition, sits in a single plane because it is bounded by only three points.

We can then render the shaded solid model by shading each planar triangle with a brightness equal to the amount of light that the triangle would see from the imaginary light source. The shading computation is made by taking the cosine of the angle between a line perpendicular to the triangle's plane, and a line from the triangle to the light source. The shade value is equal to one (the maximum) when the angle is 0°, and it is equal to zero (the minimum) when the angle is 90°. For angles of 91° to

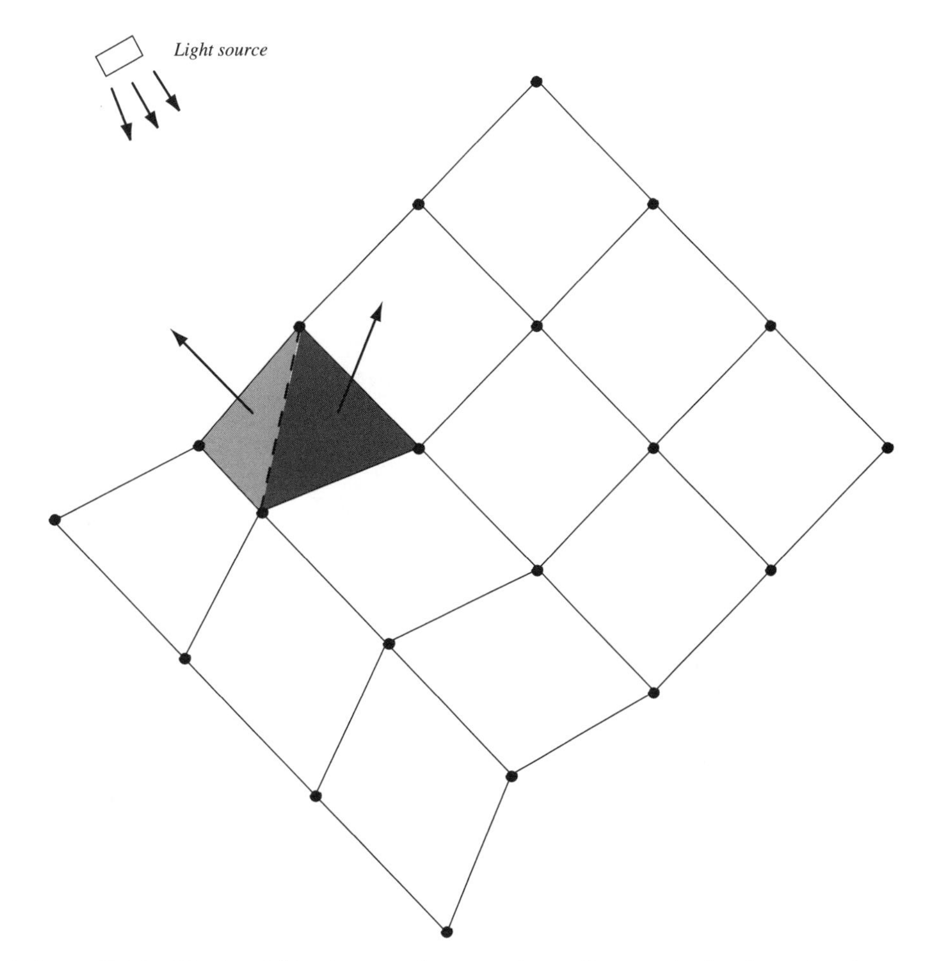

Figure 7.20 *Pixels within the area between four adjacent pixels of an elevation model do not always point in the same direction. Dividing the square regions into two triangles solves the problem because each triangle is planar.*

359°, the shade value is also zero. This is because these angles indicate a light source that is behind the triangle's surface. This geometry is shown in Figure 7.21.

Terrain rendering provides a tool to visualize the features of Earth's surface from any angle and with any sun-angle, from sunrise to sunset. Most importantly, it can be done on a computer rather than from an aircraft. Figure 7.22 illustrates several renderings of the same terrain area. Of course, elevation data of any origin can be processed in the same manner. The Magellan spacecraft, for instance, generated terrain elevation data for much of the surface of the planet Venus. Other spacecraft have mapped portions of the planet Mars and other extraterrestrial bodies. Other sources of elevation and depth data are also available. Material sample depth data can be created using the techniques discussed earlier in this chapter.

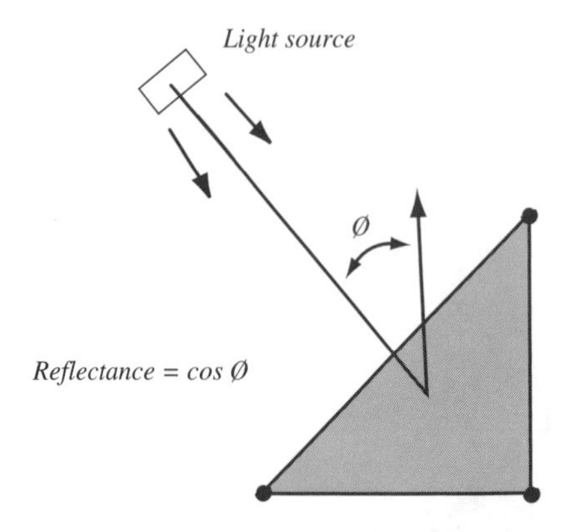

Figure 7.21 *The calculation of a light source's reflectance from a triangular surface.*

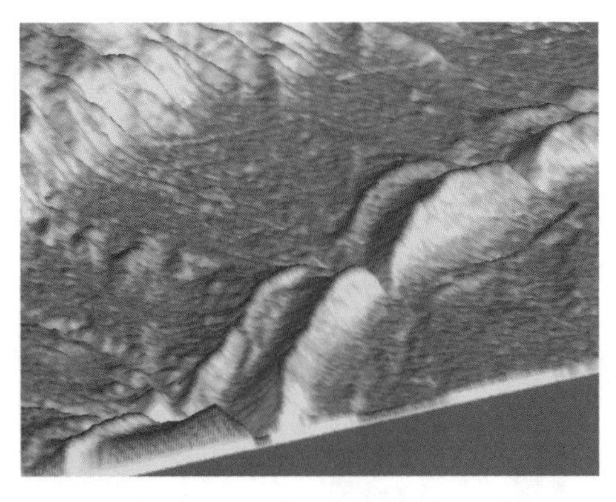

Figure 7.22a *A shaded solid rendering of elevation-modeled terrain at a −60° elevation angle, looking northwest.*

Figure 7.22b *20° elevation angle, looking northwest.*

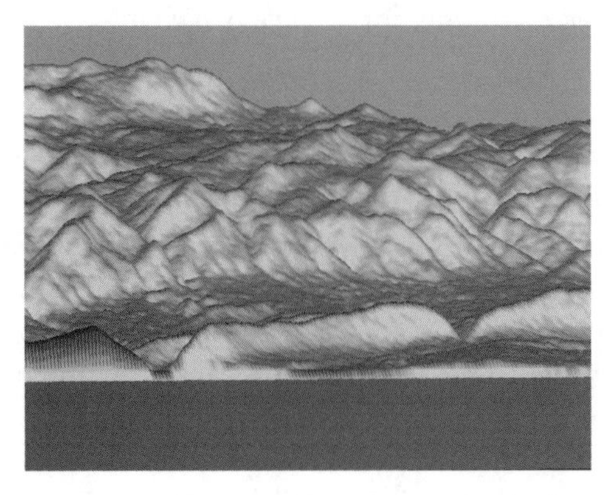

Figure 7.22c *60° elevation angle, looking west.*

Figure 7.22d *20° elevation angle, looking west.*

Visualization of Abstract Data Sets

The same techniques used to render models of physical objects, such as tomographic data, mechanical parts design, and elevation data, can be applied to visualize nonphysical *abstract data sets*. The term "abstract data sets" refers to data sets that have no spatial, physical significance. In other words, they are not viewed with any sort of physical relevance. In fact, abstract data does not even originate in a form intended to be visually perceived.

Data sets with many variables are the most difficult to interpret in their raw form. They have become the best candidates for image synthesis visualization operations. There has been much interest in the synthesizing of images from numerical scientific and financial data. In the study of a phenomenon, both the science and finance disciplines, as well as others, can create large volumes of data with many variables. The goal is often to find mathematical connections between the data variables that might lead to an understanding of their interrelationships. Tying several variables together and displaying them in a unique way can provide a valuable visualization tool for analyzing and interpreting data trends.

To visualize numerical data, we need to convert it from a tabular form to a visual image form. The image must accurately represent the relationships of the original numerical data, but can do so with pictures. These pictures can include three-dimensional views of geometrical forms, colors, textures, and so on. The visual form of the data can take advantage of the brain's ability to parallel-process image components, rather than sequentially processing each numerical data line. A case can be made for visualization techniques by looking at a common device that most of us use daily—the common wristwatch.

The contemporary wristwatch comes in two varieties, one that displays time numerically and one that displays time visually. The digital watch displays the time of day as numerical data with a digital readout. The traditional analog watch, with hour and minute hands, represents the time visually. Even though the digital watch literally tells us the time, the analog watch makes time easier to interpret because it allows us to visualize the current time relative to a target time, as shown in Figure 7.23. The brain instantly sees relative time as a pie-shaped wedge between the current hour-hand position and the target hour-hand position. With the digital watch, we must compute the difference between the current time and a target time by subtracting one from the other. In this respect, the analog watch is far superior to the digital watch because it improves our ability to interpret quickly the relevance of the time-of-day data.

As a simple multivariable data-visualization example, let's look at the presentation of some numerical data in a marketing report. Let's say that the report is required to display the average annual household expenditure for a particular product for each state in the United States. This information might be useful for showing which states are good target markets for a new product introduction. We

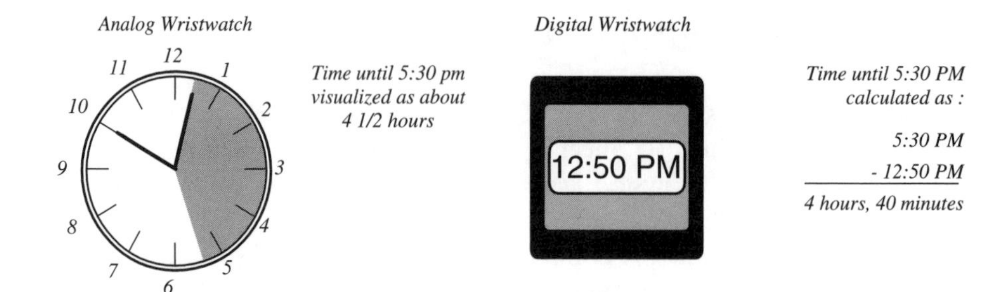

Figure 7.23 *On an analog wristwatch, the relative time between the current time and a target time is visualized as a pie-slice. With a digital watch, the relative time must be calculated as the difference of the current and target times.*

could easily gather the data from, perhaps, the latest Census Bureau reports. It could then be tabulated and entered, as a table, into the report. To help clarify the data, we could go a step further and order it by expenditure amount, from the state with the largest expenditures to the state with the smallest expenditures.

Certainly, this effort would achieve our goal of displaying the required data. But we can go another step and add significant additional value to the reader's interpretation of the data. We can do this by presenting the same data visually. For example, we could create a three-dimensional map of the United States rotated back to appear almost flat. Then we could show the outline of each state with a height based on the state's expenditure figure, as shown in Figure 7.24. In areas of the country where the states appear with lower heights, the expenditure levels are low; in high-height areas, the expenditure levels are high. This visual form of the data would help the reader spot geographic trends such as regional expenditure habits. By presenting the data visually, we enable the reader to perceive the data's essence instantly without the tedium of numerical analysis.

Further, we could color-code the state outlines to represent an additional data variable. For instance, the color could span the colors of the rainbow to represent the historical trend of each state's expenditure figure. If a state experienced an upward spending trend over the past year, its color would appear in the red tones to show that it is a "hot," growing market. If a state experienced a downward

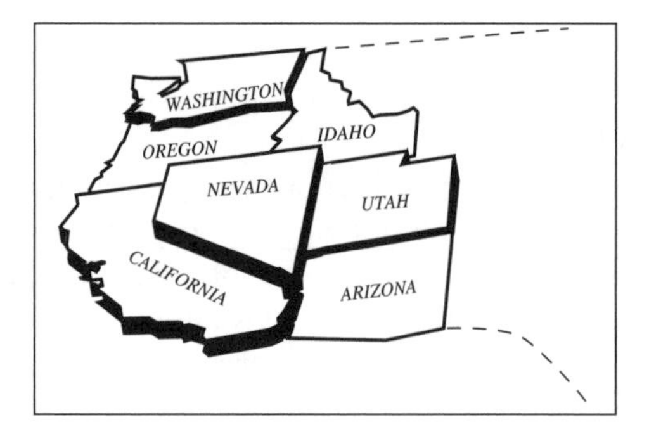

Figure 7.24 *A visual representation of relative expenditure data for a product is shown as the height of each state's outline for the western United States. The relative data can be more quickly interpreted visually than by viewing a tabulation of numbers.*

trend, its color would be in the blue tones to show that it is a "cold," declining market. This way, the reader could visually see the data's trend in addition to the current absolute expenditure numbers. Our final image would vividly and instantly illustrate three distinct data variables—average annual expenditures, geographical expenditure locations, and upward or downward trends of the expenditure data.

Figure 7.25 shows an abstract data-visualization example involving four variables. Complex data sets, ranging from three or four variables to 20 or more variables, can be visualized using the same tools of three-dimensional object rendering and the attributes of color, shading, texture, and so on. By combining the variables and representing them through visual cues rather than strict numerical tabulations, we can help the observer perceive subtle patterns more easily.

Image synthesis techniques are playing a big role in the evolution of data visualization methods. Data representing the physical or nonphysical properties of objects and events can be presented visually in compelling new ways to yield increased understanding and interpretation.

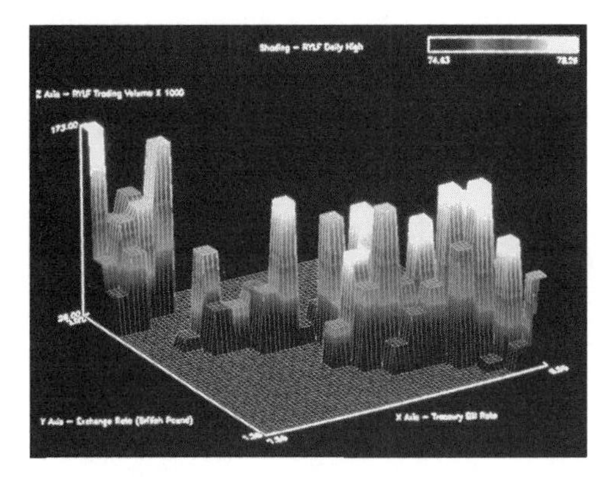

Figure 7.25 *This three-dimensional graphic portrays the relationships among four independent financial variables. The* x*-axis is the Treasury Bill rate, the* y*-axis is the British Pound exchange rate, the* z*-axis is the trading volume for a particular stock, and the shading is the stock's daily high. Viewing this graphic provides an observer a quick evaluation of the stock's performance with respect to the Treasury Bill and exchange rate indicators. (Image courtesy of Visual Numerics, Inc.)*

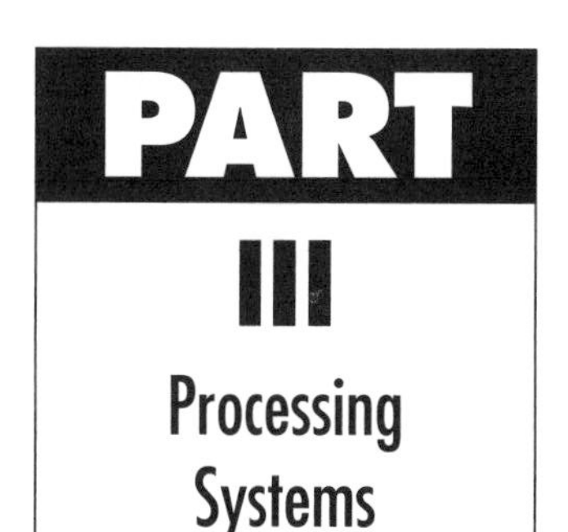

8 Image Origination and Display

In the following three chapters, we will investigate the basic hardware and software of digital image processing systems. While the range of commercially available systems and components is broad, a distinct set of functionality must exist in any system. This part of the book explores the functions and the limitations of the primary digital image processing system building blocks.

In this chapter, we will concentrate on the topics of originating and displaying images. We will also examine the standard video formats. Chapter 9 continues with a discussion of the hardware necessary for creating and handling a digital image. Chapter 10 covers the hardware and software needed to process digital images. That discussion also reviews the hardware and software implementation techniques used to carry out the image processes and operations discussed in Part II of this book.

First, though, let's look at the evolution of digital image processing systems.

A Brief History of Digital Image Processing Systems

Since the 1960s, digital image processing system configurations have evolved through several phases, as illustrated in Figure 8.1. Initially, systems comprised large mainframe computers with added devices that provided the image digitization, storage, and display functions. These systems were largely custom-built, costing hundreds of thousands of dollars. These systems relied on the mainframe computer for all processing tasks. The 1970s brought the development and availability of commercial off-the-shelf digital imaging components, allowing more cost-effective systems to be built. These systems began using minicomputers as their processing backbones. In the late 1970s, fully integrated, high-power systems became available that packaged minicomputers along with the associated imaging components. Additional special-purpose image processors were added in these systems to improve their digital image processing performance dramatically.

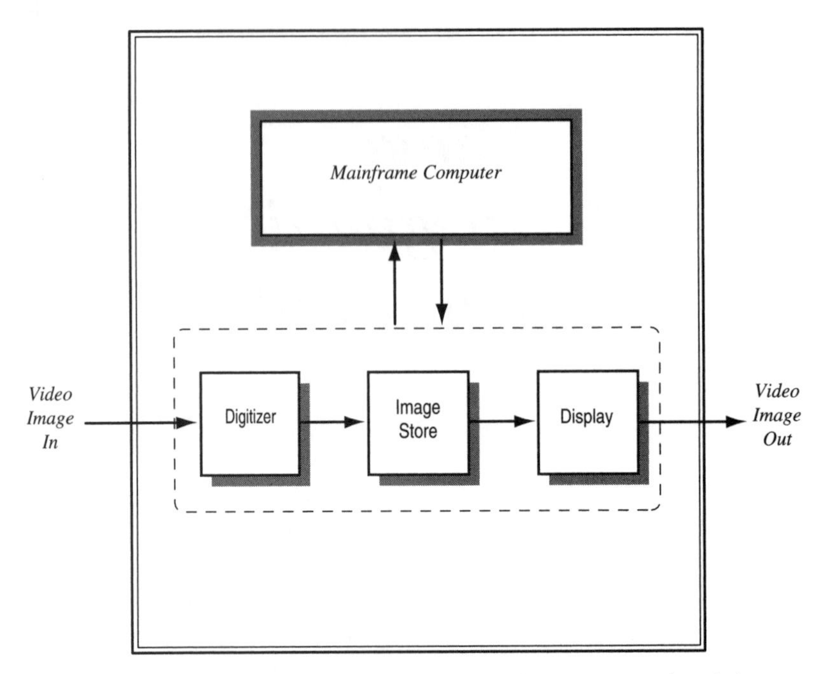

Figure 8.1a *The digital image processing system of the 1960s placed the entire processing burden on the mainframe host computer.*

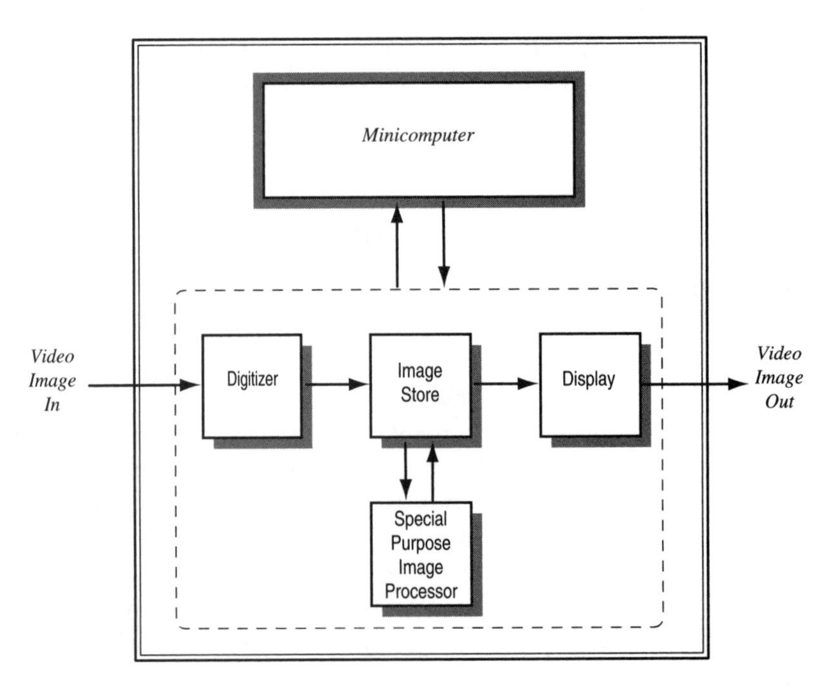

Figure 8.1b *The late 1970s saw fully integrated systems with some special-purpose digital image processors.*

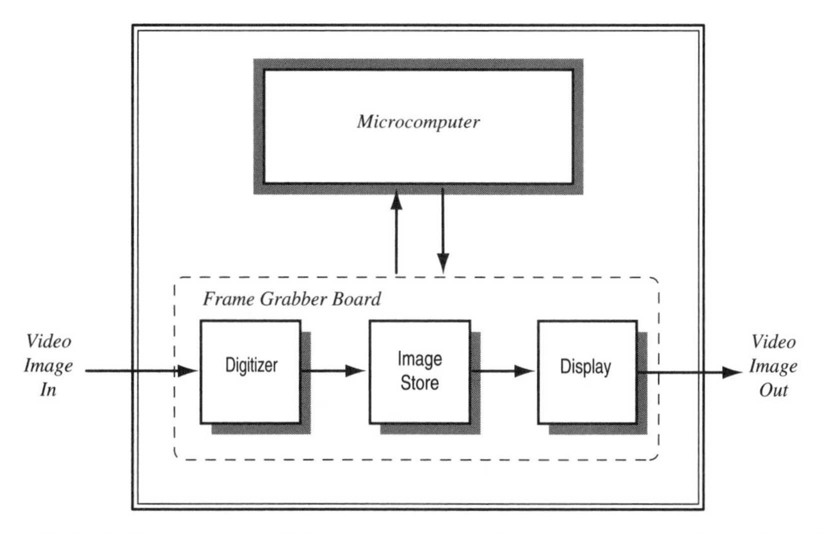

Figure 8.1c *Microcomputer digital image processing systems emerged in the early 1980s.*

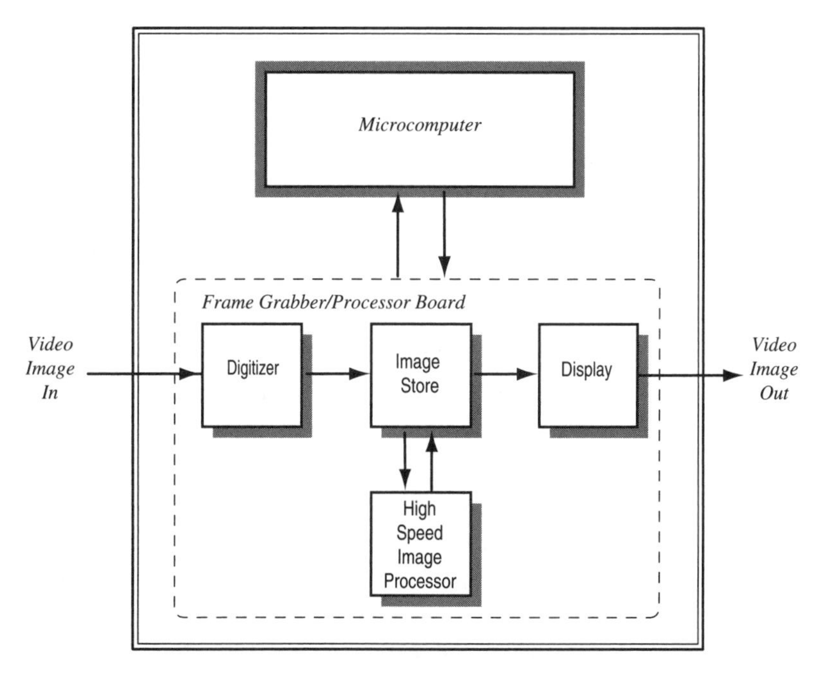

Figure 8.1d *The late 1980s brought special-purpose processors to the microcomputer digital image processing environment.*

At the same time, the microcomputer, based on single-chip microprocessor architectures, was evolving into an inexpensive general-purpose computing engine. The early 1980s introduced microcomputer systems and a variety of plug-

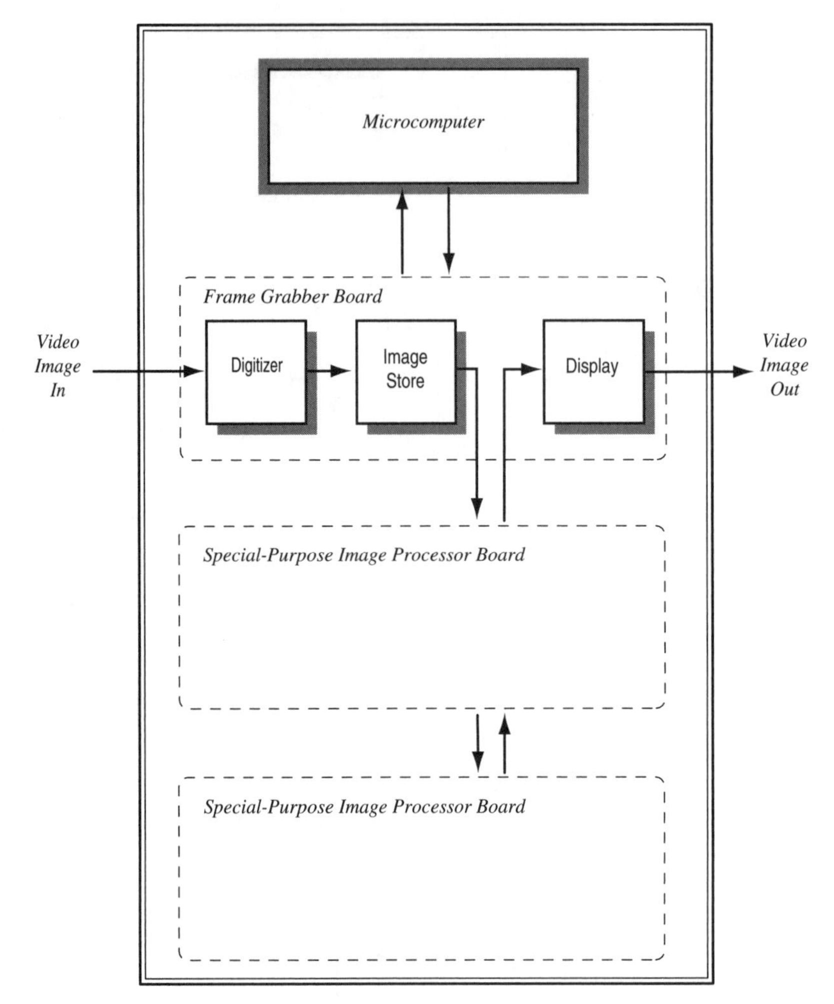

Figure 8.1e *The 1990s see increased digital image processing power and functional modularity.*

in imaging peripheral boards that provided the digitization, storage, and display functions on a single, printed circuit board. The image peripheral boards became known as *frame stores*, or *frame grabbers*, because they provided all the functions necessary to grab an image frame from a live video source. While the frame grabber provided the imaging-related functions, these systems relied on the microcomputer for all digital image processing tasks. In the mid- and late-1980s, frame grabbers became more powerful, offering significant digital image processing capabilities.

Meanwhile, microcomputer system architectures diverged in the mid-1980s to satisfy the differing needs of high-end and low-end markets. *Graphics workstations* evolved for the high-end markets to handle primarily engineering design and scientific number-crunching tasks. These systems used the 32-bit UNIX operating system almost exclusively, and were optimized for computer graphics computing needs. As a result, they tended to make good digital image processing engines. *Personal computers*, or *PCs*, based on the IBM PC and the Apple Macintosh, evolved to satisfy the business and home marketplaces. Based on simpler 16-bit operating systems, like PC/MS-DOS and Microsoft Windows, PCs provided a simple, cost-effective mechanism for general computing tasks. Frame-grabber designs proliferated to meet both graphics workstation and PC architectures. Toward the end of the 1980s, many special-purpose digital image processing boards were also available for most platforms. Modularity of the designs allowed mixing and matching of digital image processing components based on the application.

The 1990s see the merger between the graphics workstation and PC architectures. Common operating systems, such as Microsoft Windows NT, and sophisticated software development tools are providing easy software code portability and uniform look-and-feel user interfaces between graphics workstations and PC platforms. Additionally, the production scales of PCs have skyrocketed because of a widespread, multidisciplinary appeal for the machines. This volume has fueled the technical growth of the PC architecture, bringing it to parity with the graphics workstation. The result is the homogenous *computer workstation*. Additional hardware bus standardizations could, someday, further homogenize the two environments.

A broad range of frame grabbers and digital image processors have become widely available to plug into virtually any workstation chassis. The current level of processing power and sophistication includes real-time implementation of all the basic image processes and operations described throughout this book. Modularity makes it easy to match digital image processing functions with application requirements. While the original digital image processing systems of the 1960s cost hundreds of thousands of dollars, prices for today's microcomputer-based systems start at under a few thousand dollars.

Video Input and Output Devices

In any digital image processing system, an input image must originate from some source and an output image must, generally, end at some display. As we mentioned in Chapter 3, an image can originate in many different energy forms. The most common form is that of light energy. However, X-ray, infrared heat, radar radio-frequency, and acoustic energy forms are also common, along with other forms. Regardless of what form the image originates in, it must ultimately be converted to an analog electrical signal that can then be converted to a digital data form.

In the case of the common light energy image, a *video camera* or *optical scanner* can conveniently convert it to the required analog signal. If an image originates in an energy form other than light, though, it is usually detected with a special sensor that converts the image directly to an analog signal. Alternately, some image forms can be converted to a light energy form first—perhaps using photographic film, for example—and then to an analog signal, via a video camera or scanner.

As an example, in the X-ray imaging process described in Chapter 7, we can convert an image to an analog signal in one of two ways. We can use an X-ray-sensitive electronic photodetector to create an analog signal during the imaging process. Or, we can capture the image to X-ray-sensitive film and then image the film with a video camera or scanner to create an analog signal. Either way, we are left with an analog signal that represents the original X-ray energy image.

In practice, a video camera is the most common tool for acquiring an image. It has become a standard way of getting real-world images into digital image processing systems. Of course, there are applications that inherently do not use a video camera as an image input device. These applications include computed tomography, most large document or high-resolution film scanning, and many spacecraft imaging systems. However, common applications generally exploit the use of video cameras because of their widespread availability and inexpensive cost. Similarly, the most practical way to display images from a digital image processing system is with a video display monitor. We are most acquainted with these devices in the form of our home video camera-recorder combination and television receiver. These consumer versions have all the functionality of any video camera and monitor. For the purposes of the following discussions, we will assume that our images originate as analog video signals, produced by a video camera (or other standard video device like a video recorder), and that all images are displayed on a video display monitor.

For digital image processing activities, usually the cameras and monitors we choose for an application will meet greater quality levels than those for consumer uses. In particular, RGB or Y/C component video devices (discussed in the following section) are typically used for color applications because of their increased color fidelity. For applications that do not require color images, monochrome cameras are used. When an application requires image acquisition with a higher spatial or brightness resolution than common video devices can supply, special high-resolution and high-speed cameras, or document and film scanners, are available.

Video Cameras

Video cameras commonly produce standardized video signals, as we will discuss in the following section. Some, however, produce nonstandard signals, particularly when operating at increased spatial or brightness resolution, or at greater frame rates. In these cases, special interface devices are necessary to interconnect the camera with a digital image processing system.

The photosensitive devices used within video cameras can be divided into two broad categories—solid-state and non-solid-state. Non-solid-state devices are the classical photosensors, such as *vidicons* and *image orthicons*, found in older cameras. They are based on vacuum tube technology. Solid-state devices are based on silicon chip technologies and are found in most contemporary video cameras. These devices evolved into cost-effective, high-quality imaging solutions through the late 1970s and 1980s. Solid-state photosensor devices include *photodiode*, *phototransistor*, and *charge-coupled devices* (*CCDs*). The CCD devices have reigned superior from a producability standpoint and are now almost exclusively used in video camera designs.

Most commonly, the photosensor device used in a video camera is a two-dimensional device called an *area photosensor*. The photosensor receives an image of a scene that is projected through a lens upon its surface. The photosensor then converts the light into a proportional analog voltage signal.

The CCD photosensor device is inherently a sampled device. It has discrete photosites, or pixels, that accumulate electrical charges based on the quantity of light hitting them. Each photosite is scanned out of the CCD device in a pixel-by-pixel, line-by-line sequence, thus creating an analog video signal. Typically, the analog voltage levels are sequenced in accordance to the RS-170 or CCIR video standard (discussed in the following section), and the appropriate synchronization signals are added. The result is a standard video signal that is compatible with other standardized devices.

For cost and system simplicity reasons, sometimes a *linear* or *line-scan photosensor* device is used in special-purpose camera designs. A line-scan photosensor comprises a single line of photosites instead of a two-dimensional array, and costs considerably less than a two-dimensional sensor. The image produced by the line-scan camera is simply an image composed of a single line of pixels. This type of camera works well with an imaging application where the imaged object moves, as shown in Figure 8.2. For instance, where parts are moving on a conveyer line, a line-scan camera can compose a two-dimensional image by taking successive line images of the parts as they move under it. In this way, the imaged object "self-scans" itself, in a vertical direction, under the camera. The line-scan camera generally produces a nonstandard video signal that requires a special interface to a digital image processing system.

Most cameras use a standard lens geometry, as shown in Figure 8.3. A simple lens can be modeled with equations that relate the imaged object dimensions, the photosensor dimensions, the distances between the object and lens, the distance between the lens and photosensor, and the focal length of the lens. While we can't model all lenses with these simple-lens equations, they do provide a good first-order approximation. The simple-lens equations are as follows:

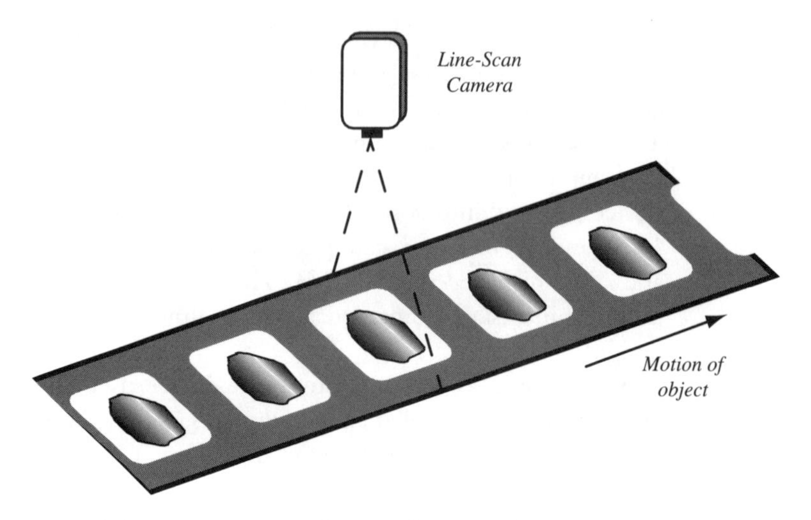

Figure 8.2 *A line-scan camera relies on the imaged object's motion to scan a two-dimensional image.*

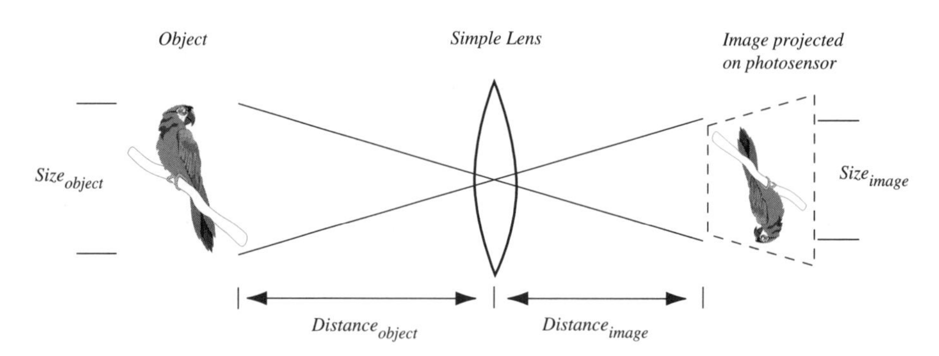

Figure 8.3 *The simple-lens object/image geometry.*

$$\text{Magnification} = \frac{\text{Size}_{\text{image}}}{\text{Size}_{\text{object}}}$$

$$\frac{\text{Distance}_{\text{image}}}{\text{Distance}_{\text{object}}} = \frac{\text{Size}_{\text{image}}}{\text{Size}_{\text{object}}}$$

$$\frac{1}{\text{Focal length}} = \frac{1}{\text{Distance}_{\text{object}}} + \frac{1}{\text{Distance}_{\text{image}}}$$

where $\text{Size}_{\text{image}}$ and $\text{Size}_{\text{object}}$ are the respective sizes of the image and object, $\text{Distance}_{\text{image}}$ is the distance between the lens and the photosensor, and $\text{Distance}_{\text{object}}$ is the distance between the imaged object and the lens.

Using these equations, the placement of the camera relative to the imaged object and the lens focal length can be chosen such that the object fills the photosensor's field of view.

Additionally, we must also consider several other issues when selecting a lens:

1. *Optical transfer function* (*OTF*)—a measure of the optical resolution abilities of a lens. As the brightness details of an object get finer and finer, the lens will depict them with less and less brightness difference, or contrast.
2. *Cosine^4 light falloff*—a phenomenon causing the reduction of image brightness in the outer field of a lens. The brightness of the image projected by a lens experiences a falloff that is proportional to the cosine of the field-of-view angle, raised to the fourth power.
3. *Geometric distortion*—a spatial distortion caused by inaccuracies in the lens shape. The spatial locations of details in an imaged object will have some error; this error tends to get worse in the outer field of a lens.
4. *Depth of field and depth of focus*—a measure of how much the distances $\text{Distance}_{\text{object}}$ and $\text{Distance}_{\text{image}}$, respectively, can vary before the image will appear out of focus. These measures are related to the lens aperture—the wider the aperture, the shallower the depth of field and depth of focus.
5. *Aperture*—the size of the light opening of a lens. Generally, as the aperture increases, diffraction distortions of light around the aperture decrease and object brightness increases, but the depth of field and depth of focus decrease.

Video Display Monitors

Video display monitors, like video cameras, commonly produce standardized video signals, as discussed in the following section. Prior to the late 1980s, nonstandard display monitors were common. They were used to display high-spatial-resolution images. With the advent of the high-resolution personal computer SVGA standards, however, nonstandard display monitors have become rare.

Off-the-shelf display monitors, manufactured for the personal computer markets, are usually standard fare for the digital image processing markets as well. Many of these display monitors provide the capability to detect the timing of a video signal automatically and adjust to its synchronizing signals. This has simplified the task of interfacing a video display monitor to an arbitrary video signal.

Selection of a video display monitor does not impact the overall quality of a digital image processing system in the same way that a video camera selection does. When selecting a video camera, the spatial and brightness resolutions of the camera directly dictate the quality of the digital image that can be acquired. A video display monitor, however, can have poorer display quality than the digital image that it is intended to display. By using geometric transformation operations, as discussed in Chapter 4, we can zoom in and out of the digital image and pan left to right in order to view the image in closer detail. Although a display monitor must meet basic quality requirements, its quality does not permanently impact the quality of the digital image, as does the camera.

Video display monitors have classically been most prevalent in the form of a *cathode-ray tube* (*CRT*). These tubes are manufactured with vacuum-tube technology, and are identical in form to those found in standard consumer television receivers. They are available in sizes from about 1 inch across to greater than 27 inches across.

A monochrome, CRT-based video display monitor operates by scanning an electron beam across the face of the CRT screen, as shown in Figure 8.4. The CRT face is coated with a phosphor-based compound that emits light proportional in brightness to the amplitude of the electron beam. The electron beam varies based on the voltage level of the video signal feeding the beam. In the case of a color CRT display monitor, the CRT is subdivided into many phosphor triads. Each triad is composed of a red, green, and blue phosphor dot. Three electron beams—one each for red, green, and blue—are scanned across the face of the CRT to excite the phosphor triads, hence emitting a color light.

The spatial resolution of a display monitor depends primarily on its ability to focus its electron beam accurately on the CRT face. The smaller the width of the electron beam, the finer the detail that the monitor can display. Additionally, if the beam is not held in constant focus over the entire CRT face, the image will appear out of focus around the edges. For color display monitors, the size of the color phosphor dots, called *dot pitch*, also affects resolution. The smaller the phosphor dots are, the greater the spatial resolution of the display.

Liquid crystal display (*LCD*) monitors are replacing the CRT display monitor, much like solid-state photosensors have replaced the vacuum-tube photosensor. LCD displays are solid-state versions of CRT displays. They are flat, consume less power, and can operate effectively in areas of high ambient light, unlike CRT dis-

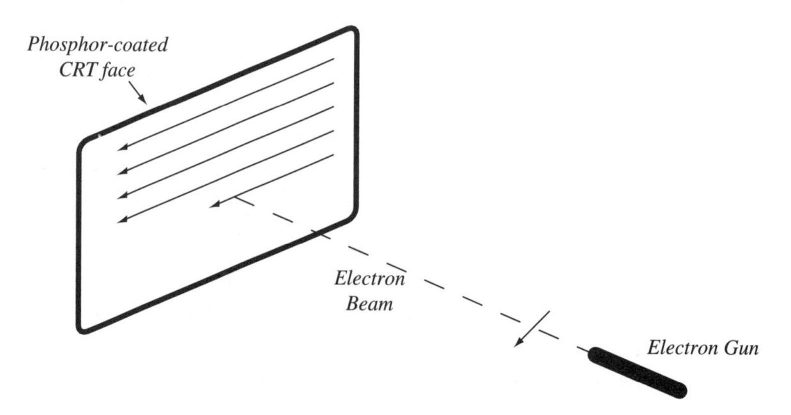

Figure 8.4 *An electron beam repeatedly scans across the face of a CRT to generate a displayed image.*

plays. LCD displays are available in transmissive and reflective modes. Transmissive LCDs are illuminated with a light source behind the device, shining through the device to the viewer. Reflective LCDs reflect light from the viewer-side of the display. Ambient light is generally the illumination source for reflective-mode LCDs. LCDs are composed of individual sites that control the amount of light that either passes through them or reflects from them. The video signal's voltage level controls the sites so that light is passed in proper proportions to create an image. The LCD is an inherently sampled device because it has a discrete array of light transmission or reflection sites, or pixels, with which to create the image display. LCD display monitors are available in sizes from less than 2 inches across to greater than 20 inches.

Standard Video Formats

Video signals to or from various devices such as video cameras, video display monitors, recorders, signal generators, and measurement equipment generally conform to standard video format specifications. This way, the signal from one device will match that of another, allowing multiple devices from independent manufacturers to be interconnected.

A video standard specifies the synchronization timing and electrical parameters for the video signal. No matter what the format, all video standards have certain consistencies. In particular, they all break an image frame into discrete horizontal lines, called a *raster*, beginning with the line at the top of the image and ending with the line at the bottom. Further, each line is sequenced from left to right. The brightness of the image across a line is represented by a voltage level; usually a low voltage represents a dark brightness and a high voltage represents a bright brightness. Synchronization signals are added to the brightness signal to signify the beginning of each line and frame. The final video signal appears as a varying voltage level that represents image brightnesses across each line in succession. After all the lines in the image frame are scanned out, the process begins back at the top of the image and repeats indefinitely.

In addition, each video standard must establish the following criteria:

1. Specific scan-rate timing.
2. Number and order of lines in the image frame.
3. Image aspect ratio.
4. Synchronization signals that indicate the beginning of each line and frame.
5. Color signal encoding (for color devices).
6. Image brightness and color signal voltage levels.

Numerous video standards have evolved throughout the world. The most common, however, are those that have been adopted as national standards for commercial broadcast television use. We will first examine the standard monochrome video format found in the United States. This examination will give us an in-depth look at the most common video standard and how it defines the various video signal aspects. We will then look at the primary European monochrome video standard, color component and composite standards, and the emerging digital video and high-definition television standards.

EIA RS-170/RS-330/RS-343/CCIR Monochrome

In the United States, the *EIA RS-170* standard defines the parameters for a monochrome (gray-scale) video signal. All monochromatic video source and destination devices conform to this standard. The RS-170 standard, produced by the *Electronic Industries Association*, embodies technical specifications that were originally defined in the late 1930s. While the intent was to standardize monochrome commercial broadcast television signals, the standard has also found use in many other video applications. The RS-170 video standard specifically prescribes a video signal's synchronization timing, electrical voltage levels, and quality measures.

Like all video signals, an RS-170 video format image frame is scanned out line by line, from top to bottom. Each line is scanned from left to right. RS-170 further specifies that the dimensions of the image frame must have an *aspect ratio* of 4:3. As discussed in Chapter 3, this means that the horizontal width of the image is 4⁄3 the vertical height. Additionally, an *interlaced scan* technique is used—the image frame is broken into two parts, one composed of the odd lines and the other composed of the even lines. First, the odd lines of the frame are scanned out, followed by the even lines. The odd-line portion of the image frame is referred to as the *odd field*. The even-line portion is called the *even field*. The interlacing scheme is used to produce an apparent update of the entire frame in half the time that a full-frame update actually occurs. The eye's integration of sequential fields gives the impression that the frame is updated twice as often as it really is. This results in a television display monitor image with less apparent flicker.

Each line of the RS-170 video signal is composed of a varying analog voltage signal that represents image brightness across the line. A synchronizing pulse, called *horizontal sync*, separates one line from the next. Further, a longer synchronization pulse, called *vertical sync*, separates the fields of the image frame. When a CRT video display monitor receives the video signal, its electron beam scans the face of the display tube. The amplitude of the video signal controls the amplitude of the beam, and hence the brightness on the CRT face. Whenever a horizontal sync signal is detected, the beam is positioned at the leftmost side of the screen and moved down to the next line location. When a vertical sync signal is detected, the beam is positioned at the top leftmost point of the screen to a line centered between the first two lines of the previous scan. This allows the current field to be displayed between lines of the previous one, thus creating the interlaced scan.

An entire RS-170 video frame is made up of 525 lines and is sequenced out every 33.33 milliseconds (mS), or every 1⁄30th of a second. This means that each field contains 262.5 lines and is sequenced out in 16.67 mS, or 1/60th of a second. By dividing the field time by the number of lines in a field, we arrive at the line time of 16.67 mS/262.5 = 63.49 microseconds (μS). The reciprocal of this line time is 15750 Hz, the RS-170 line scan frequency standard. The determination of line time plays an important role in digitizing: it tells us how fast to sample and digitize the analog video signal to yield the desired number of pixels per line in our resulting digital image.

Figure 8.5 illustrates the timing and electrical voltage levels dictated by the RS-170 specification. The signal's vertical sync interval occurs over the first nine line times. Eleven "no-video" lines follow the vertical sync period, which is then followed by 242.5 active video lines. The total of 9 + 11 + 242.5 = 262.5 line times compose an entire field. Within each line, there is a horizontal sync interval

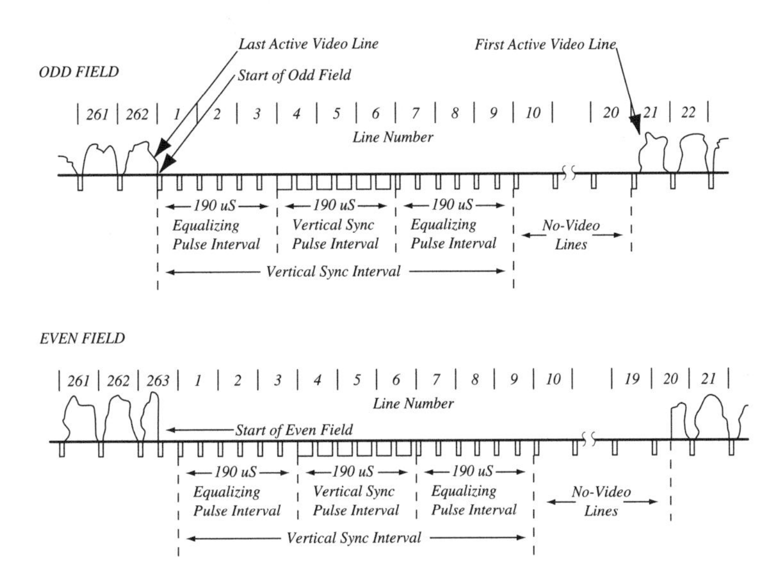

Figure 8.5a *The RS-170 standard video format—vertical frame timing.*

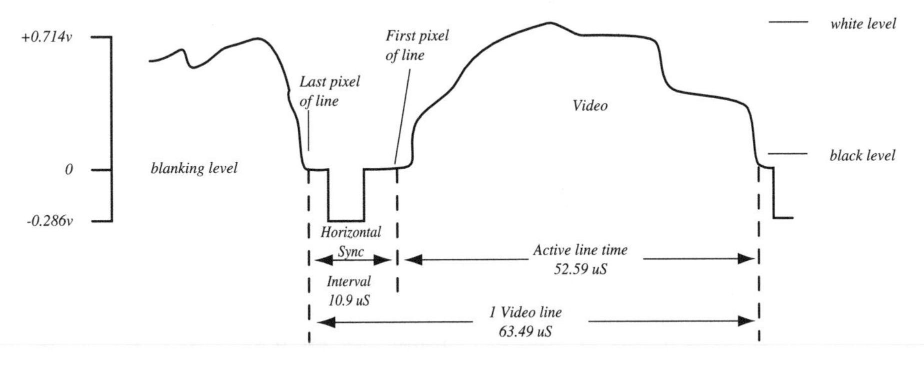

Figure 8.5b *Horizontal line timing and electrical signal voltage levels.*

of 10.9 mS duration. This leaves 52.59 µS of active line time for the visual portion of the video signal. By sampling and digitizing pixels within this active line time, we can obtain our end product: the digital image.

The RS-170 specification also gives us information about the electrical voltage levels of the video signal for the video and sync portions. The overall range is a 1-volt (v) swing from −0.286 v to +0.714 v. The portion of the video swing between +0.143 v and +0.714 v is used to represent the image brightness between black, through the grays, to white. The zero-volt level is called the *blanking level*, where the video is considered "blacker than black." Sync pulses go from O v down to −0.286 v.

The RS-170 specification defines 242.5 active video lines per field, which is equal to 485 lines per frame. With an aspect ratio of 4:3, this yields $485 \times \frac{4}{3} =$ 646.66 square pixels per line. As discussed in Chapter 3, square pixels are preferred for virtually all digital image processing operations. A slightly smaller image size, such as 640 pixels × 480 lines, is generally used to portray square-pixel digital images derived from an RS-170 video source.

The accuracy with which the analog voltage amplitude represents an image brightness along the horizontal line dictates the ultimate brightness resolution in the RS-170 video signal. Usually, this resolution is equivalent to 8 bits when quantized. Greater brightness resolution can be obtained by using a higher-performance camera or other video source.

There are other specifications that extend the RS-170 standard. The EIA RS-330 specification defines additional video signal electrical performance characteristics for the RS-170 signal. The EIA RS-343A specification defines video signals of higher resolution containing between 675 and 1023 lines per image frame. The RS-330 and RS-343A each use the base RS-170 specification with tighter tolerances and modified timing waveforms to provide additional signal characterization.

The RS-170 standard video timing can be summarized as follows:

- 525 lines/frame
- 30 frames/second or 60 fields/second
- 2:1 interlaced scan
- 4:3 aspect ratio
- 1-volt video signal amplitude

The *CCIR* video standard (named for the standardizing body, *Comité Consultatif International des Radiocommunications*), the equivalent to the United States RS-170 video standard, is found primarily in European countries. The standard specifies a 625-line image frame with a frame rate of 40 mS, or $\frac{1}{25}$th of a second. Like RS-170, it specifies an interlaced scan and a 4:3 aspect ratio. The specific timing for the synchronizing signals are, for the most part, similar to the RS-170 standard; they are simply scaled to the slower frame rate. The CCIR standard's frame rate of 50 frames/second was chosen, as was RS-170's 60 frames/second rate, to be identical to the respective national electrical power frequencies. The original intent was to lock the video frame rates to the power line frequencies to minimize television display interference caused by the power distribution systems.

The CCIR standard video timing can be summarized as follows:

- 625 lines/frame
- 25 frames/second or 50 fields/second
- 2:1 interlaced scan
- 4:3 aspect ratio
- 1-volt video signal amplitude

Although the world has gone to color video standards for commercial broadcast television applications, the RS-170 and CCIR video standards and their derivatives live on in many monochrome video devices. In particular, many industrial applications that do not require color images use monochrome cameras. Monochrome cameras tend to be less expensive and exhibit higher spatial resolution than equivalent color cameras. Digital image processing systems almost always accept monochrome video signals from RS-170 and CCIR sources as their preferred mode. Even when color video images are required, the video signal timing remains similar to RS-170 and CCIR systems, as we will discuss later.

RGB/HSB Component Color

Many systems, when dealing with color images, simply use three RS-170 type signals, one for each of the three additive primary color components—red, green, and blue (RGB). These systems accept and generate RGB-compatible component video.

Digressing slightly, two primary methods for formatting color video signals exist—*composite color* and *component color*. A composite color signal is a single physical signal; both the color and brightness information are combined into one signal, as shown in Figure 8.6. A component color signal, on the other hand, is made up of multiple physical signals. RGB component video is one form of a component color video signal. Generally, component video signals are best for color video sources and display devices, because composite color systems usually introduce various color and brightness distortions.

When using an RGB component video source, a digital image processing system will store the image as an RGB image and process each color component image separately. This makes a color processing system nearly identical to a gray-scale system. The only difference is that the color processing system acquires and displays all three component images simultaneously. The digital image processing algorithms work the same way, but are simply applied independently to all three images.

Alternately, as discussed in Chapter 3, the RGB component images can be digitally converted to hue, saturation, and brightness (HSB) component images prior to applying any digital image processing operations. For many applications, pro-

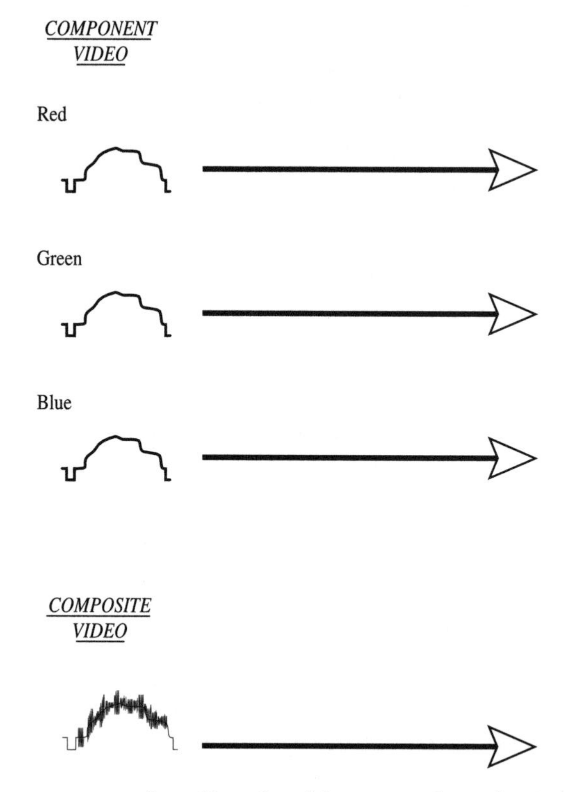

Figure 8.6 *A component color video signal is conveyed as physically separate signals, whereas a composite color video signal is conveyed as a single physical signal.*

cessing HSB component images provides superior results. Following the processing, the HSB component images are converted back to RGB component images in preparation for display.

NTSC/PAL Composite Color

Standards were created with the advent of color commercial broadcast television to define the representation of color video signals. Really, the RS-170 and CCIR standards were adapted for the color task. The basic timing of each of these standards was maintained so that, at the time, the existing base of consumer monochrome television receivers could still receive the color video signals in a monochrome form.

The color system adopted in the United States is known as *NTSC*, which stands for the originating organization, the *National Television Systems Committee*. It was developed in the late 1940s. The NTSC standard, also known as *EIA RS-170A*, modifies the RS-170 standard to work with color video signals by adding color information to the existing monochrome brightness signal. The NTSC signal is a composite color signal because it is created by combining color and brightness information together into a single signal.

The color signal is encoded by taking the three RGB component signals and combining them into intermediate color and brightness signals. The color signals are combined using phase and amplitude modulation techniques. The result is a single color subcarrier signal, called the *chrominance signal*, which represents the hue of the color as a phase and the saturation of the color as an amplitude. The chrominance subcarrier signal is added to the remaining brightness signal (now called the *luminance signal*), which is the same as the original RS-170 brightness signal. Additionally, a color reference signal, called the *color burst*, is added at the start of each video line, following the horizontal sync signal. The result is an NTSC, or RS-170A, signal. It should be noted that several other intricacies, including slightly varied signal timing, are also added in the NTSC standard to create an effective color encoding technique. Figure 8.7 compares the NTSC and RS-170 video signals.

At the receiving end, the NTSC signal is decoded by first separating the luminance and chrominance signals. The chrominance signal is then demodulated using the color burst as the demodulation reference signal. The demodulated color signals are finally combined with the luminance signal to form RGB component signals.

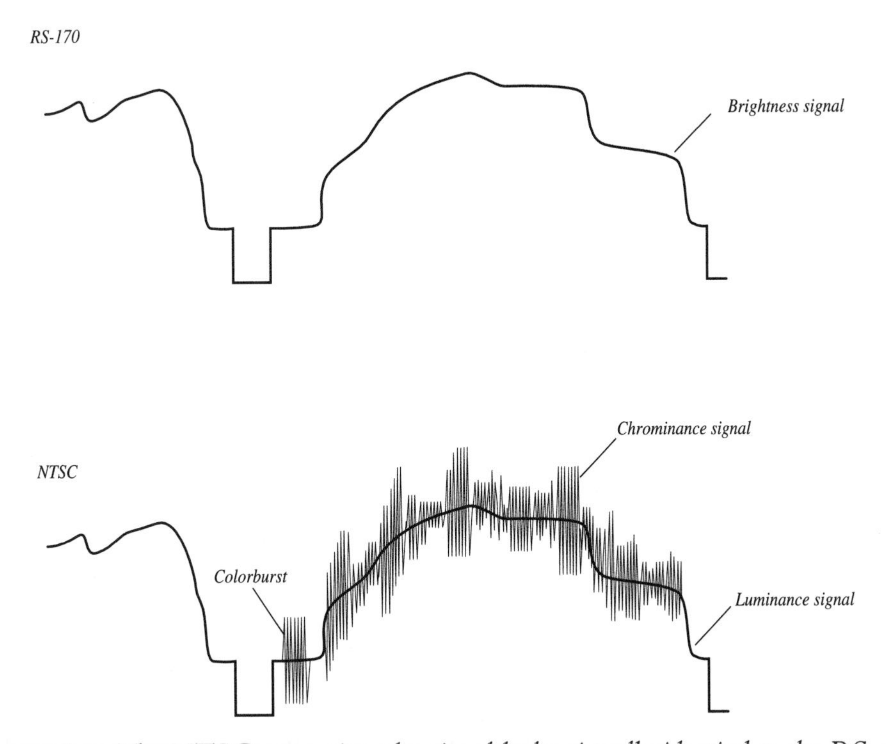

Figure 8.7 *The NTSC composite color signal looks virtually identical to the RS-170 signal with the addition of the color subcarrier.*

The European CCIR video standard was likewise adapted to a color video standard called *PAL*, for *Phase Alternation Line*. The scheme, developed in the early 1960s, uses a color encoding and decoding method that is similar to the NTSC standard.

France and most Middle-Eastern and Eastern-European countries use a third, less popular, color commercial broadcast television standard called *SECAM*, for *Séquentiel à mémoire*. Developed in the late 1950s, SECAM uses techniques that are similar to NTSC and PAL to create a color composite video signal.

Y/C Component Color

The *Y/C component color* standard conveys the color video signal as two distinct components: the *luminance*, or *Y*, signal and the *chrominance*, or *C*, signal. The Y signal is identical to a standard RS-170 monochrome video signal. The C signal is identical to the chrominance subcarrier signal defined in the NTSC standard. The only difference between the Y/C video signal and the NTSC signal is that the Y/C signal physically separates the Y and C components, rather than combining them into a single composite color signal. There is also a similar PAL-compatible Y/C format that conforms to the timing and signal parameters of the PAL composite color signal.

Keeping the Y and C signals separate creates a middle-of-the-road quality solution between RGB component video and NTSC composite video signals. The result is a video quality with color and brightness resolution that are superior to those of NTSC.

The Y/C video standard evolved from the video tape recorder industry as a way to create improved quality for professional video uses. It is also known as *S-Video*, *Super-Video*, and *S-VHS*.

CCIR Recommendation 601-1 Digital Video

The *CCIR Recommendation 601-1 digital video* standard is a digital component version of the NTSC and PAL video signals. It has the same timing as the NTSC or PAL standard signals, but differs from them in two significant ways. First, the signals are entirely digital. The synchronization pulses and video brightness and color signals are all conveyed digitally. Second, the signals are component signals, rather than composite signals. The color information is represented by individual digital RGB component signals or other similar component signals. CCIR-601-1 is intended for use within the commercial broadcast television industry for video transport between studios, control rooms, and transmission facilities.

The CCIR-601-1 standard has several versions that support the NTSC 525-line and PAL 625-line video standards. The data transfer can be physically handled using either parallel or serial data transmission formats, depending upon the application.

HDTV

High-definition television (*HDTV*), standards began evolving in the early 1980s and were finalized in the early 1990s. Similar in nature to the preceding video standards, HDTV standards increase the number of lines in the image frame, widen the aspect ratio, and increase the image frame's effective horizontal resolution.

The United States HDTV standard video timing can be summarized as follows:

- 1125 lines/frame
- 30 frames/second or 60 fields/second
- 2:1 interlaced scan
- 16:9 aspect ratio

Although it will take years for the commercial broadcast television industry to fully exploit HDTV, industry is already making widespread use of HDTV for applications requiring higher spatial resolution. The equipment costs, however, are still high. As the commercial television broadcaster and consumer demands increase over time, HDTV costs will drop.

Other Video Formats

The preceding video standards have all evolved from the commercial broadcast television industry. Because of the television industry's high-volume consumer and professional markets, a large variety of conforming video equipment is available at costs far below those meeting unique requirements. As a result, it has always been advantageous for the digital imaging markets to use existing video standards and equipment, when possible, rather than to design special-purpose devices.

Other standards have emerged, nonetheless. In particular, the personal computer industry has created several video display standards. These standards are known as the *Super Video Graphics Array* (*SVGA*) video modes. An industry consortium called *VESA*, for *Video Electronics Standards Association*, controls the SVGA video standards. The SVGA standards provide for spatial resolutions of 1024 pixels × 768 lines, 1280 pixels × 1024 lines, and beyond. SVGA standard frame rates include up to 100 image frames per second. These standards offer very high-resolution display capabilities along with a flicker-free appearance.

Some special, nonstandard video formats have also emerged. Often, these nonstandard formats stick with the use of the synchronization timing established by the RS-170 or CCIR standards. An example of this is found in digital video cameras. These cameras conform to the RS-170 timing standard, but transfer their image as a digital data stream rather than as an analog voltage signal. As a result, the video signal is higher quality. The video signal is different from a strict analog RS-170 signal, yet conforms to its timing parameters.

Other cameras stray far away from the RS-170 standard timing. Instead, they require special digital image processing system interfaces that are capable of receiving their special video signals. Use of nonstandard video devices precludes their arbitrary mixing and matching. Rather, they must be carefully matched to guarantee interoperability between the devices. This can be an insignificant issue, however, if the application is well-defined and the qualities of the nonstandard video are required.

9 Image Data Handling

Any digital image processing system must be capable of performing three primary image data handling functions—image digitization, image storage, and image display. These functions comprise the components of a typical frame grabber. *Image digitization* converts an image from the analog electrical form—from a camera or similar device—to the digital form. *Image storage* provides short-term and long-term storage of image data for processing, display, and archive purposes. The image storage function also provides interconnectivity between the short-term image store and image data processor functions. *Image display* converts a digital image back to the analog electrical form for display on a video display monitor. Figure 9.1 is a block diagram illustrating how these functions fit together.

In this chapter, we will discuss these primary elements of a digital image processing system. The individual hardware and software components comprising these functions are also described.

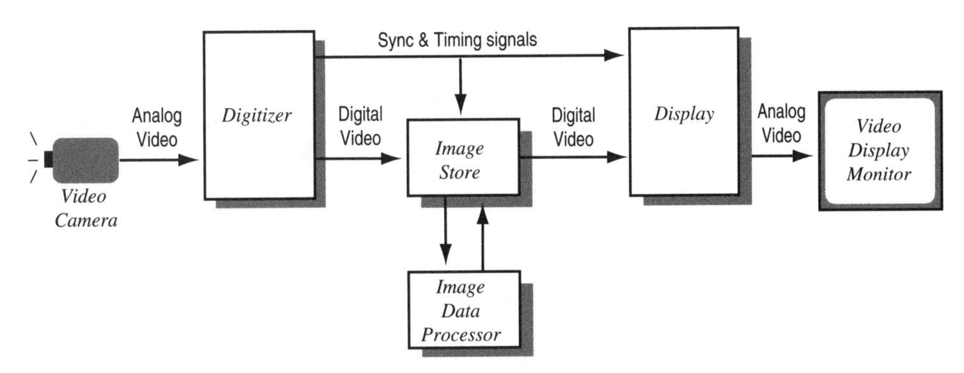

Figure 9.1 *The image handling functions of a digital image processing system: image digitization, storage, and display.*

Image Digitization

Before we can process an image digitally, we must first acquire it in a digital form. This means that an originating analog video signal must be converted to digital image data. Image digitization provides this elementary function.

The image digitization process can be divided into two principal functions, image sampling and image quantization. The latter is also referred to as analog-to-digital conversion. In addition, several ancillary functions are employed to pre-process the analog video signal prior to the quantization process. Figure 9.2 shows a block diagram of the subcomponents of the image digitization process.

In Chapter 3, we discussed the fundamental issues of sampling and quantization and how they relate to the spatial and brightness resolution of an image. We also showed how spatial aliasing can result from undersampling an image. It is in the hardware circuitry of the image digitization process that these issues must be addressed.

Sampling

In sampling a video signal to create a digital image, the first step is to select the rate at which sampling will occur. This rate is usually based on the desired spatial

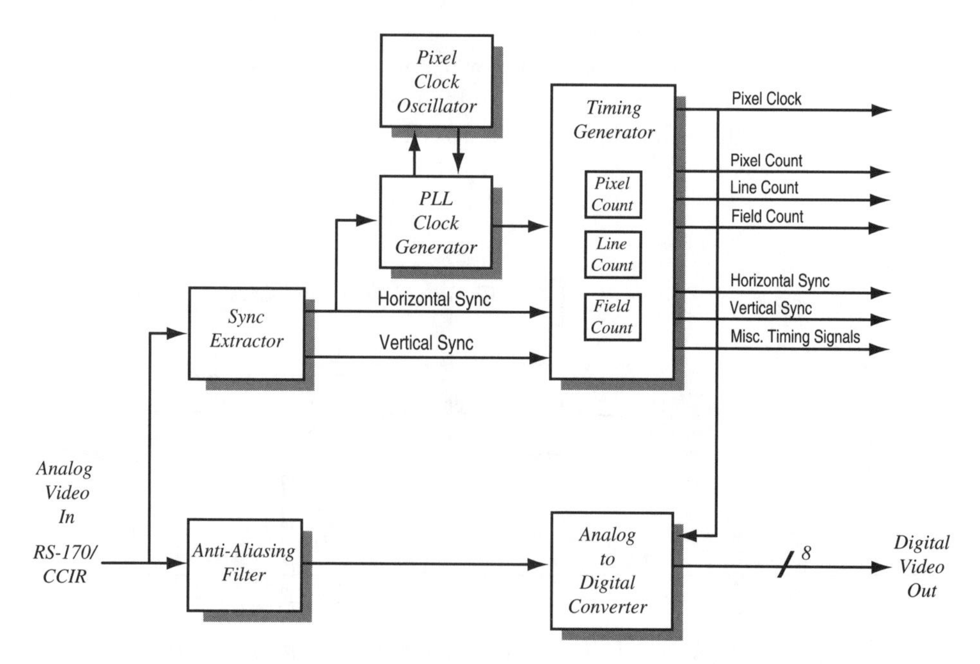

Figure 9.2 *The image digitization process.*

resolution of the resulting image. The inherent line resolution and signal quality in the video signal may also play a role in determining the appropriate rate. Let's look at the common case of sampling an RS-170 monochrome video signal. Incidently, sampling a CCIR signal is similar, only with different timing, pixel counts, and line counts.

In our earlier RS-170 timing discussion, several important facts emerged. First, an RS-170 video signal has 485 active lines of visual information. Second, the active line time—that part of each line containing visual information—is equal to 52.59 μS. The RS-170 signal was also defined to have an aspect ratio of 4:3. With this information, we can compute the sample rate needed to create a digital image.

With the 4:3 aspect ratio and number of visual lines, we compute the number of pixels per line as $485 \times 4/3 = 646.66$ pixels. These will be square pixels, meaning that they cover a square unit of image area. The pixel sampling rate can be computed as the active line time divided by the number of active pixels. This is $52.59\ \mu S/646.66 = 81.33$ nS per pixel, which is equivalent to a sampling frequency of $1/81.33\ nS = 12.30$ MHz. In practice, a few lines and pixels are trimmed from the RS-170 image size to yield a final image of dimensions 640 pixels $\times$ 480 lines.

To create the image, we acquire individual pixels of the image by sampling the active portion of each video line at the equal intervals of the pixel sampling rate, or 81.33 nS. This period is also commonly referred to as the *pixel clock period* or *pixel time*, and is the critical pacing factor to the design of the image digitization system. During each pixel time, the sampled pixel must be quantized, using an analog-to-digital converter, and then stored in the image store memory. Once we have acquired all the pixels in the image, the entire digitized image will be located in the image store.

To follow the flow of the digitization process, let's refer to the RS-170 timing diagram presented earlier in Chapter 8. The process begins with the vertical sync signal. The vertical sync pulse indicates the start of a field, which we'll assume is the even field. The video line counter is reset to its starting point. The nine lines of the vertical sync interval are allowed to pass. Then, the following 11 "no-video" lines pass. When the 21st horizontal sync pulse is detected, we arrive at the first active video line. At this point, the video signal is at the upper left corner of the visual portion of the image frame. The video pixel counter is reset. The horizontal sync interval of 10.9 μS is allowed to pass; then, digitization begins.

At the rate of 81.33 nS per pixel, 640 pixels are sampled, quantized, and stored in memory from the video line. At the end of the line, digitizing stops and the process awaits the next horizontal sync pulse. When the following horizontal sync pulse is detected, the process repeats, line by line, throughout the field. When line number 261 arrives, all 240 visual lines of the field have been digitized (lines 21 through 260), and the process halts for the remainder of the field. The following vertical sync pulse indicates the start of the odd field of video. The digitization process repeats for the odd field, completing the acquisition of an entire image frame, 640 pixels × 480 lines.

If we want to acquire only a 320-pixel × 240-line image from an RS-170 video source, we use all of the above timing, but store only every other pixel and only the lines of one field of the video frame. Likewise, we can acquire other spatial resolutions depending on the application's requirements.

Sync Extraction and Timing

The horizontal and vertical sync signals found in the incoming video signal (see the RS-170 timing illustration in Figure 8.5) establish the digitization process timing. A sync extractor circuit detects the voltage levels in the video signal, watching for a transition into the sync signal range, below the "blacker-than-black" level. The duration of the sync signal is then evaluated to determine whether it is a horizontal sync signal or a vertical sync signal. A short sync pulse indicates a horizontal sync signal, while a long pulse indicates a vertical sync signal.

The horizontal and vertical sync signals are used to generate a variety of timing signals. The timing signals coordinate the digitization process, as well as image storage and display activities. First, a pixel clock signal is generated by using an oscillator running at the desired pixel clock rate, 12.30 MHz for an RS-170 video signal. A *phase-locked loop* (*PLL*), circuit locks the pixel clock to the horizontal sync signal, as shown in Figure 9.3. The purpose of phase-locking the pixel clock is twofold. First, it guarantees that the first pixel of each video line occurs at precisely the same time relative to the horizontal sync pulse. This results in the first pixel of each line being placed at precisely the same spatial location relative to the left side of the video image frame. Without this, horizontal jitter in the image can occur, degrading its visual quality.

The second reason for using a phase-locked pixel clock is to keep the timing of the pixel clock and other digitization timing signals constant, even when a disturbance occurs in the video signal. A dropout can occur, especially with video

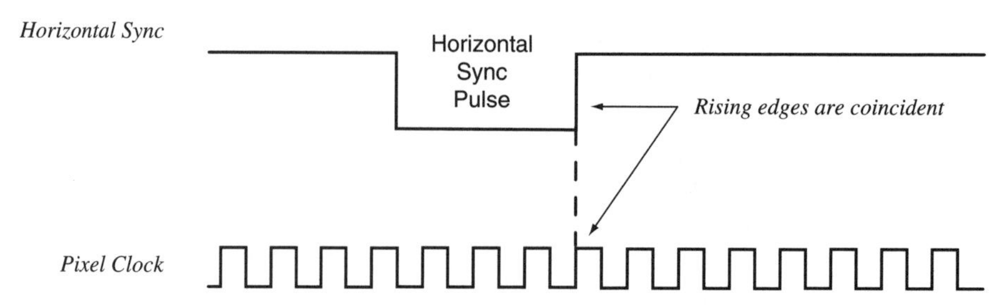

Figure 9.3 *Phase-locking the pixel clock to the horizontal sync signal eliminates image jitter and fills in when disturbances occur in the video signal.*

sources such as video recorders or transmission links, where the video disappears for some period of time. Dropouts may last for only a few or several hundred microseconds, interrupting a critical horizontal or vertical sync signal. This can destroy the sample timing of the digitization process. A PLL "free runs" during times when the video signal disappears. It is like a mechanical flywheel whose momentum fills in for missing horizontal and vertical sync signals by keeping the pixel clock running, timed with the horizontal sync pulse last detected.

The stable, PLL-generated pixel clock is used along with the raw horizontal and vertical sync signals to create several timing signals for image digitization, storage, and display. Of most importance are a pixel count, a line count, and a field count. For an RS-170 video signal, the pixel counter counts from 0, at the beginning of a line, to 639, at the end. Similarly, the line counter counts from 0, at the top of the image frame, to 239, at the bottom. The field counter simply counts from 0 to 1, indicating whether the current field is the even or odd field.

These counts keep track of where we are in the video frame during digitization. They are coordinated by the horizontal and vertical sync signals. Whenever a horizontal sync pulse occurs, the pixel counter resets to 0. This indicates that the video signal is at pixel 0, the left side of the image frame. The horizontal sync pulse is also used to increment the line counter to the next count, indicating that the video signal has moved down one line. Whenever a vertical sync pulse occurs, the line counter resets to 0, signifying that the video signal is at the top of the image frame. The vertical sync pulse also increments the field counter, indicating the start of a new field. Every time the field counter counts from 0 to 1 and back to 0, an image composed of two fields has been acquired.

Some special-purpose cameras, such as high-speed, low-speed, or line-scan cameras, do not convey horizontal and vertical sync signals within their video signal, as prescribed by the RS-170 standard. In these cases, separate sync lines with nonstandard timing are often used to couple the digitization hardware to the camera. This way, the digitizer can stay perfectly synchronized with the camera.

Anti-Aliasing Filtering

To avoid spatial aliasing effects, as described in Chapter 3, the video signal is filtered prior to sampling and quantization by the analog-to-digital converter. The goal of an anti-aliasing filter is to eliminate all spatial frequencies from the video signal that exceed one-half the digitization sampling rate. Our sampling rate for an RS-170 video signal is 1/81.33 nS = 12.30 MHz. This means that any video frequency components above 6.15 MHz will be aliased to a new, erroneous frequency, adding unwanted degradation to the acquired digital image.

We implement an anti-aliasing filter as a low-pass filter circuit with a cut-off frequency at the Nyquist rate (one-half the sampling rate). The perfect "brick-wall" filter would have absolutely no effect on the video signal for frequencies below the Nyquist rate, and would totally eliminate frequencies above the Nyquist rate. In practice, though, a filter like this is physically impossible to realize. Instead, we will lose a little of the video signal composed of "good frequencies" below the Nyquist rate, while some of the "bad frequencies" exceeding the Nyquist rate will remain.

Figure 9.4 shows the frequency response curves for a perfect "brick-wall" low-pass filter and a practical low-pass filter. When designing an anti-aliasing filter, the

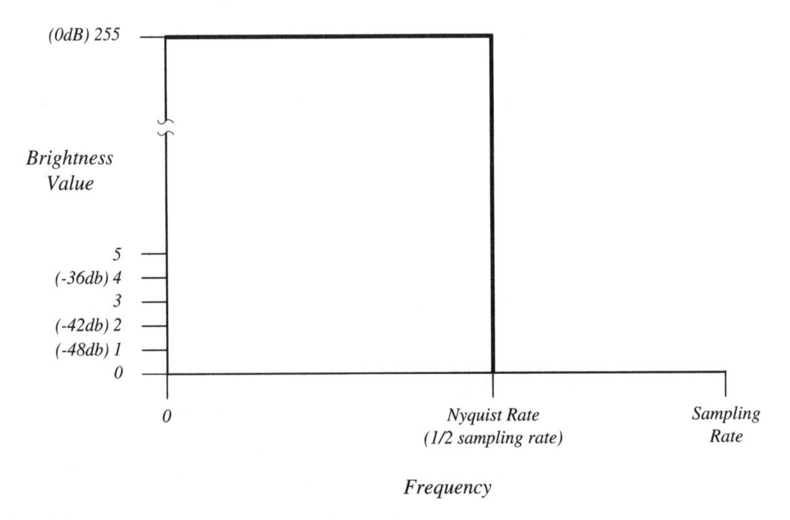

Figure 9.4a *A "brick-wall" anti-aliasing filter removes all frequency components above the frequency given by the Nyquist criterion.*

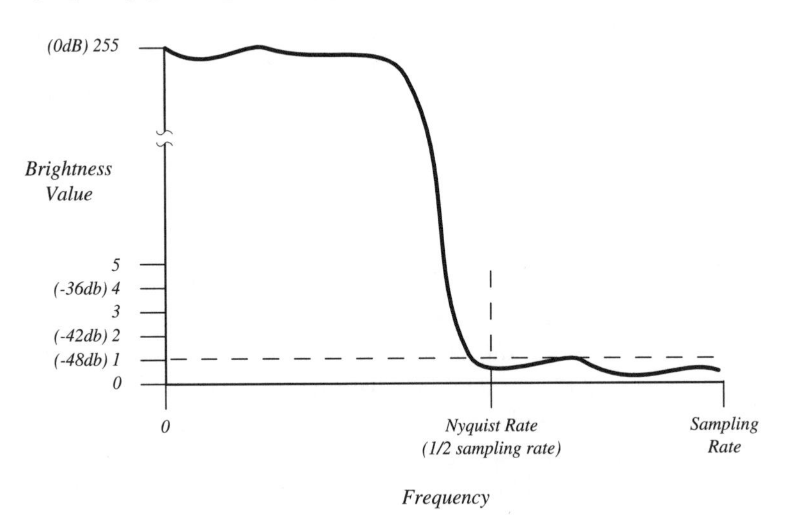

Figure 9.4b *A practical anti-aliasing filter attenuates frequencies above the Nyquist rate to a level that is not resolvable by the quantizer.*

goal is to minimize the attenuation of frequencies below the Nyquist rate while attenuating the frequencies above the Nyquist rate to a point where they cannot be resolved by the quantization resolution. For instance, when quantizing to a brightness resolution of 8 bits, we can only resolve one part in $2^8 = 256$. This means that each brightness step is 1/256th of the total brightness range. If we can attenuate the brightnesses of the frequencies above the Nyquist rate by a factor of 256, the quantization process will not be able to detect the remaining brightness information at these frequencies.

Filters are generally specified as having attenuation characteristics measured in *decibels* (*dB*). The decibel scale is logarithmic, where 6 dB equals 1 bit of quantization resolution. Decibel equivalents are listed in Table 9.1.

When describing the anti-aliasing, low-pass filter for an RS-170 video signal, we would say that it must attenuate the signal by 48 dB at 6.15 MHz, the Nyquist rate.

For some applications, an anti-aliasing filter may not be necessary because the video source may be devoid, or close to devoid, of frequency components above the Nyquist rate of the digitizer. This can occur with certain video cameras and other video sources where they may not have the inherent capability to produce a signal above about 6 MHz. In these cases, the anti-aliasing filter is effectively in the video signal source itself.

Analog-to-Digital Conversion

Following the anti-aliasing filter is the *analog-to-digital converter* (*A/D*), the heart of the digitization process. The A/D converter carries out the sampling and quantization processes. It must be capable of converting the video signal's voltage level to a digital value in less than 1 pixel time. Typically, video rates require the use of a high-speed A/D known as a flash A/D converter, shown in Figure 9.5. A flash

Table 9.1 Decibel Equivalences

GRAY LEVELS	QUANTIZATION BITS	DECIBELS
16	4	24
32	5	30
64	6	36
128	7	42
256	8	48
512	9	54
1,024	10	60
2,048	11	66
4,096	12	72

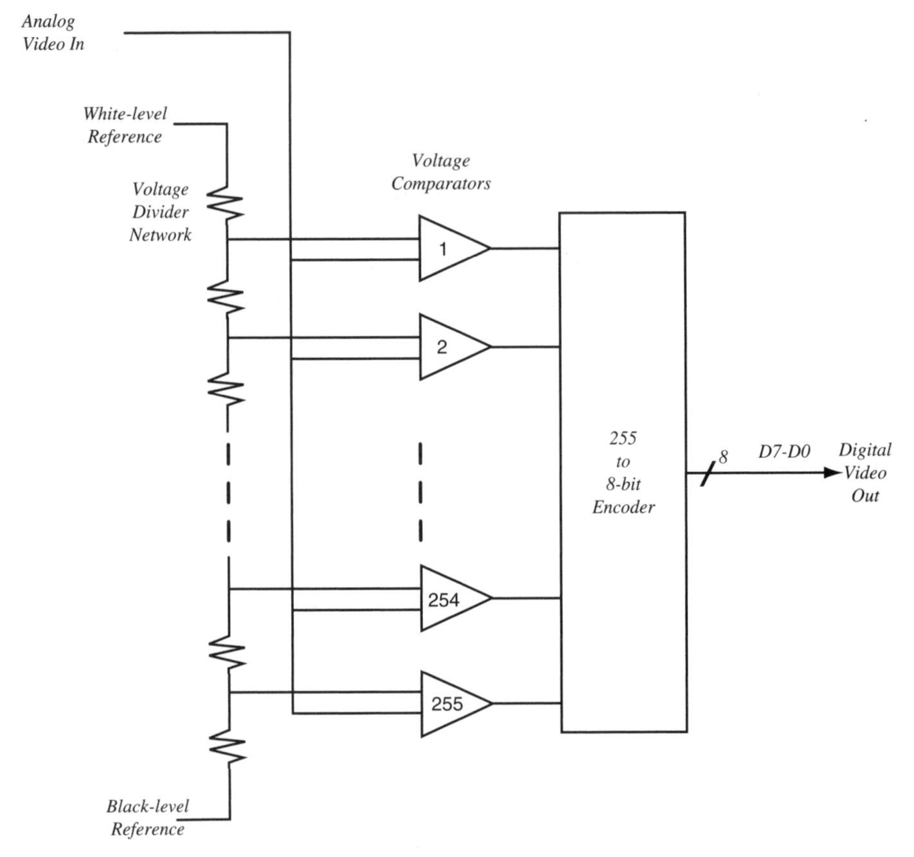

Figure 9.5 *A flash analog-to-digital converter typically used for video-rate digitization.*

A/D is a single-chip device consisting of individual voltage comparators, one for each brightness level to be detected. In the case of 8-bit quantization, the flash A/D has 256 comparators. As the required brightness resolution increases by 1 bit, the number of comparators in the flash A/D doubles.

The flash A/D conversion process works by comparing the video input signal against two reference voltage levels. The two reference levels represent a *black-level reference* and a *white-level reference*. In an 8-bit A/D device, a resistor voltage-divider network creates 256 individual voltage levels, spanning the range between the black-level and white-level references. Each of the 256 voltage levels feeds a comparator, with which the incoming video signal is compared. Digital logic looks for the highest active comparator to determine the digital value representing the video voltage level (the brightness) of the video signal. An A/D conversion occurs once every pixel time—which is, of course, equal to the image sampling rate.

Flash A/D converters evolved out of United States military electronics programs in the late 1970s. They have been an essential ingredient in the development of low-cost digital image acquisition products. Many current A/D converter devices include some, or all, sync extraction, analog-to-digital conversion, and color decoding functions.

Color Decoding

Often, we may need to use composite color video signals as input to a digital image processing system. We can digitize composite color signals by first breaking them into individual video component signals. This is done by a *color decoding* process that decodes the color signal to an RGB form. In the case of an NTSC color signal, the RGB form is simply three RS-170 signals that are subsequently sampled and quantized as described earlier. Similarly, a PAL color signal is decoded to three CCIR video signals.

A color decoder that decodes an NTSC signal into RGB signals is composed of a luminance/chrominance separator followed by a color demodulator and matrix, as shown in Figure 9.6. The luminance/chrominance separator is basically two filters, one that allows only the luminance portion of the signal to pass and another that allows only the chrominance portion to pass. The result is two outputs, luminance and chrominance (Y and C). The Y and C signals are identical to the Y/C color component form of a color video signal, discussed in Chapter 8. The chrominance signal is demodulated into two intermediate color signals (I and Q) called *color vectors*. The two color vector signals are finally transformed through the color matrix to form red, green, and blue color component signals. Each of these signals is equivalent to a monochrome RS-170 signal.

Sometimes a digital image processing application requires gray-scale images but only has access to a color video source. In these cases, only the color separator portion of the above color decoder is used to remove the chrominance portion of the signal. By removing the chrominance portion, the quality of the remaining luminance signal is improved. The luminance signal is identical to an RS-170

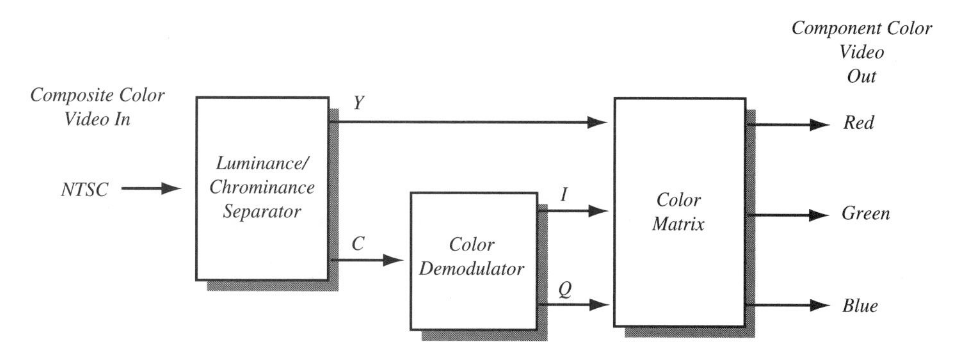

Figure 9.6 *The color decoding process converts a composite color video signal to individual color component signals.*

monochrome signal. The function of removing the color information from a color video signal is referred to as an *NTSC* (or *PAL*) *color filter*, or a *chrominance filter*.

Image Display

To view a digital image on a video display monitor, we must convert the digital image back to an analog video signal. This conversion process is referred to as *analog signal reconstruction*. It is the electronic reformulation of an analog video signal from a stored digital image, and is not in any way related to the digital cross-sectional image reconstruction operation used in computed tomography (see Chapter 7). The image display process is the reverse of the image digitization process. Generally, image display is a simpler task than image digitization, because many of the necessary timing signals are already available from the digitization process.

The image display process centers around the digital-to-analog conversion function. Additionally, several ancillary functions are used to post-process the analog video signal prior to its display on a video display monitor. Figure 9.7 shows a block diagram of the subcomponents of the image display process.

Digital-to-Analog Conversion

The digital image data is first converted to an analog voltage signal. This is done using a *digital-to-analog converter* (*D/A*). Like the A/D converter, the D/A con-

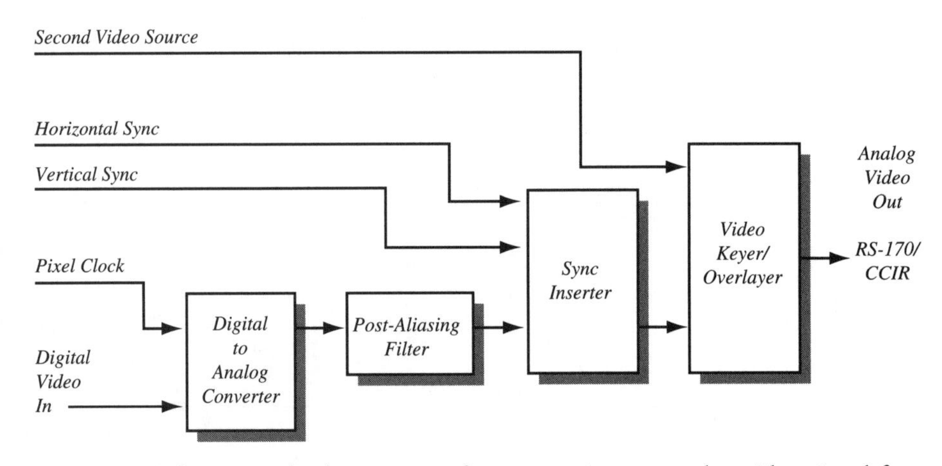

Figure 9.7 *The image display process of reconstructing an analog video signal from digital image data.*

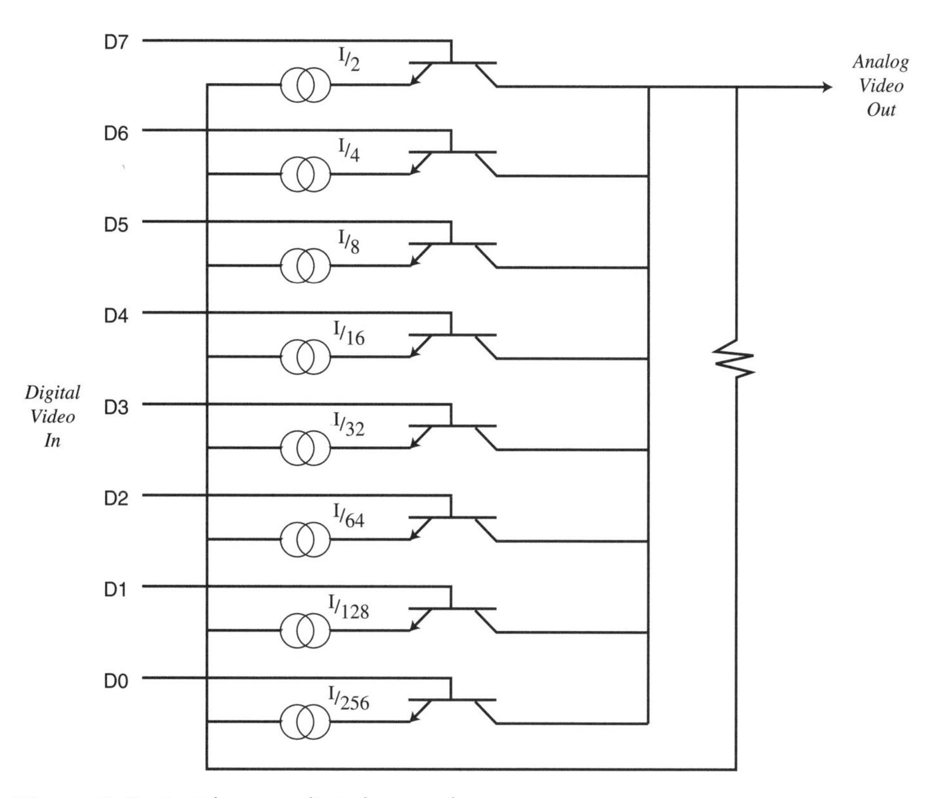

Figure 9.8 *A video-rate digital-to-analog converter.*

verter must be capable of converting a pixel's digital data to an analog voltage level in less than one pixel time. High-speed video D/As are used for this purpose.

A D/A converter uses a current-summing network to create a voltage level output that is proportional to the input digital value, as shown in Figure 9.8. The number of precision current sources is equivalent to the number of bits in the digital input value. Each current source in the network provides a current that is one-half the preceding current source's value, looking at the bits in descending order. This corresponds with the fact that each bit in a digital value is weighted as one-half the value of the preceding bit's value. When each of the bits of the digital input value drives its corresponding current source, a total current is generated that is proportional to the digital value. A final resistor divider in the network acts to convert the current to a voltage that becomes the output voltage-level signal. The D/A conversion occurs once every pixel time, which is equal to the image sampling rate.

High-speed video D/As, like flash A/Ds, evolved out of military electronics programs in the late 1970s. Many current D/A converter devices include some, or all, digital-to-analog conversion, sync insertion, keying and overlaying, and color decoding functions.

Post-Aliasing Filtering

The voltage-level signal produced by the D/A converter contains artifacts, making it unsuitable for display. In particular, it contains tiny stairstep patterns often combined with high-frequency spikes and ringing. The stairstep patterns occur because each time the D/A does a conversion, it jumps almost instantaneously from one voltage level to another. Similarly, the fast speed of the signal slewing from one voltage level to another introduces spikes and ringing into the signal. Because these artifacts were not present in the original input image, we want to remove them before displaying the output image.

The stairstep patterns actually cause an aliasing phenomenon. Frequency components that are higher than the Nyquist rate are created by the sharp stairstep edges. We can remove the stairsteps by using a low-pass filter circuit called a *post-aliasing filter*, as shown in Figure 9.9. In fact, this filter can be similar, or even identical, to the anti-aliasing filter described earlier. The goal, again, is to remove frequency components from the video signal that exceed the Nyquist rate.

For some applications, a post-aliasing filter may not be required, because the video display monitor may not be capable of displaying the high-frequency components generated by the D/A converter. This is common with some video display monitors, especially less expensive ones. In these cases, the video display monitor effectively contains the post-aliasing filter.

Sync Insertion

Like any standard video signal, our reconstructed analog video signal must contain the required synchronization signals. This is so that standard video devices, such as a video display monitor, can understand the horizontal and vertical timing of the signal.

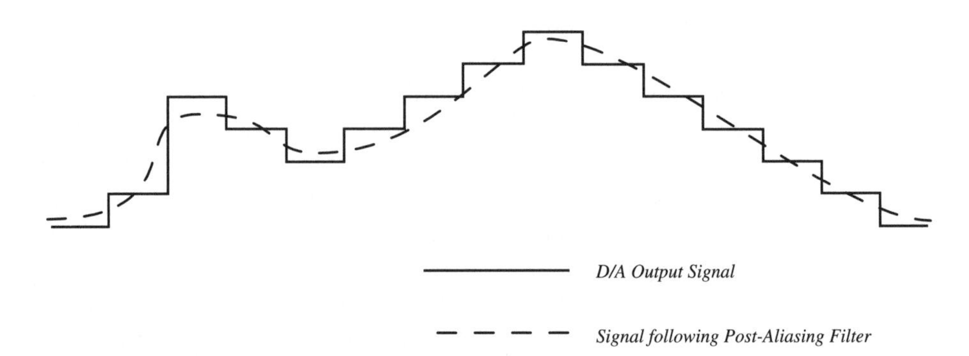

Figure 9.9 *The raw output from a D/A converter contains stairsteps at the transitions between pixels. The post-aliasing filter removes these unwanted signal artifacts.*

To accomplish this, we must insert a sync signal—both horizontal and vertical portions—into the outgoing analog video signal generated by the D/A converter. This mixing is typically done with analog circuitry that forces the video signal to the correct RS-170 sync voltage levels when appropriate, as shown in Figure 9.10. The resulting video signal becomes entirely RS-170-compatible.

The sync signals extracted from the incoming video signal can be inserted into the outgoing video signal, or sync signals from another video source or an internal sync generator can be used. The important issue is that the digital image data must be sequenced from the image store memory to the D/A converter in step with whatever sync source we use.

It is also possible to encode and mix the sync signal digitally prior to the D/A conversion process. Then, the signal leaving the D/A has sync signals that are

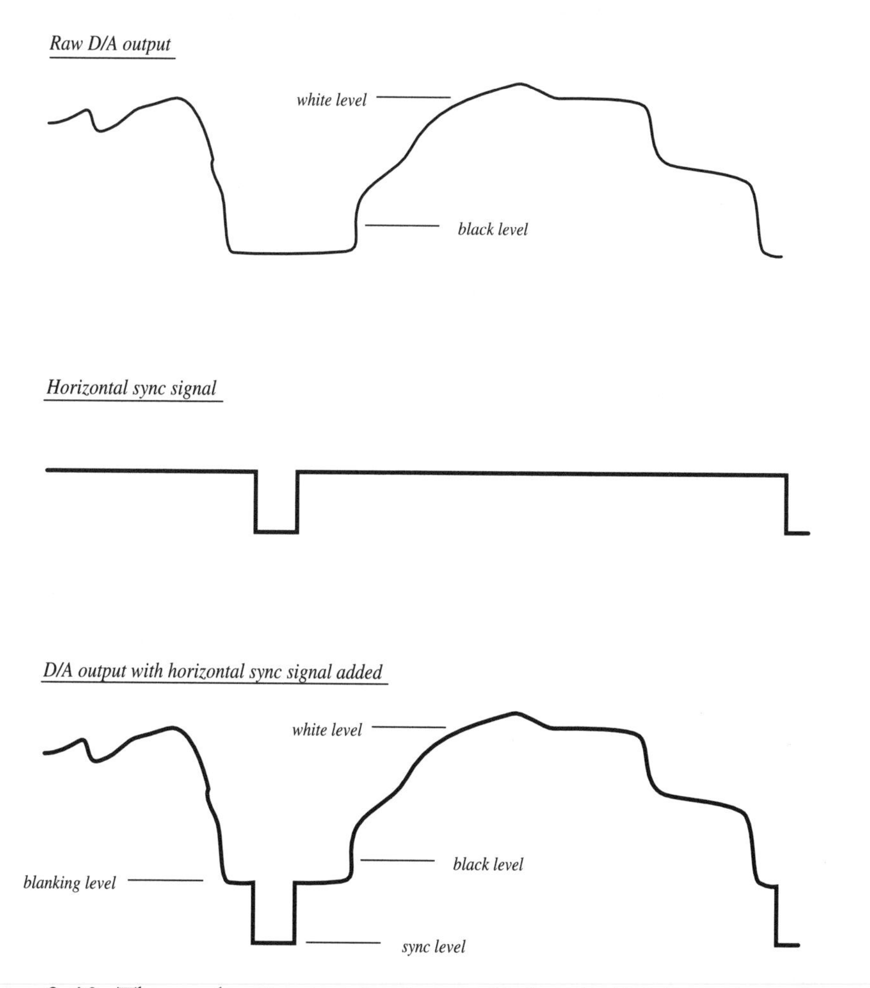

Figure 9.10 *The synchronization signals are added to the D/A's analog video signal output, making the resulting signal RS-170-compliant.*

already mixed with the video signal. This approach, however, eats up the video signal's dynamic range and results in a reduced brightness resolution. Because the sync signal occupies nearly one-third of the video amplitude, about one-third of the 8-bit brightness range of the D/A converter's output is wasted to create the sync signal. Fewer than 200 gray levels remain for the image brightness, rather than the full 256 levels. This will result in reduced image brightness resolution. If a 9-bit or greater D/A is used, then the image brightness resolution can be maintained. However, the D/A converter cost increases.

Keying and Overlaying

Sometimes we may want to mix the video signal with another video signal before displaying it. *Video keying* and *video overlaying* provide two ways to mix the two video signals. It is important to synchronize the two video signals, one from the D/A and sync inserter with one from a second source. This means that their synchronization signals must be precisely matched in time. Generally, the sync signal is extracted from the second video source and inserted into the D/A video signal. This way, both video signals have the identical sync signals, and are therefore perfectly synchronized. The video keyer and overlayer are composed of an analog switch that switches between the two video signals based on the content of the video signals. Figure 9.11 shows the block diagram of a video keyer and overlayer.

The video keying function switches between the two video sources based on the content of the D/A video signal. It works by defining a particular brightness or color, or range of brightnesses or colors, in the D/A video source as being the *color key*. The video keyer looks for the color key in the D/A video signal. When

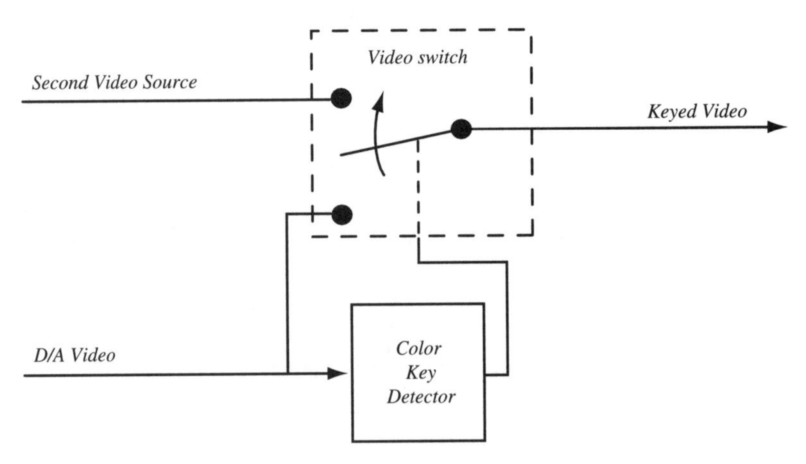

Figure 9.11a *Video keying switches in a second video source whenever the color key is detected in the D/A video signal.*

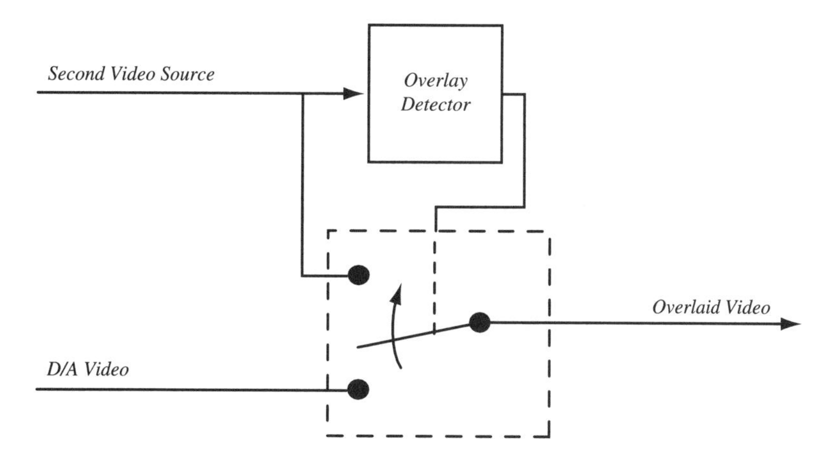

Figure 9.11b *Video overlaying switches in a second video source whenever the background color is* not *detected in the second video source signal.*

the color key is detected, the video signal from the second source is switched in, in lieu of the D/A video signal. The result is a video signal containing the second video source within the D/A video signal wherever the D/A video source is the same brightness or color, as the color key. This allows the D/A video to control the mixing of the two video signals, as shown in Figure 9.12.

Video overlaying is exactly the reverse of video keying. This process switches between the two video sources based on the content of the second video signal. It works by defining a particular brightness or color in the second video source as a *background color.* The video overlayer looks for the background color in the second video signal. Whenever it is *not* detected, the video signal from the second source is switched

Figure 9.12a *A D/A video signal where an inset portion of the background appears with a distinct brightness. The color key is set to the same distinct brightness.*

Figure 9.12b *The second video source of a satellite weather image to be substituted wherever the color key brightness is detected in the D/A video signal.*

Figure 9.12c *The resulting video keyed signal.*

in, in lieu of the D/A video signal. The result is a video signal that has the second video source appearing on top of the D/A video signal wherever the second video source is *not* the same brightness or color as the background color. This allows the second video signal to control the mixing of the two video signals, as shown in Figure 9.13.

Video keying is used to mix two video signals when the intelligence of the mixing control is in the D/A video signal. As an example, let's look at a case where an original image contains several objects of interest upon a background of uniform brightness. Let's say that we want to replace the background of the image with a particular pattern. This is accomplished by making the second video source portray the desired background pattern and by setting the color key to the background brightness found in the original D/A video image. Whenever the video keyer detects the background

Figure 9.13a *A D/A video signal that is to be overlaid with the contents of the second video source.*

Figure 9.13b *The second video source of graphical information on a constant background brightness. The background color is set to the same background brightness. Wherever the background color is* not *detected in the second video signal, the graphical information is substituted in place of the D/A video signal.*

Figure 9.13c *The resulting video overlaid signal.*

brightness (the color key), the second source's background pattern is switched into the resulting video signal. Whenever the background brightness is not detected, the original image passes through unaltered. The video keying function is very similar in its action to the image compositing function discussed in Chapter 4.

Video overlaying is used to mix two video signals when the intelligence of the mixing control is in the second video signal. This case typically arises when we want to overlay graphics material, such as text or lines on top of the D/A video signal. The graphical overlay may be in the form of image annotations or other pertinent information regarding the image's content. We do the video overlay by making the graphics image, with a constant background brightness, the second video source. The background key is then set to the same background brightness as in the graphics image. Whenever the video overlayer detects the background brightness (the background key), the original D/A image passes through unaltered. Whenever the background brightness is not detected, the second video source (the graphics image) is switched in as the resulting video signal and appears on top of the original D/A video signal.

Color Encoding

Reconstructed color images are in the RGB form of three component video signals. These can be combined into a composite color signal by using a *color encoding* process. For instance, an NTSC color signal can be created from three individual RS-170 signals. Likewise, a PAL color signal can be created from three CCIR video signals.

A color encoding circuit that converts RGB signals into an NTSC composite signal is made up of a color matrix and modulator followed by a luminance/chrominance mixer, as shown in Figure 9.14. The color matrix first transforms the red, green, and blue color component signals into two intermediate color vector signals. The color vector signals are then modulated into a single chrominance signal (C) and a luminance signal (Y). The Y and C signals are identical to the Y/C color component form of a color signal. Finally, the Y and C signals are mixed to create the resulting NTSC composite color signal.

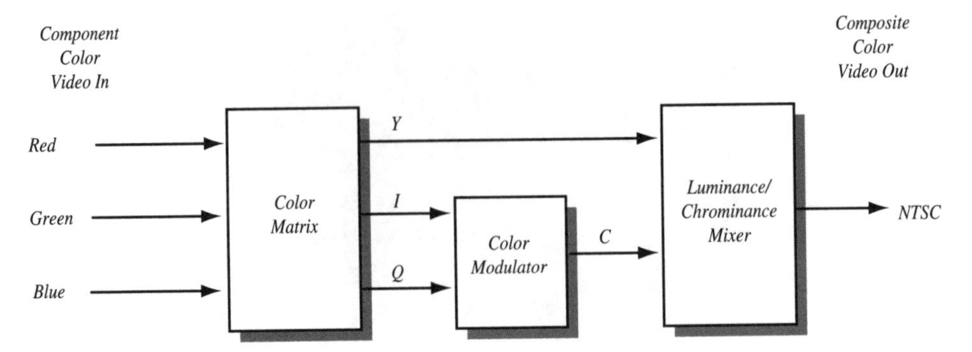

Figure 9.14 *The color encoding process converts individual color component signals to a composite color video signal.*

Image Storage

We have now interfaced a common analog video signal with the digital world. The signal enters the digital image processing system through the image digitization function and exits through the image display function. Between these two functions, all remaining data handling and processing hardware is digital.

Entire digital image frames from an RS-170 video source arrive at the digitizer output at the rate of one every 1/30th of a second. Sometimes, required digital image processing operations can be applied to the image in this amount of time. When this is possible, we refer to the processing operation as a *real-time processing* case. In the real-time case, the image data can flow directly from the digitizer, through the digital image processor, and back out to the display function. Often, however, we need to store the digital image for some time. This is true when (1) processing is too complex to be carried out in real-time, (2) the processing operation requires access to all pixels in the image frame before processing can begin (such as geometric transformations), or (3) we need to freeze the image for subsequent display analysis.

For these reasons, most digital image processing systems provide a way to temporarily store a digital image, called *working image storage.* Further, most systems also provide a means for *permanent image storage.* In this section, we will examine the hardware and software used for both the working and permanent digital image storage functions.

Working Image Storage

Working image storage in a digital image processing system is commonly referred to as the *image store.* It is composed of memory devices to store the digitized video data produced by the image digitizer. For an RS-170 video signal, image data arrives at the image store at a rate of 1 pixel every 81.33 nS. Image data also leaves the image store, at the same rate, to feed the image display circuitry. Additionally, the image data processor must have time to access the image data within the image store for processing. To meet these requirements, the image store architecture uses special organization, data buffering and control techniques.

Here, we will explore memory devices suitable for an image store design, the image data flow into and out of the store, and some image store architectures.

Image Store Memory Types

An image store is composed of *random access memory* (*RAM*). RAM memories are semiconductor read/write memory devices. We can randomly process any data in a RAM memory device at any time. There are two primary forms of RAM memories—*static RAM* (*SRAM*), and *dynamic RAM* (*DRAM*). Static RAMs are the simplest memory devices. They are composed of an array of storage elements, each of which is addressable. As an element is addressed, its contents can be read out of the device or written into the device. Often multiple bits of information are grouped together with the same address and are read or written simultaneously.

Some common static RAM memory data sizes are 256 Kbits and 1 Mbits. These particular devices are available in the following configurations:

256 Kbit devices	—	256K × 1-bit
		64K × 4-bit
		32K × 8-bit
1 Mbit devices	—	1M × 1-bit
		256K × 4-bit
		128K × 8-bit

A static RAM device has several data input/output and address signals, a chip enable, data input/output selection, read/write control, and power lines. The contents of a static RAM can be read by placing the address of the desired memory location on its address lines, enabling the chip, selecting the data output mode, and setting the read/write control line to the read state. A period of time later, called the *access time*, the data appears on the data lines. Similarly, the contents can be written by placing the data on its data lines, placing the address of the desired memory location on the address lines, enabling the chip, selecting the data input mode, and setting the read/write control line to the write state. The data is written into the device within the access time period. Static RAM access times run from about 10 nS to over 100 nS.

Dynamic RAMs differ from static RAMs in several ways. First, dynamic RAMs can be manufactured in data sizes that are about 16 times greater than static

RAMs. They also cost considerably less and consume less power per bit. This gives dynamic RAMs a large advantage over static RAMs for bulk storage applications, such as multiple digital image storage. The memory element circuitry used in a dynamic RAM is much smaller than that of a static RAM. The dynamic RAM storage element is based on a capacitor to store the state of each memory location, whereas static RAMs use several transistors to do the same thing. Unfortunately, the dynamic RAM's capacitor memory element requires all elements of the device to be regularly *refreshed*, which requires additional external circuitry and time. Also, the dynamic RAM has slower access times and an additional period called *cycle time* that slows down the data-reading and writing process.

Some common dynamic RAM memory data sizes are 1 Mbits, 4 Mbits, and 16 Mbits. These particular devices are available in the following configurations:

1 Mbit devices	—	1M × 1-bit
		256K × 4-bit
4 Mbit devices	—	4M × 1-bit
		1M × 4-bit
		512K × 8-bit
		256K × 16-bit
16 Mbit devices	—	16M × 1-bit
		4M × 4-bit

A dynamic RAM device has more signals than a static RAM device. These include multiple data input/output and address signals, a chip enable, data input/output selection, read/write control, row and column address selection, and power lines. The contents of a dynamic RAM can be read by first enabling the chip, selecting the data output mode, and selecting the read state, followed by sequencing the row and column halves of the location address to be read. The data appears on the data lines following the access time period. There is a waiting period following the access time before the next read or write operation can begin. The total period to complete an entire read or write cycle is the cycle time.

The contents of a dynamic RAM can be written by placing the data on its data lines, enabling the chip, selecting the data input mode, and setting the write state, followed by sequencing the desired row and column addresses. The data is written into the device within the access time period. Again, prior to a subsequent read or write operation, the cycle time period must be observed. Additional refresh cycles must be added periodically to refresh all memory locations with the device. Dynamic RAM access times run from about 60 nS to more than 100 nS, and cycle times run from about 100 nS to more than 200 nS. This means that random access to data within a dynamic RAM device can take a minimum of 100 nS, dictated by the cycle time.

A variety of addressing modes have been developed by DRAM manufacturers to speed up cycle times. Among them, techniques referred to as *read-modify-write mode*, *page mode*, *enhanced page mode*, *nibble mode*, and *static column decode mode* are available. These alternate modes come with various addressing order and timing constraints, but can significantly reduce read and write cycle times in some applications.

Additionally, there are two types of dynamic RAM devices produced especially for video data storage applications. They are known as *video random access memory* (*VRAM*) and *dynamic field store memory*. Both of these devices are unique in the way that they sequence data into and out of the device. Video RAM devices are composed of typical dynamic RAMs with the addition of large internal input and output data buffers. These buffers have their own additional data path into and out of the memory device, called the video data path, and can be written and read at video data rates. Once either internal data buffer is full, the dynamic RAM array is internally written or read with a wider data size and lower data rate. In this way, the internal cycle time of the video data path becomes hidden in the background, making the video data path's effective cycle time about 30 nS. The higher input and output rates of this additional data path match the requirements of video-rate applications.

A common video RAM memory data size is a 1 Mbit device. It is available in the following configurations:

1 Mbit devices	—	256K × 4-bit
		128K × 8-bit

Dynamic field store memory devices are specifically intended to sequentially store a field of RS-170- or CCIR-timed video data. They have fully self-contained refreshing and addressing circuitry which makes their use relatively simple. Video data flows into and out of a dynamic field store memory device sequentially and random access is not available. Therefore, these devices are excellent for storing, freezing, and delaying image data for processing purposes, but they have limited value for general image data processing needs. The most common dynamic field store memory data size is also a 1 Mbit device.

Two-Port Image Store Access

The image store design must meet tough data speed requirements. As we discussed earlier, an RS-170 video signal has a rate of 1 pixel every 81.33 nS. This means that every 81.33 nS, we must write the pixel's data from the image digitizer to the image store memory or read the pixel's data from the image store to the image display circuitry. Also, we must provide additional read and write accesses to the image store so that the digital image processor can process the stored image. Other standard video formats can produce image data at even faster rates, requiring an image store to operate at yet higher speeds.

The primary design mechanism in an image store design is a *two-port memory access* scheme. The idea is to provide two totally independent data paths into and

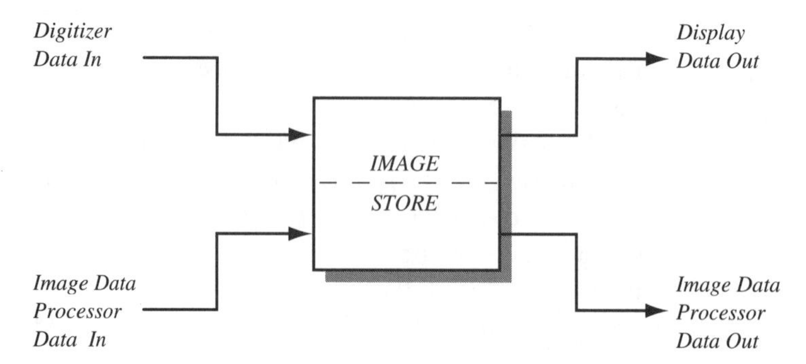

Figure 9.15 *The two-port image store. The digitizer/display and image data processor data flows each follow independent paths through the image store.*

out of the store, as shown in Figure 9.15. The first input/output data path supports the image data flow from the digitizer and to the display circuitry. The second path supports the digital image data processor data flow.

The typical way to implement a two-port memory design is by *time-slicing* access to the memory. We create two separate data paths, both in and out of the memory, and alternately switch between the two to the memory devices. One chunk of time is dedicated to getting image data into the image store from the digitizer and getting it out for display. The second chunk of time is dedicated to the image data processor's access to stored image data.

The most common approach to time-slicing the data accesses to the image store is the *pixel interleave* method. It provides equal memory access time to the image store for each of the two data paths. To accomplish this, two complete access cycles to the image store are provided every pixel time, as shown in Figure 9.16. The first cycle is a write cycle that writes data into the image store from the digitizer. The same data is passed to the image display circuitry for display. When the incoming image data flow stops to freeze an image in memory, the write operation is replaced with a read operation that reads image data from the memory to the display circuitry. The second cycle is a read or write cycle that belongs solely to the image data processor. This means that every 81.33 nS, the incoming digitizer and outgoing display image data are handled *and* the image data processor gets access to any random pixel for processing. The result is 100 percent access time for both data flows.

Image Store Memory Architectures

We must consider two primary issues when designing an image store memory. Required data size is the first consideration. To handle an RS-170 video image, the store must have the capacity to store 640 pixels × 480 lines = 307,200 pixels. If

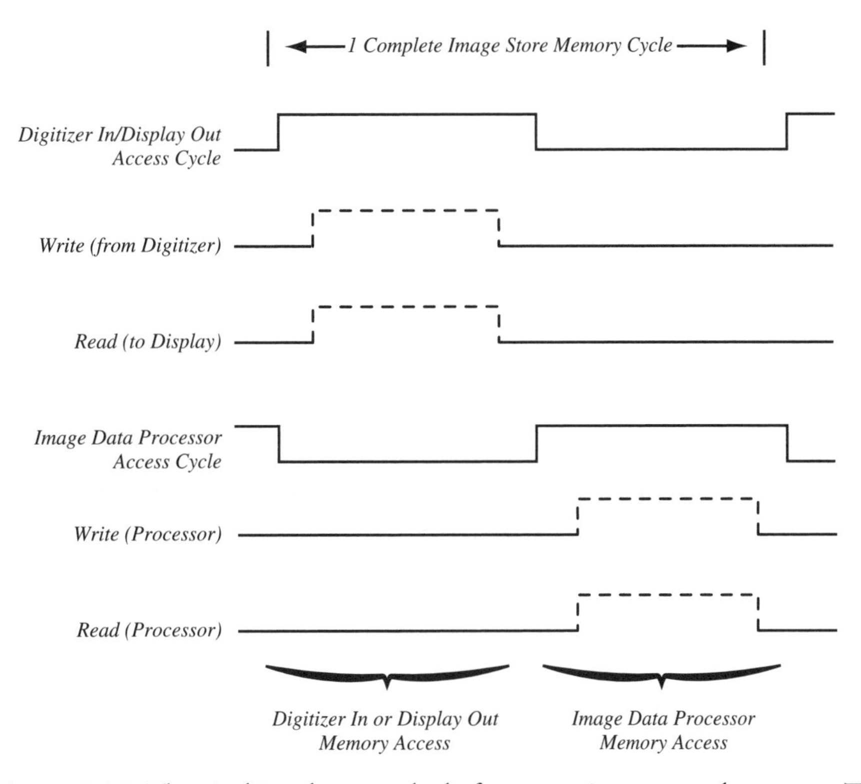

Figure 9.16 *The pixel interleave method of two-port image store data access. This method provides alternating image data processor access after every digitizer/display access.*

pixels are quantized to the typical brightness resolution of 8 bits, the equivalent memory size is 307,200 bytes, which is equal to 307,200 × 8 bits = 2,457,600 bits. For comparison, an image with the dimensions of 1024 pixels × 768 lines requires 1024 × 768 = 786,432 bytes of storage. An image with the dimensions of 320 pixels × 240 lines requires 320 × 240 = 76,800 bytes. For our analysis here, we will assume an image resolution of 640 pixels × 480 lines × 8 bits gray-scale.

The second design consideration is the required *data bandwidth* necessary to accommodate the volume of image data in and out of the image store. The data bandwidth is the quantity of image data per unit time that must be transferred to and from the image store's memory. It can be computed as the number of bits per pixel divided by the pixel time. Our RS-170 data bandwidth can be calculated as 8 bits/81.33 nS = 98.4 Mbits/second. Because we need two memory access cycles per pixel (for pixel interleave access) to the image store memory—one for the digitizer input or display output, and one for the image data processor—the image store must have twice the data bandwidth of the video signal alone. This means that our image store needs a data bandwidth capacity of 98.4 Mbits/second × 2 = 196.7 Mbits/second.

We can compute an image store memory's data bandwidth using its *memory cycle time* and *memory data width.* The memory cycle time is the entire amount of time required to complete a read or write cycle to the memory device. In the case of static RAM devices, the cycle time is the same as the access time. For dynamic RAMs, the cycle time is always about twice that of the access time. The memory data width is the number of data bits that can be transferred to the memory each read or write cycle. As an example, if an image store memory has a cycle time of 40 nS and a data width of 8 bits, its data bandwidth is 8 bits/40 nS = 200 Mbits/second. If its cycle time is 160 nS and its data width is 32 bits, its data bandwidth is the same—32 bits/160 nS = 200 Mbits/second. This means that an image store memory can be slow if its data width is large. If the data width is small, however, the cycle time must be fast, as shown in Figure 9.17.

Let's consider an actual image store architecture. To store a standard 640 pixel × 480 line × 8 bit RS-170 image frame, we require about 2.5 Mbits of memory. Our first response might be to use a single 4-Mbit dynamic RAM device—say, the 512K × 8-bit device. It would be a physically small, single-device solution, and it certainly has enough data storage capacity to hold the image. Unfortunately, however, it will not work. To see why, we must compute the image store's data bandwidth and compare it against our required data bandwidth of 196.7 Mbits/second.

The 512K × 8-bit dynamic RAM has a data width of 8 bits and a read/write cycle time of about 130 nS. This a data bandwidth of 8 bits/130 nS = 61.5 Mbits/second. It falls short of the data bandwidth requirement of 196.7 Mbits/second by a factor of about 3 to 1. There is no way that the single-device 512K × 8-bit dynamic RAM solution can work in this application.

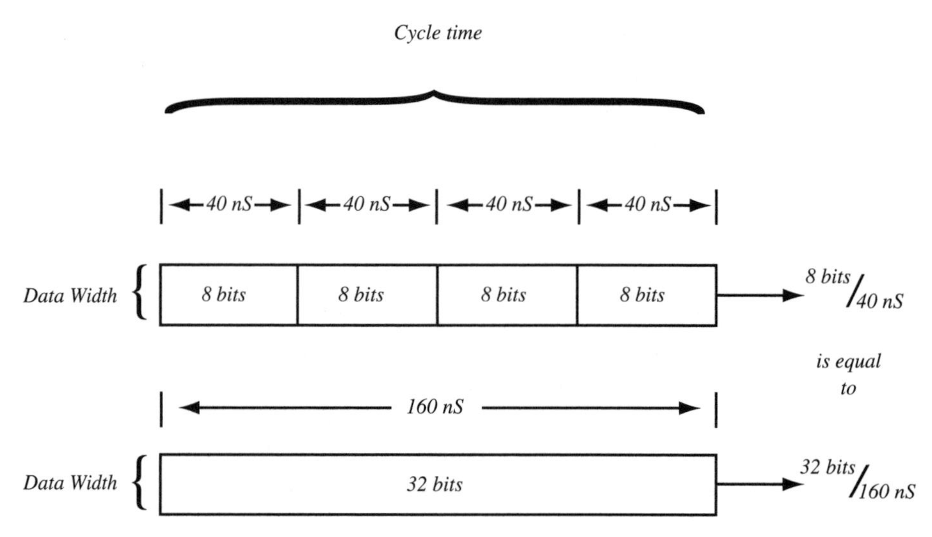

Figure 9.17 *Image store data bandwidth is a function of memory cycle time and memory data width.*

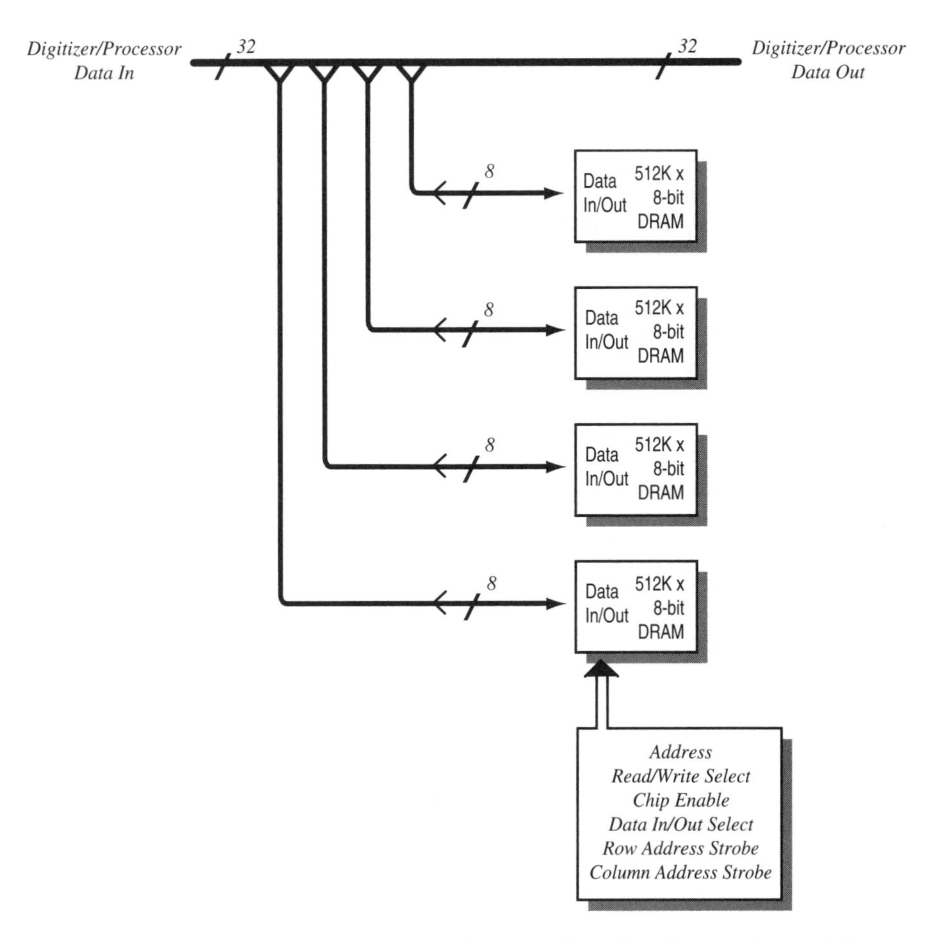

Figure 9.18 *An RS-170 image store architecture based on four 512K × 8-bit dynamic RAM memory devices. This design has the capacity to store six images of the dimensions 640 pixels × 480 lines.*

To achieve a design that will work, we must use faster RAM memories or expand the memory's data width. Either option increases the image store's data bandwidth. The first solution is to use a faster RAM device type, like static RAMs or video RAMs. The second option is to stick with dynamic RAMs and increase the data width of the memory. If we wish to stay with dynamic RAMs, we can use either four 512K × 8-bit devices, four 256K × 16-bit devices, or 16 256K × 4-bit devices. The 512K × 8-bit architecture is shown in Figure 9.18. Either choice expands the data width to 32 bits and gives us more than the required amount of data storage, enough for more than six images. The data bandwidth becomes 32 bits/130 nS = 246.2 Mbits/second.

Because the data bandwidth is 32 bits, we will read or write 32-bit chunks of image data to and from the image store. This means that incoming image data from the digitizer must be buffered into groups of 4 pixels before they are written to the image store. Outgoing pixels come out of the image store in 4-pixel chunks and must also be buffered and then sequenced to the display circuitry. Accesses by the image data processor are also made in 32-bit data chunks. So, we are doing two accesses to the image store every 4 pixel-times, alternating between a digitizer input or display output access, and an image data processor access. This timing is shown in Figure 9.19.

Alternately, we can implement the image store using static RAM devices. We can use three 128K × 8-bit devices with an access time of 40 nS, or less, as shown in Figure 9.20. The data capacity is 128K × 8 bits × 3 = 3.0 Mbits. The data bandwidth of this scheme can be calculated as 8 bits/40 nS = 200 Mbits/second. This is probably the best match for required data size, memory speed, and physical space. Cost, however, will be higher than a dynamic RAM approach. Similarly, we can use 10 32K × 8-bit devices or 10 64K × 4-bit devices. In these static RAM schemes, the image store's data width is 8 bits, and therefore, the image store alternates every one-half pixel-time (rather than every 2 pixel times) between digitizer input or display output and image data store accesses. This timing is shown in Figure 9.21.

Video RAMs merge dynamic RAM technology with internal pixel data buffering. They typically provide a 30 nS access/cycle time on the video data path because of the internal data buffering, as discussed earlier. The standard dynamic RAM data path, although slower, is left available for exclusive access by the image data processor. This means that we can use either three 128K × 8-bit devices or four 256K × 4-bit devices. The 256K × 4-bit architecture is shown in Figure 9.22. The video data-path bandwidth is 8 bits/30 nS = 267 Mbits/second, and the image data-processor data-path bandwidth is 8 bits/130 nS = 61.5 Mbits/second. The primary benefit of the video RAM architecture is that no external data buffering is necessary, as in the dynamic RAM implementations.

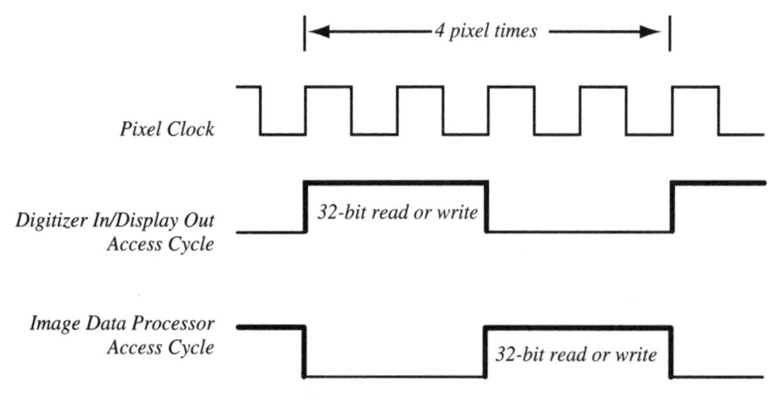

Figure 9.19 *Because four pixels are accessed from the image store at a time, the digitizer/display and image data processor data flows alternate once every two pixel-times.*

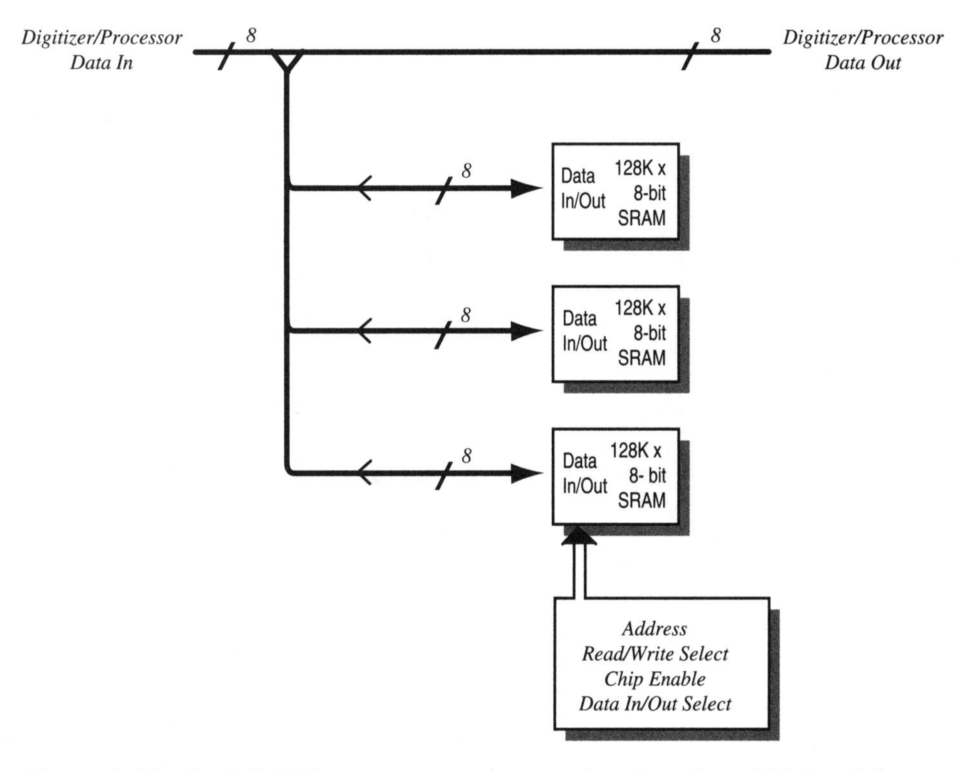

Figure 9.20 *An RS-170 image store architecture based on three 128K × 8-bit static RAM memory devices. This design has the capacity to store a single image of the dimensions 640 pixels × 480 lines.*

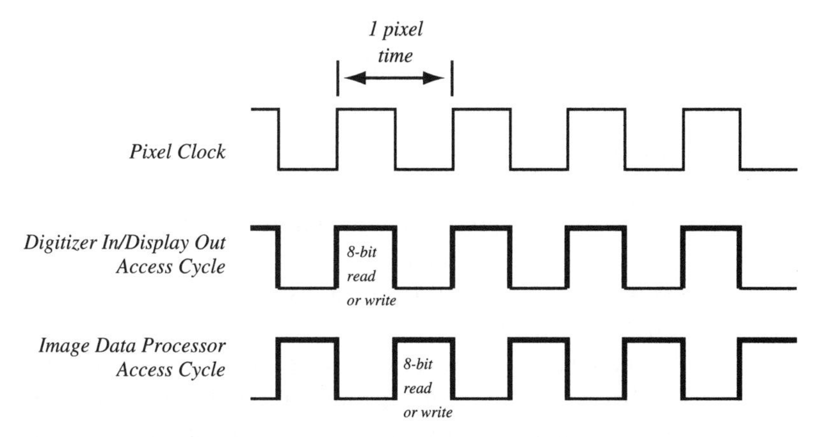

Figure 9.21 *Because a single pixel is accessed from the image store at a time, the digitizer/display and image data processor data flows alternate once every one-half pixel-time.*

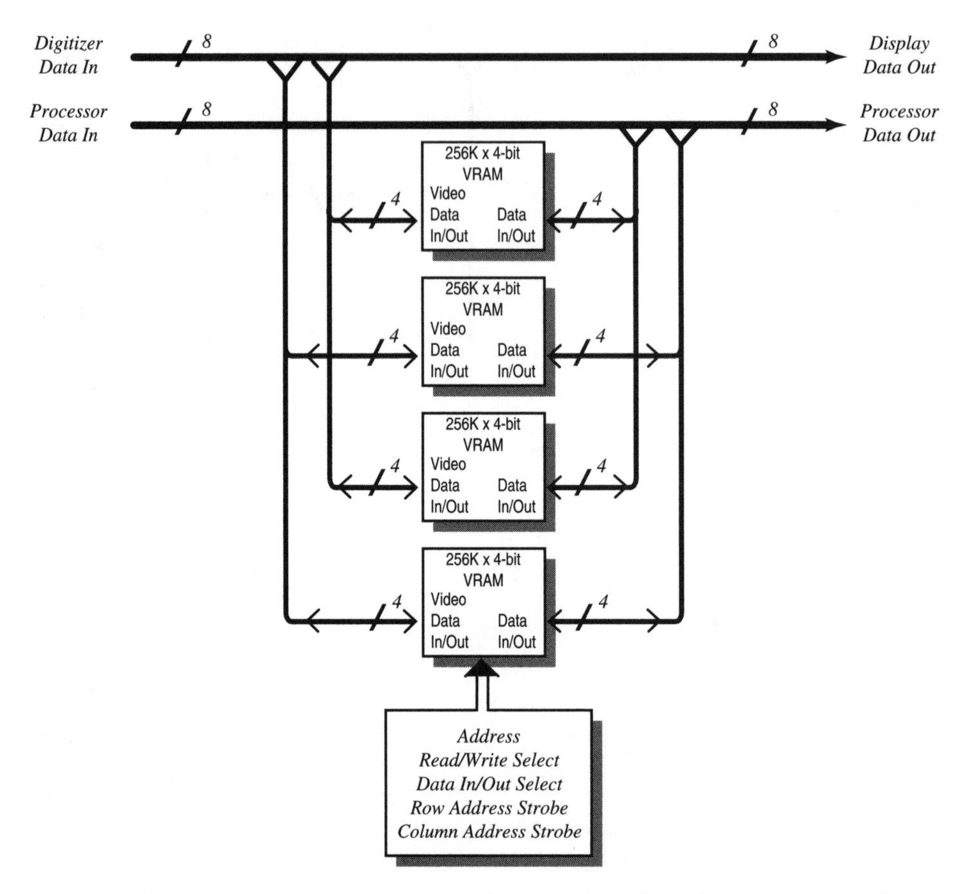

Figure 9.22 *An RS-170 image store architecture based on three 256K × 4-bit video RAM memory devices. This design has the capacity to store a single image of the dimensions 640 pixels × 480 lines.*

Permanent Image Storage

Once we have processed a digital image, we may want to store it on a permanent or semipermanent basis. We may also want to transport the image to another computer or digital image processing system. These operations are generally done by transferring an image's data from working storage (within the digital image processing system) to a permanent digital storage medium or digital transport link.

Because there are many different computing platforms, as well as many varieties of electronic data storage and transport links, it is necessary to package a digital image into a standardized format before archiving or transferring it. Further, the typical image's data size can be quite large, as we saw in Chapter 6. Image data formatting often calls upon various image compression mechanisms to reduce the data quantity being handled.

Physical Storage Media and Transport Links

Many forms of physical storage media and electronic data transfer mechanisms are available. Because a typical image can range in size from a fraction of a megabyte to tens of megabytes, high-density media and high-bandwidth data links are generally desirable. For motion image sequences, the data requirements skyrocket, making these forms of media and links absolutely essential.

Appropriate storage media for digital images begins, at the low end, with *floppy disks*. Floppy disks can store 1 to 2 megabytes of digital data. With image compression, a number of moderately sized images can be accommodated per disk. Floppy disks are a magnetic storage medium and are physically portable between digital image processing systems. *Hard disks* are also a magnetic storage medium, but with vastly greater storage capacity than floppy disks. Hard disks can store more than a gigabyte of data. These disks come in two varieties: fixed, which permanently resides in one system, and removable, where the disk media can be exchanged between systems. Both floppy and hard disks have random data access abilities, meaning that any desired data on the disk can be accessed in about the same time as any other data. Hard disk data transfer rates are very high, while floppy disk transfer rates are somewhat slower.

Magnetic tape is another magnetic medium for image data storage. It is similar to audio and video recording tape, but meets higher standards for data storage applications. Data stored on magnetic tape is not randomly accessible. Rather, it is sequentially accessible, which can add substantial image file access time when the tape must be searched for a particular image.

Optical data storage media also offers high-density data storage. The *compact disc–read only memory* (*CD-ROM*) is a derivative of audio compact disc technology. A CD-ROM can store more than 500 megabytes of image data and is portable between systems. A CD-ROM is a read-only medium, meaning that data cannot be written to it except when it is first manufactured—a process of thermo-plastic pressing. It is, however, a very inexpensive medium, compared with magnetic hard disk technology. As a result, CD-ROMs are primarily used for the mass distribution of large digital image collections.

Writable varieties of CD-ROM-like media have evolved since the CD-ROM's introduction. *Write-Once Read-Many* (*WORM*) optical disks are similar to CD-ROMs but they can be written only once, using an optical process. Following the data writing, they become a read-only medium, just like a CD-ROM. *Magneto-optical disks* are a read/write variation of optical disks. They are similar to a floppy disk but provide considerably higher storage capacity and data transfer rates.

Transport links for transferring digital image data between digital image processing systems include several mechanisms. The most rudimentary method is the common *voice-grade telephone line*. A modulator converts digital data to audio signals and a demodulator performs the inverse process. Together the *modulator/demodulator*, or *modem*, provide the interface between a digital data stream and the audio voice-grade telephone link. Data transfer rates can exceed 2,000 bytes/second.

Switched digital telephone circuits can provide considerably faster data rates. The *integrated services digital network* (*ISDN*) provides such a digital data link. Data rates of 16 Kbytes/second are available as a dial-up service between ISDN-serviced locations. Alternately, dedicated telephone circuits can provide transfer rates from 2 to 180 Kbytes/second, and greater. Dedicated links are most appropriate when nearly continuous high-rate data transfers are required.

Local area networks (*LANs*), using protocols such as *Ethernet* and *Token-Ring*, are shared network links that allow systems to transfer data between each other. Data rates in excess of 100 Kbytes/second can be achieved between arbitrary systems in practical LAN environments. The *Fiber Distributed Data Interface* (*FDDI*) and emerging *asynchronous transfer mode* (*ATM*) technologies can further boost intersystem data transfer rates by a factor of ten or more. Of course, direct, dedicated, system-to-system links can be established between digital image processing systems that can provide uninterruptable, maximal-speed transfers of image data.

Numerous direct and networked microwave and satellite data transfer links are also available and are cost-effective for long-haul image transport. These types of data links are also important for data transport to or from physically remote locations.

Packaging Digital Image Data

Regardless of the mode, image archive and transport is typically done by wrapping an image's data into a standard *image file interchange format*, transporting or storing the image file, and later unwrapping the original image data from the image file and transferring it back to the working storage (of the same digital image processing system or another). This process is illustrated in Figure 9.23. Standard image file formats make it possible for many different digital image processing systems to read and write the same image files by using a commonly understood format. This provides easy permanent storage and interchange of images between systems.

Image file formats define the data structures for organizing image data within a file. Often, compression techniques are employed, as discussed in Chapter 6, to reduce the data size of the resulting image file. While image file formats typically make reference to image compression schemes, they usually do not define them. Rather, the algorithms for image compression are defined and controlled by standards external to the image file format.

Image file formats can be divided into two principal groups—*raster image file formats* and *vector image file formats*. Raster file formats consider an image to be composed of multiple sequential lines, each containing a uniform number of pixels. Vector file formats, however, consider an image to be composed of many primitive graphical elements such as lines, circles, characters, and so on. For digital image processing activities, we use raster file formats almost exclusively. This is

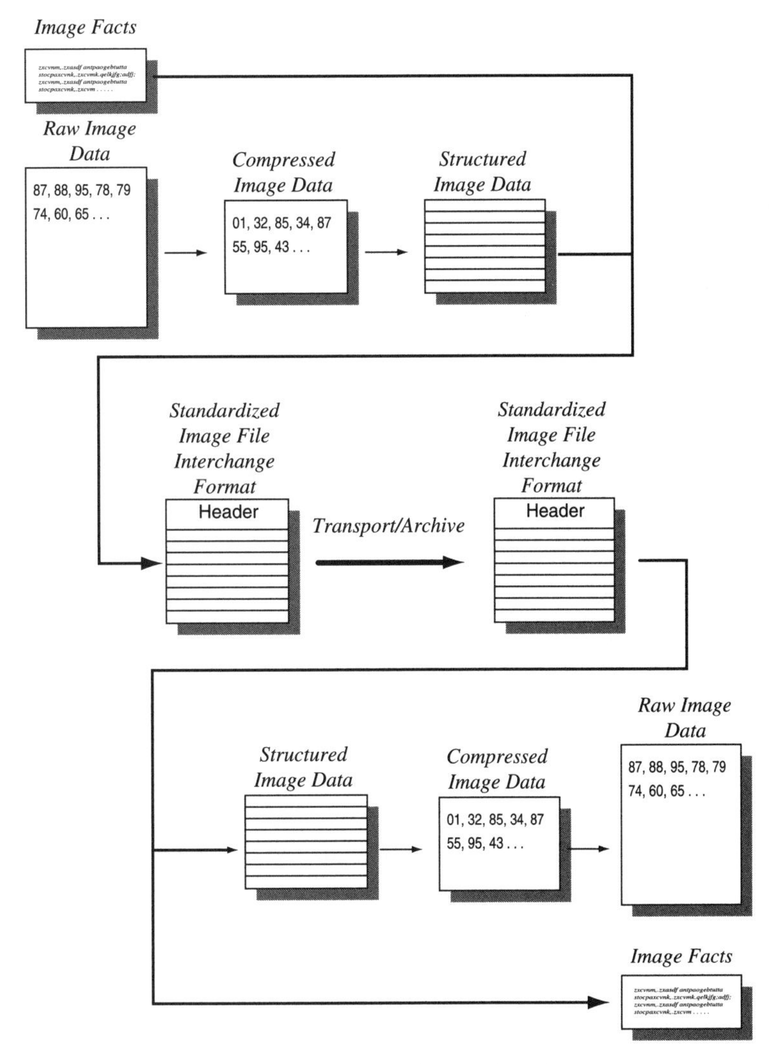

Figure 9.23 *The process of packaging image data for storage or transport.*

because our digital images have an inherent raster structure of sequential lines of pixels.

There are numerous raster image file formats, most of which have been created by software vendors to support image storage and interchange between multiple software products. Several of these formats are widely used, and have therefore become de facto standards. Regardless of the format, an image file is generally separated into two parts—a file header and image data. The header portion contains several fields of pertinent data regarding the following image data. Header fields usually include, at the minimum, the following information:

1. Horizontal image width (in pixels)
2. Vertical image height (in pixels)
3. Format of image data (gray-scale, full color, etc.)
4. Number of bits per pixel
5. Image compression technique used

The header must provide all the information necessary to reconstruct the original image data and its organization from the stored image data. Following the file header is the image data. The image data may be compressed and organized in a variety of ways. Using the information in the header, the original digital image can be unambiguously recreated.

As an example, we can package a 640 pixel × 480 line × 8-bit gray-scale image file into a simple image file format, as follows:

Header
Width [2 bytes] = 640
Height [2 bytes] = 480
Format [1 byte] = Gray scale
Number of Bits Per Pixel [1 byte] = 8
Image Compression [1 byte] = None

Image Data
[307,200 bytes of image brightness data]

A simple format like this typically assumes that image data is organized line by line from the top of the image to the bottom and pixel by pixel from the left side of the image to the right.

For digital image processing applications, the Tag Image File Format (TIFF) has become a ubiquitous, versatile, and well-controlled standard. While it is not controlled by a national or international standards body, its keepers are open to industry scrutiny and advice. The TIFF format provides unique portability and the ability to grow over time and respond without partiality to industry needs.

Another image file format that is important to digital image processing applications is the Photo CD format. This format specifically handles images on CD-ROM media. It is a proprietary format that is intended to provide the well-controlled storage and exchange of color photographic-quality images.

While many other image file formats also play a role in image storage and transport between imaging systems, the TIFF and Photo CD formats are the most commonly used in practice. In the following section, we will examine the Tag Image File Format in some depth to get a feeling for the aspects of an image file format. We will then overview the Photo CD format. In depth information on these and other image formats can generally be obtained from the originating vendors.

Tag Image File Format (TIFF)

One of the most common digital image file formats found in digital image processing applications is the *Tag Image File Format* (*TIFF*). This format was jointly developed by Aldus Corporation and Microsoft Corporation to specifically format large arrays of raster image data originating from scanners and video frame grabbers.

The TIFF image file format was designed to be portable between computing platforms. It was also intended to be extensible. These features allow its growth over time by providing the means for adding new provisions as required by industry or individual users. The two originating organizations continue to maintain and extend the TIFF standard.

The TIFF image file format is highly flexible. As a result, it is somewhat complex. Basically, a TIFF image file is composed of several entries, each of which has a tag and some associated data. Each entry's tag indicates the purpose of its associated data. The tags and their data are used to define an image's format and size, as well as many other parameters including the image data itself. The file is made up of a header followed by a mix of *Image File Directories* (*IFD*), *IFD Entries*, and data. The TIFF file organization is shown in Figure 9.24.

The TIFF file header is simple, made up of only 8 bytes of data. It comprises the following information:

1. 2 bytes—Byte order indicator for 16-bit and 32-bit values found in the rest of the file
2. 2 bytes—TIFF version number (not revision number)
3. 4 bytes—Address of first (and often only) IFD

The *byte order indicator* has ASCII values of either "II" or "MM" to indicate whether bytes are ordered by least significant byte first (as used in Intel processors) or most significant byte first (as used in Motorola processors). This way, portability of the image file is maintained between virtually all computing platforms.

The *TIFF version number* has the value "42" to indicate the original, and probably only, version number. It is not to be confused with the TIFF revision number, which does change as extensions are made to the standard.

The *address of the first IFD* is the location of the first IFD relative to the beginning of the file. Generally, a TIFF image file contains a single IFD.

An IFD contains several IFD Entries, as shown:

1. 2 bytes—Number of image file directory entries in IFD
2. 12 bytes—First Directory Entry (0)
3. 12 bytes—Directory Entry 1
4. 12 bytes—Directory Entry 2

   ⋮

5. 12 bytes—Last Directory Entry
6. 4 bytes—Address of next IFD

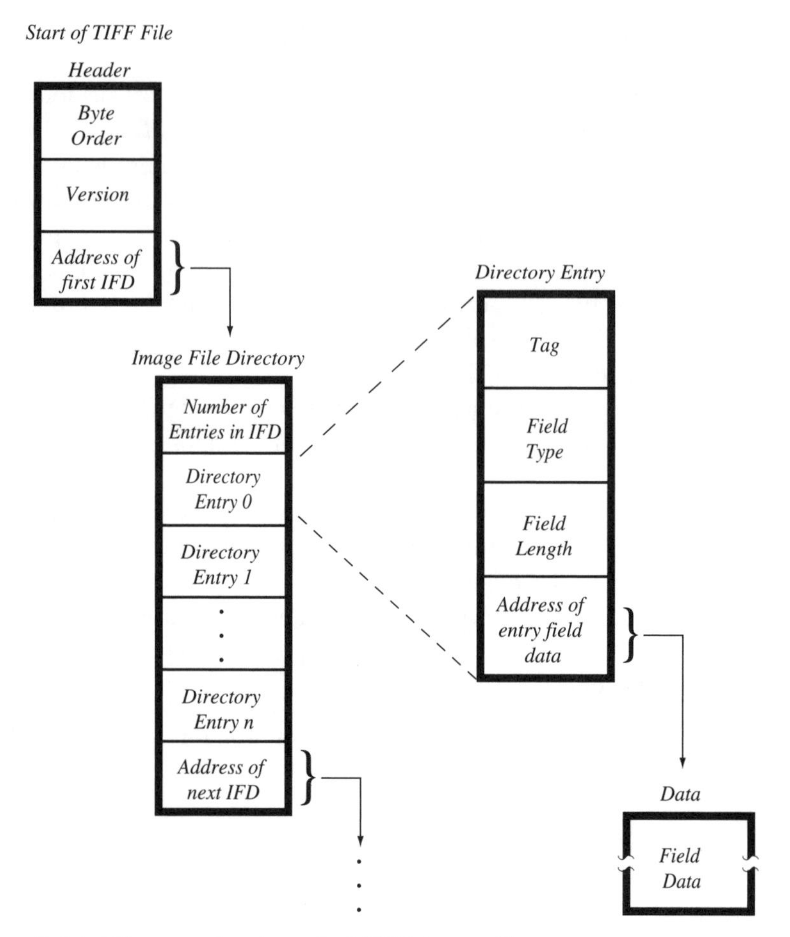

Figure 9.24 *The organization of a TIFF image file.*

The *number of IFD entries* indicates how many entries are listed in the current IFD.

Each IFD entry then follows. At the end of the IFD is the *address of the following IFD*. Generally, an image file contains only a single IFD.

Each Directory Entry contains the following information:

1. 2 bytes—Entry tag number
2. 2 bytes—Entry field type
3. 4 bytes—Entry field length
4. 4 bytes—Address of entry field data

The *entry tag number* is a tag that indicates the meaning of the following data. The data can be various image formatting information or a strip of actual image data.

The *field type* defines the type of data. Types include BYTE (8-bit unsigned integer), ASCII (8 bits containing a 7-bit ASCII character), SHORT (16-bit unsigned integer), LONG (32-bit unsigned integer), and many others.

The *field length* is the number of data items contained in the field data.

The *address of field data* is the location, relative to the start of the file, of the data associated with the tag. The data can appear in any order in the file as long as the address correctly points to it.

There are numerous tags that define the parameters, and the data, of the image stored in the TIFF file. Table 9.2 lists several commonly used tags for storing a gray-scale image.

ImageWidth, *ImageLength*, and *BitsPerSample* define the number of pixels in the image, lines in the image, and bits per pixel, respectively. *Compression* defines the type of compression used to store the image. Typical compression types include run-length coding, Lempel-Ziv-Welch (LZW), and Joint Photographic Experts Group (JPEG), as described in Chapter 6. *PhotometricInterpretation* defines whether the image is quantized with black = 0 and white = 255, or the reverse (with the gray levels in between either ascending or descending, respectively).

The *ImageDescription* and *DateTime* tags define the image's verbal description and its creation date and time. *ResolutionUnit* describes the units of measure represented by the pixels. The measure can be in inches, centimeters, or no units. The *Xresolution* and *Yresolution* tags represent the number of pixels in the x and y directions, respectively, per unit of resolution.

Finally, the image data is conveyed. Typically, the image data is broken into strips representing small portions of the overall image. This way, an imaging system can quickly get to small portions of a large image for display purposes. First, the *RowsPerStrip* tag defines how many image lines exist in each strip. Then the

Table 9.2 Common TIFF Gray-Scale Image Tags

TAG NAME	TAG NUMBER
ImageWidth	256
ImageLength	257
BitsPerSample	258
Compression	259
PhotometricInterpretation	262
ImageDescription	270
StripOffsets	273
RowsPerStrip	278
StripByteCounts	279
XResolution	282
YResolution	283
ResolutionUnit	296
DateTime	306

StripOffsets and *StripByteCounts* tags state the locations and sizes of the strips of image data.

So, a TIFF image file is an array of entries, each having a tag defining the entry's purpose and an address of where the entry's data is located in the file. The format's only data placement requirements are that there must be a header at the top of the file and that the IFD entries must be in tag-number sequential order. Otherwise, image format information and data can be located anywhere within the file, as long as the entry addresses properly point to them.

A number of *private tag-numbers* are available, allowing the user to customize the TIFF format for specific nonstandard uses. The TIFF format's flexibility and platform portability make it ideal for most digital image processing archive and interchange needs.

Photo CD

The *Photo CD* (*PCD*), image file format was developed by Eastman Kodak Company. It was created to provide a mechanism for handling digital color photographic images. The Photo CD image file format is a proprietary format of Eastman Kodak and is exclusively maintained by the company. It is available through license to companies that wish to embody the format into their digital image processing products.

One of the most important parts of the Photo CD image file format is its use of a calibrated color space. The components of a color image are handled independently as luminance and two chrominance components rather than as red, green, and blue components. Unlike other image file formats, the Photo CD format specifically references its color representations to a tightly controlled, albeit proprietary, standard. As a result of this control, applications that use the Photo CD format for interchange can expect the true colors of the original image to be faithfully represented. This is not always true for many other image file formats.

Several forms of the Photo CD image file format exist to meet the demands of different application markets. The baseline format, called Photo CD Master, is intended to handle full-quality 35mm photographic format images. Several extensions to the baseline format serve the professional photographer, consumer, print, and medical imaging markets, as shown in Table 9.3.

Each of these market-specific Photo CD versions trades off image spatial resolution for the number of images that can be stored per CD-ROM. The number of images per CD-ROM listed above can vary widely based on the resolutions of the images stored on a particular CD-ROM.

The Photo CD image file format uses proprietary image compression techniques to store images in a compact form. For the Photo CD Master format, 35mm photographic images are scanned at a resolution of 3,072 pixels × 2,048 lines. Then, a reduced resolution image, called the base image, is created that has

Table 9.3 Photo CD Image Formats

FORMAT NAME	IMAGE QUALITY	# OF IMAGES/CD-ROM
Photo CD Master	35mm photo format	~100
Pro Photo CD Master	35mm, 120, and 4" × 5" photo formats	~25
Photo CD Portfolio	Television and higher	~800
Photo CD Catalog	Very low	~6,000
Photo CD Medical	X-ray, CT, MRI, etc.	~200
Print Photo CD	35mm, 120, and 4" × 5" photo formats	~25

1/16th the spatial resolution of the original image. The base image is stored along with Huffman-coded difference images that record the differences among the base image, the original image, and an intermediate-sized image with 1/4th the spatial resolution. This way, the entire spatial resolution of the original image need not be stored.

Using the base image and the two difference images, an image with any spatial resolution can be recreated with high quality. Additionally, two further reduced resolution images are stored, one with 1/4th and one with 1/16th the resolution of the base image. These thumbnail images are used to preview an image quickly without recreating the full resolution version. The spatial resolutions stored for each Photo CD Master image are shown in Table 9.4.

The Pro Photo CD Master format adds a base × 64 (6144 pixels × 4096 lines) Huffman-coded difference image to provide increased spatial resolution for professional photographic applications.

The Photo CD image file format is especially useful for graphic art applications where photographic quality color images must be efficiently stored and accurately reconstructed. Further, the Photo CD format provides a convenient way for professionals and consumers alike to get digital images into their computers for processing.

Table 9.4 Photo CD Master Spatial Resolutions

IMAGE NAME	SPATIAL RESOLUTION
Base/16	192 pixels × 128 lines
Base/4	384 pixels × 256 lines
Base	768 pixels × 512 lines
Base × 4	1,536 pixels × 1,024 lines (stored as a compressed difference image)
Base × 16	3,072 pixels × 2,048 lines (stored as a compressed difference image)

Other Image File Formats

Many other image file formats have also been created, several of which are standards in particular fields or on particular computing platforms. Because interdisciplinary image file interchange is generally limited, this is typically not a problem. However, when an image must be exchanged between two varied disciplines, a common interchange format must be determined or created from scratch.

Many image file formats have evolved, because often a particular application has unique needs for file organization or compression. If an existing image file format standard is not appropriate for a particular application, an imaging system manufacturer may choose to create a customized format. Additionally, a proprietary format gives the manufacturer a trade secret that can keep others from using the format to process image data. This can help protect a company that wishes to maintain proprietary control over the use of images created with its equipment.

10 *Image Data Processing*

At the heart of all digital image processing systems is the capability to process a digital image. The speed at which digital image processing operations are executed determines the time-efficiency of the entire system. In the previous two chapters, we explored how images physically originate with video cameras and how they are displayed with video display monitors. We then looked at digitally handling an image—digitizing the analog video signal, storing it in an image store, and later reconstructing it back to an analog signal for display. These functions provided the basic image paths into and out of the digital image processing system.

The final function, yet to be discussed, is the *image data processor*. It is this digital image processing function that implements all of the techniques discussed in Part II of this book. In this chapter, we will explore the hardware and software of the image data processor. With this discussion, the digital image processing system is complete.

Image Data Processor Interface

As we have discussed earlier, the flow of image data through a digital image processing system is as follows:

1. The analog video signal enters the system.
2. The analog video signal is digitized to a digital image.
3. The digital image is stored in the image store.
4. The digital image is processed by the image data processor.
5. The processed digital image is reconstructed back to an analog video signal.
6. The analog video signal exits the system for display.

A frame grabber performs the image data handling functions—steps 1, 2, 3, 5, and 6—as discussed in Chapter 9. The image data processor handles step 4.

Image data processors come in a variety of hardware and software configurations. Regardless of the particular configuration, though, the image data processor must be physically connected to the image store. This way, the processor can access a stored image's data.

As discussed in Chapter 9, the most common image store design uses two memory access cycles that constantly alternate in an interleaved manner. The first cycle moves image data in and out of the store from the digitizer and to the display circuitry. In the second cycle, the image data processor reads and writes image data for processing. These two image store access cycles form a two-port memory access scheme.

Along with the two access cycles, two physical paths must also be provided to and from the image store. Each path must provide for the transfer of image data, a pixel address (pixel and line locations within an image), and a read or write signal. The two data flows can independently access pixel locations within the image store every time their access cycle is available.

Figure 10.1 is a block diagram of a two-port image store showing how it links with the image data processor and image data handling functions. On the digitizer/display data path, a pixel value is written to the image store every pixel time. The same pixel value is passed around the image store to the display circuitry, as well. We can freeze the image in the image store memory by simply disabling the write operation during the digitizer/display cycle. When this is done, the digitizer write cycle is replaced by a

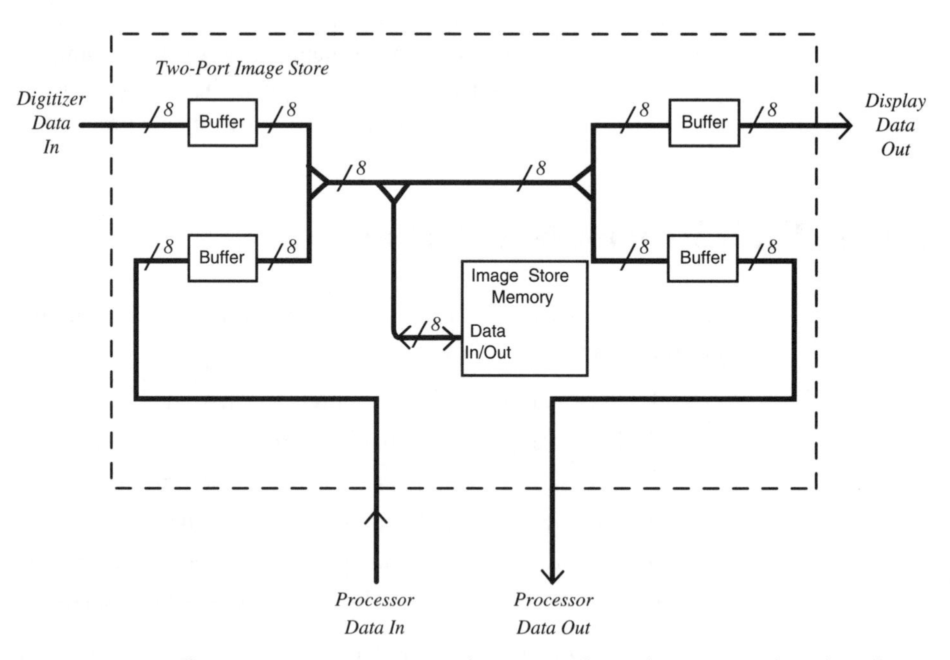

Figure 10.1 *The two-port image store, showing independent image data-handling and image data-processing access paths.*

display read cycle so that the frozen image will be continuously displayed. In this case, during the digitizer/display cycle, a pixel value is read from the image store for display. Because a video signal is sequential, the digitizer/display write and read accesses to the image store are done sequentially, pixel by pixel.

On the image data processor data path, a pixel value can also be read or written to the image store every pixel time. Generally, the image has been frozen in the image memory when processing begins. Unlike the digitizer/display accesses, the image data processor does not always access pixels in sequential order. Rather, the processor accesses pixel values at random locations, based on the operation that it is carrying out. A *pixel address generator* determines the location of the pixel to be read from, or written to, every time the processor accesses the image store.

The image data processor can also access image data in line with the image store's outgoing data flow to the display circuitry. This type of processing is called *pipelined processing*. Instead of the image data processor accessing pixel values from the image store, processing them, and returning them back to the store, pixel values are grabbed from the image store's outgoing data stream. They are then processed on the fly and returned to the data stream, to the display circuitry. This processing must be done in real time, at the video pixel clock rate, or the outgoing video image data stream will be interrupted, thus interrupting the image being displayed.

We can use pipelined processing techniques to implement any digital image processing operation. However, they are particularly well suited for operations that can be implemented sequentially, pixel by pixel. Otherwise, large amounts of image memory may be required within the pipelined processor circuitry. Pipelined processing is the most costly way to process image data because of its required operation speed—the video pixel clock rate. But, when we require high-speed, real-time processing for an imaging application, pipelined processing is an especially effective technique.

Image Data Processing Hardware

All digital image processing systems have a *host computer* that controls and oversees all system functions. The host computer can be a simple, embedded controller, but is most generally a computer workstation with significant computing capacity. The host computer usually provides the interface between the system and the user.

There are three general methods for implementing an image data processor. The first and most rudimentary method uses the host computer to carry out all digital image processing tasks. The second method uses an internal accelerator processor that has direct access to the image store's data. The third, and highest performance method, uses pipelined, special-purpose digital image processing hardware. Each method is a mix of hardware and software components, and each has varying levels of speed, flexibility, and cost trade-offs that must be considered for a given application. Let's look at each of these image data processor configurations.

Host Computer Processor

The simplest way to process a digital image is to use solely the host computer's processor, as shown in Figure 10.2. In this configuration, the image to be processed is in the form of an image file. The image is loaded into the host computer's system memory and displayed using its video display monitor. The host computer processes the image using its *central processor unit* (*CPU*). The processed image can be saved as a new image file for future use or for transport to another digital image processing system.

This system configuration relies on external equipment to digitize the image. This can be ideal for many applications, especially those requiring that many independent processing tasks be applied to the same set of stored images. However, there is often a need to acquire images locally from a live source. By adding a frame grabber consisting of image digitization, storage, and display functions, the host computer can acquire and display images, as well as process them. In this configuration, shown in Figure 10.3, the host computer instructs the frame grabber to digitize the image when required. At this point, the digitized image is in the frame grabber's image store. The host computer accesses the image data from the image store, processes it using its CPU, and returns the resulting image data to the image store. The resulting image is displayed on a video display monitor driven by the frame grabber's display circuitry.

This host computer/frame grabber configuration allows the digital image processing system to acquire images from a live video source, rather than being confined to pre-acquired image file data. Further, processed images are displayed on their own video display monitor, freeing the host computer's display for user interaction.

Accelerator Processors

When using the host computer approach, the host computer's CPU must perform all digital image processing tasks. This means that all image data must be moved from the image store, across the host computer's data bus to its CPU, and back again, for processing. The addition of an *accelerator processor*, as shown in Figure 10.4, can improve this situation in several ways. First, image data can be accessed

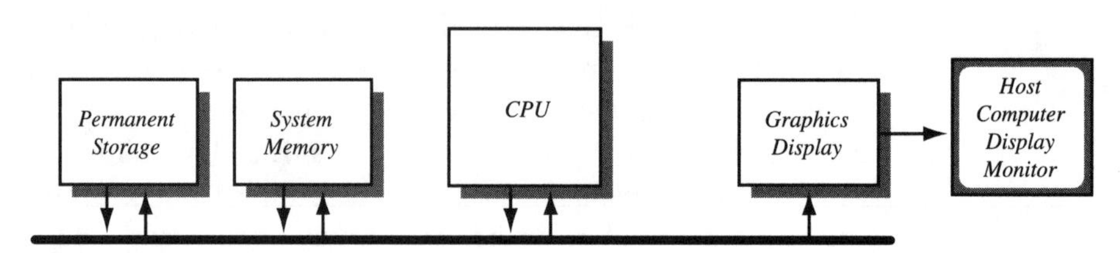

Figure 10.2 *A digital image processing system containing only a host computer for image data processing.*

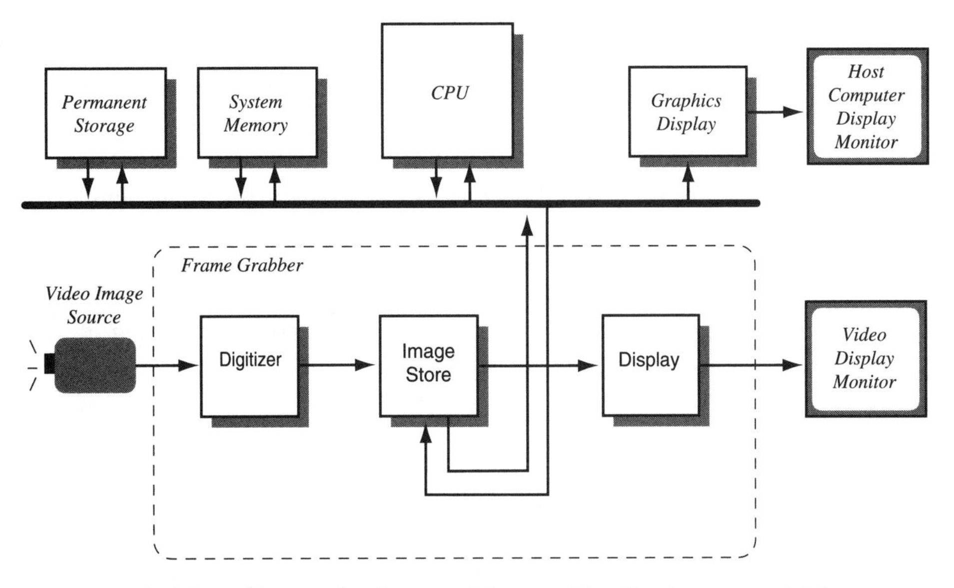

Figure 10.3 *The addition of a frame grabber provides live-image acquisition, storage, and display to the system.*

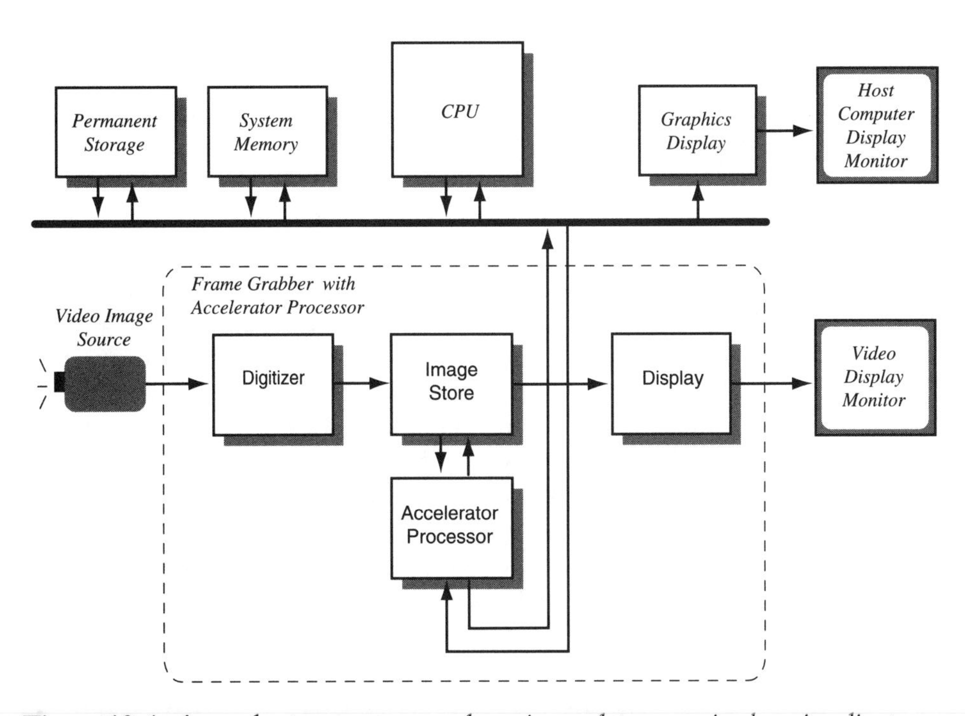

Figure 10.4 *An accelerator processor speeds up image data processing by using direct image store access, thus relieving the host computer to handle higher-level system activities.*

directly from the image store by the accelerator processor, eliminating the movement of image data over the host computer's data bus. Also, the host computer's CPU is freed from the digital image processing tasks, making it available for other tasks, such as user interaction or other processes.

An accelerator processor is generally a high-speed microprocessor that is specifically designed to process digital signals, including digital images. These microprocessors are called *digital signal processors* (*DSPs*). The accelerator processor coexists with a frame grabber, working in unison. Generally, the software to implement a particular operation is downloaded from the host computer to the DSP. The DSP is then instructed to start its processing. It accesses the image store directly, reading image data as required. The resulting processed image data is written directly back to the image store. As the processed image is written back into the image store, the image is displayed on the frame grabber's video display monitor.

Numerous DSP devices are available. Many have specific instructions that make them optimal for digital image processing operations. These processors are widely used in many fields, including telecommunications, automotive, and consumer applications. As a result, they are inexpensive and their hardware and software development resources are good. Additional specialized DSP processors are available and often better for specialized high-power digital image processing tasks. Because of their limited use, however, the specialized processors are generally more costly and their development resources are more limited.

Special-Purpose Hardware Processors

When a digital image processing operation's execution speed is critical or the operation is used frequently, pipelined processor techniques are generally required. As we discussed earlier, the pipelined approach processes image data at video image data rates as it departs the image store. As a result, it is the most costly way to process image data. However, the benefits of real-time processing power make these processing techniques essential to many applications, such as factory automation, medical imaging, and high-speed document processing. Figure 10.5 shows the pipelined processor configuration.

Special-purpose processors are generally available as hardware board-level modules. These modules allow a digital image processing system to be configured with only the functions necessary for a particular application. This way, only the operations requiring high-speed or repetitive use can be embodied in the more costly hardware, leaving the remaining operations for software implementation on the host computer or an accelerator processor.

Some digital image processing operations, however, can be inexpensively implemented using special-purpose hardware processing circuitry. An example of

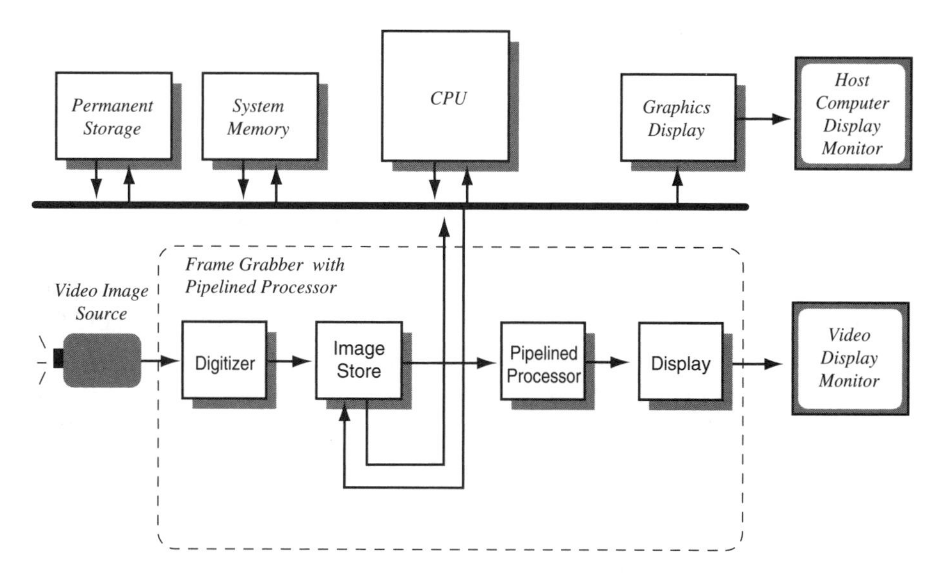

Figure 10.5 *The addition of a pipelined processor in the display out path of the image store provides real-time image data processing.*

this is the hardware for pixel point processes. As a result, real-time point processing is found in virtually all frame-grabber designs.

Let's look at several digital image processing operations that are commonly implemented using special-purpose processors.

Single-Image Pixel Point Processor

As we discussed in Chapter 4, pixel point processes operate independently on each pixel of the input image to create a resulting output image. The general equation for point processes is of the form

$$O(x,y) = M[I(x,y)]$$

where $I(x,y)$ represents the input image pixel at the location (x,y), $O(x,y)$ represents the output image pixel at the same location (x,y), and M is the mapping function. It is implied that all pixels of the image are mapped, through the mapping function, to the output image.

The mapping function takes on the characteristics of a mathematical equation that converts the brightness of an input pixel to a resulting output pixel brightness. The function can be potentially complex. However, instead of making this computation for every input pixel based on its brightness and the mapping function, there is a simpler way of implementing the process.

All pixels, quantized to 8 bits, have a brightness value between 0 and 255. This means that all possible input and output pixel values can only be between 0 and

255. Because the mapping function maps input pixel brightness values to output values, there are only 256 possible output values, corresponding to the 256 possible input pixel values. Instead of running every input pixel's brightness value through the point process mapping function to arrive at an output pixel value, we can use it to drive a simple *look-up table* (*LUT*).

To do this, we can use the point process mapping function 256 times to compute the output values for the 256 possible pixel input values. The results are then loaded into the look-up table. From then on, the point process is reduced to a look-up procedure—an input pixel value drives the LUT, resulting in the preloaded output pixel value. When implementing the LUT version of the point process, no matter how complex the mapping function is, there are only 256 possible input brightness values to map to one of 256 output brightness values.

As an example, Figure 10.6 shows a simple mapping function for a binary contrast enhancement point process. For the locations 0 through 127, the LUT is loaded with the value 0. For the locations 128 through 255, the LUT is loaded with the value 255. Any input pixel value between 0 and 127 yields a LUT output value of 0, and any input value between 128 and 255 yields a LUT output value of 255. Hence, the LUT carries out the point process. Further, more complex mapping functions can be handled the same way. Once the 256 LUT values have been computed, the point process becomes a simple look-up procedure.

The LUT point process can be implemented in software on the host computer or on an accelerator processor, or in hardware using a simple pipelined processor circuit. In the first two approaches, the 256 output values for the LUT are calculated and stored in the processor's system memory. Then, each input pixel is read from the image store and used to address the LUT. The resulting output value from the LUT is written back to the image store. This process continues over the entire image.

As a pipelined processor, the LUT is implemented as a 256-location RAM memory device, where each location stores an 8-bit data value, as shown in Figure 10.7. The LUT sits in the outgoing image data path between the image store and the display circuitry. The image data from the image store drives the LUT address

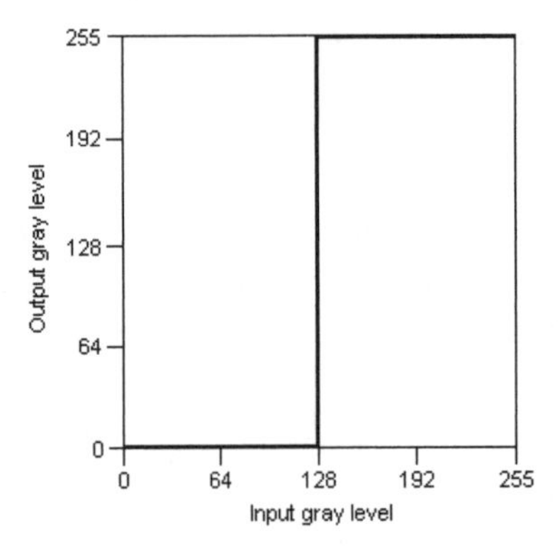

Figure 10.6 *A pixel point process mapping function for binary contrast enhancement, using a threshold brightness of 128.*

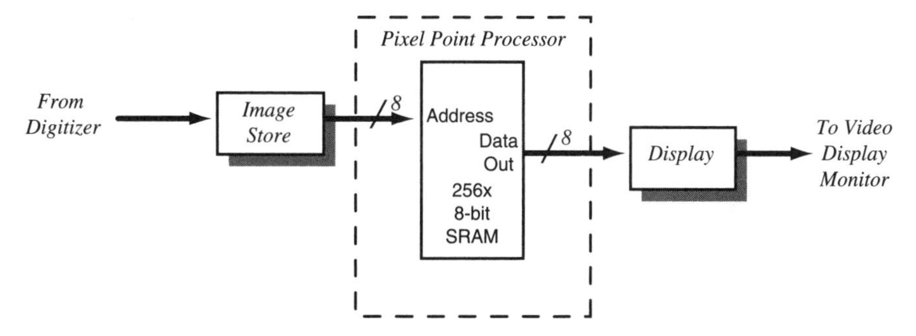

Figure 10.7 *Single-image pixel point processor LUT hardware.*

lines, while the LUT's data output lines produce the resulting output pixel values. Again, the host processor must calculate the 256 output values for the LUT and write them into the LUT memory device. From then on, the image data flows through the LUT, performing the point process in real time.

Often, pipelined LUTs are included in digital-to-analog converter devices, called *RAMDACs* (for *random access memory digital-to-analog converters*). They are loaded by the host processor just as described above. In color systems, RAMDACs can be used to map gray-scale pixel values to RGB color values in much the same way. This technique can be used for operations such as brightness slicing, to highlight a feature in a noticeable color like red.

Multiple-Image Pixel Point Processor

The difference between single-image and multiple-image pixel point processes is the number of input images used in the operation. Because dual-image point processes are the most common, we will examine their implementation here.

Dual-image point processes operate independently on two input pixel brightnesses, one from each of two input images, to create a resulting output image. The general equation for a dual-image point process is of the form

$$O(x,y) = I_1(x,y) \# I_2(x,y)$$

where $I_1(x,y)$ and $I_2(x,y)$ represent the input image pixels at the location (x,y), $O(x,y)$ represents the output image pixel at the same location (x,y), and the # symbol is used to denote the combination function. It is implied that all pixels of the image are combined, using the combination function, to the output image.

The combination function can include virtually any function that accepts two pixel values as input and creates a resulting output pixel value. Some common combination functions are $+$, $-$, $\times$, $\div$, AND, OR, and EXclusive-OR. Mathematical combinations that are significantly more complex can, however, be useful in some applications.

The dual-image point process can be implemented through direct computation or through a LUT approach. The computational method simply computes, pixel by pixel, the mathematical combination of corresponding pixel values from the two input images, creating an output image. For simple combination functions, the computational technique can be the most efficient if implemented on a host computer or accelerator processor. The dual-image LUT approach is similar to that used in the single-image point process. The LUT must be 256 × 256 = 65,536 locations and must be able to store 8-bit data values at each location. The LUT combination technique is superior when more complex combination functions are involved or when the technique is implemented as a pipelined processor.

In the LUT approach, each of the two input pixel values can be any value between 0 and 255. Therefore, the combination of the two pixel values can produce exactly 256 × 256 = 65,536 possible resulting pixel values. Of course, the resulting output pixel brightness value must also be between 0 and 255. Each one of the 65,536 possible LUT locations must be loaded with the pixel value obtained by using the combination function to combine the two corresponding input pixel values.

As an example, let's say our application requires the average of two input images. We can implement this by simply adding corresponding pixels together, pixel by pixel, and dividing each result by 2. This combination function is expressed as

$$O(x,y) = [I_1(x,y) + I_2(x,y)]/2$$

where each pixel of the input images is processed through the equation. The resulting output pixel values are the average of the two input pixel values. Or, we can load a 65,536 location × 8-bit LUT to do the same. For instance, for input pixel values of 28 and 187, the LUT location addressed by (28 × 256) + 187 = 7,355 is loaded with the value (28 + 187)/2 = 108. Hence, when the image number 1 input pixel brightness is equal to 28 and the image number 2 input pixel brightness is equal to 187, the output value from the LUT is equal to 108, which is the average of the two input pixel brightnesses. Similarly, the remaining 65,535 LUT locations must be loaded with their corresponding output values.

Like the single-image point process, the dual-image point process LUT can be implemented in software on the host computer or on an accelerator processor, or in hardware using a pipelined processor circuit. As a pipelined processor, the dual-image point process LUT is implemented as a 65,536-location RAM memory device, where each location stores an 8-bit data value. Generally, image number 1 data is fed from the incoming video data path, prior to the image store, and image number 2 data is fed from the frozen output of the image store. This way, sequential image frames can be averaged, differenced, or otherwise combined to form a resulting output. The output image data is passed from the LUT to the display circuitry. Figure 10.8 shows the pipelined processor implementation of the dual-image point process LUT.

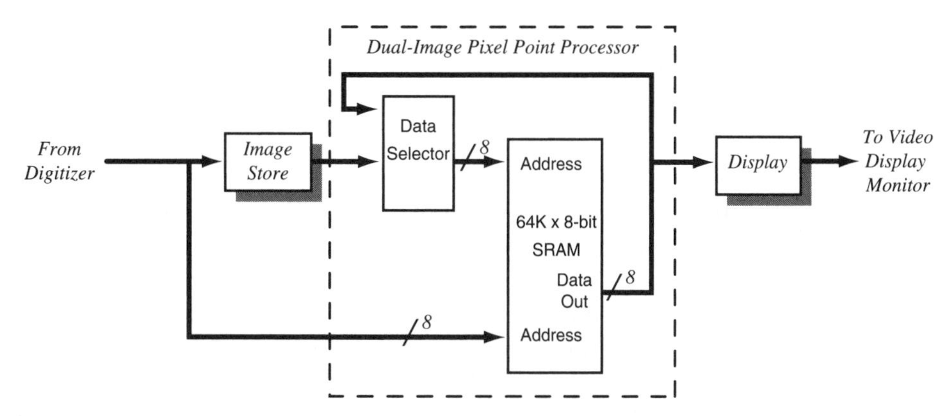

Figure 10.8 *Dual-image pixel point processor LUT hardware with input data selector.*

Additionally, the output of the dual-image point process LUT can be routed back, through a data selector, to the input of the LUT. This way, it can be continuously combined with the incoming video image data, providing a way to average an unlimited number of successive image frames.

Pixel Group Processor

Pixel group processes operate on several input pixel values, to create an output pixel value. The input pixels are generally a square block of pixels surrounding a center pixel that is being processed. The dimensions of the input pixel group are usually 3×3, although larger groups are also used in many applications. Group processes are used to implement spatial convolution operations as well as nonlinear filtering operations, such as the median filter operation and morphological operations. In this discussion, we will focus on 3×3 group process implementations.

As discussed in Chapter 4, we implement the spatial convolution process by multiplying each input pixel value by a defined weighting value and summing the results. The ensemble of weighting values is called the convolution mask. For each input pixel, the process requires nine multiplication operations followed by nine addition operations.

We can implement the spatial convolution process in software using the host computer or an accelerator processor, or using a hardware pipelined processor circuit. The host computer and accelerator approaches are implemented in software. As a pipelined processor, the 3×3 group process is implemented by using two line store memory devices, nine multiplier devices, and a summing network of nine adders. The multipliers and adders are available as single-chip devices called *convolvers*, which are specifically intended for digital image processing applications.

The pipelined spatial convolution group processor sits in the outgoing image data path between the image store and the display circuitry. As image data flows from the image store, it runs through the two line stores. Therefore, at any given time, image data is available from the current image line, as well as from the previous two image lines. The multipliers are simultaneously fed with the appropriate nine pixel values, three from each line store and three from the current video line's data. The nine multipliers each compute the multiplicand of their input pixel value and the corresponding weighting value. The nine results are summed, producing the output pixel value, as shown in Figure 10.9. This process occurs once every pixel time, providing a real-time flow of resulting image data to the display circuitry.

We can modify this pipelined processor circuit for morphological processing. The nine multipliers and adders are replaced by a pixel-value sorting and selection circuit, as shown in Figure 10.10. For the erosion operation, we must find the smallest pixel value of all of the input pixel values and select it as the minimum-valued output pixel. For the dilation operation, we need to find and select the largest pixel value, which becomes the maximum-valued output pixel.

For the median operation, a similar pixel-value comparing scheme compares the nine input pixel values against one another and places them into ascending

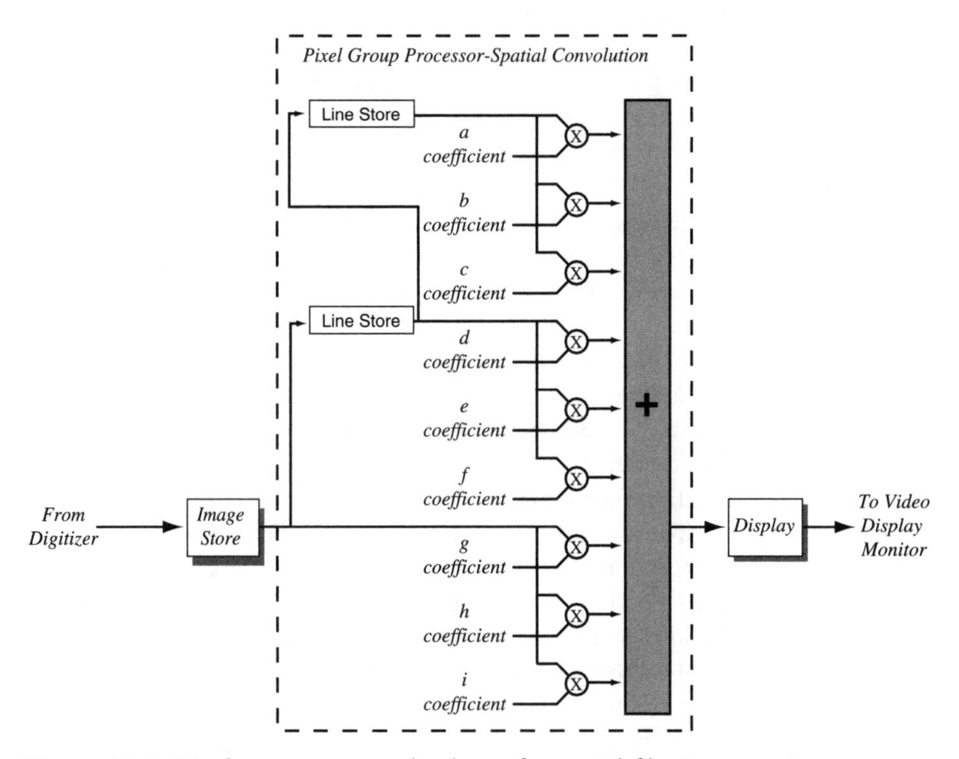

Figure 10.9 *Pixel group processor hardware for spatial filtering operations.*

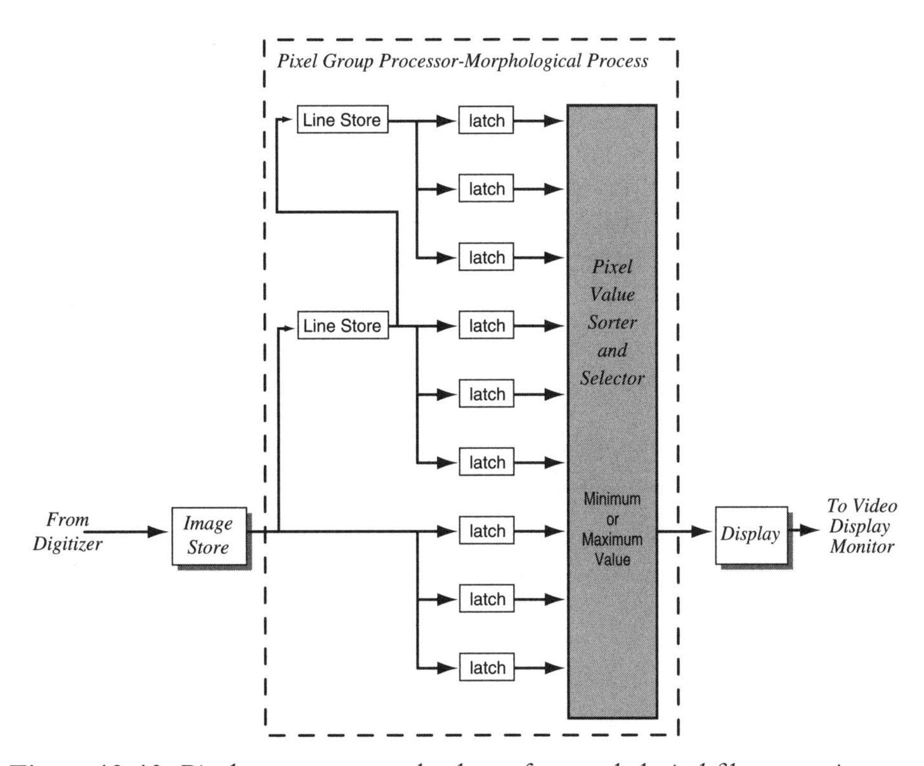

Figure 10.10 *Pixel group processor hardware for morphological filter operations.*

order. The middle value is found and selected, becoming the output median pixel value.

Other Processors

A number of other commonly used digital image processing operations are also regularly implemented using specialized hardware circuits. The following special-purpose processors are available in digital image processing systems:

- *Image Compressors and Decompressors*—Image compression and decompression hardware is common because of its widespread use. Further, many systems regularly compress and decompress images every time they are stored, retrieved, or transported to other systems, making compression facilities essential for adequate throughput.

  Image compression hardware is available to implement common compression algorithms, such as those discussed in Chapter 6. For applications requiring lossless compression, these processors generally cannot operate at full video data rates. However, many good lossy compression processors can easily provide video-rate image compression and decompression. Lossy compression processors are finding applications in television distribution industries such as cable television and direct-broadcast satellite, where image data compactness is essential.

- *Color Space Converters* (RGB-to-HSB and HSB-to-RGB)—Because color digital image processing systems often process images in the HSB color space, real-time color conversion from RGB to HSB, and back again, is often needed. Single-chip devices provide this conversion and can be embedded directly in the image data path for real-time conversion. Generally, the RGB-to-HSB conversion takes place between the digitizer and image store, while the HSB-to-RGB conversion happens between the image store and display circuitry. All image data processing occurs on the image data between the two functions, in HSB color space.
- *Histogram Generators*—For systems involved primarily in image enhancement and restoration tasks, a real-time histogram generator can greatly improve system throughput. The histogram generator is composed of a 256-location RAM memory, where each location stores the number of pixels in the image that have the same corresponding brightness value. As the image flows from the digitizer to the image store, the appropriate RAM locations are incremented based on each pixel's value. At the end of the incoming image frame, the RAM contains the 256 counts representing the brightness histogram of the image now residing in the image store. The histogram is then used to determine the appropriate enhancement or restoration parameters to be used in subsequent processing.
- *Geometric Processors*—For systems involved in geometric transformation operations, such as image resampling, perspective correction, or warping, special geometric processors are useful. These processors provide the necessary pixel address generation and interpolation computation speeds for real-time interaction between the user and resulting image transformations.

Image Data Processing Software

A digital image processing system is implemented using a mix of hardware and software components. As we have discussed, the hardware portion of the system is made up of at least a host computer, and can additionally comprise an accelerator processor and special-purpose processing hardware. Each of these hardware components requires a software program to make it carry out the digital image processing operations required by an application.

Digital image processing software can be separated into four primary layers—user interface, software application, function library, and device driver, as shown in Figure 10.11. Each layer has a distinct level of processing involvement. At the highest level, the *user interface* provides the look and feel of the digital image processing system to the user. This is the portion of software that allows the user to view the image, set the processing parameters, and generally interact with the system. The next level down is the *software application*. This software contains the system's paradigm. It defines how and what the user can do to an image, and sets up the operating environment. The *function library*, sometimes called the *engine*, contains the core of the system's software. This is where the actual digital image processing operations reside. The function library software is generally highly optimized code so that operations run efficiently. The lowest software level contains the *device driver* software. Device drivers interface the function library software with the system's image data handling and processing hardware. This is the software that knows how to communicate with the frame grabber, accelerator processor, and special-purpose processing hardware. Together, all four software layers operate with one other and with the system hardware to form the digital image processing system.

Device Drivers

The host computer must be able to communicate with the specific image data handling and processing hardware contained in the digital image processing system. Such hardware includes frame grabbers, accelerator processors, and special-purpose processors. Device drivers provide this communication function, and are necessary in all systems except those comprising only a host computer.

Devices driver software resides and runs in the host computer. This software handles hardware control functions such as frame-grabber commands to adjust spatial

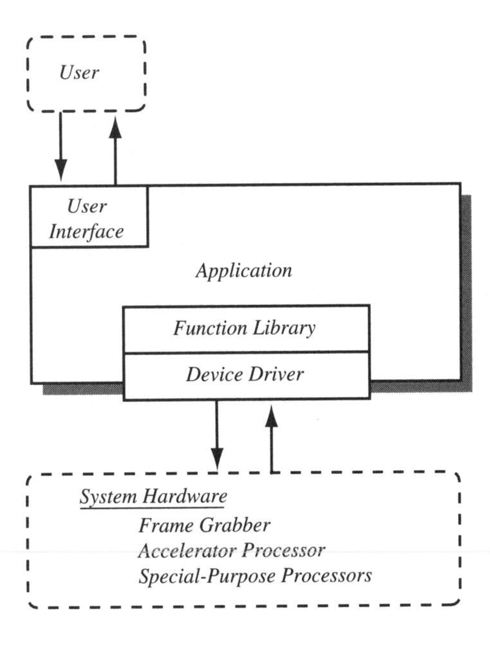

Figure 10.11 *The four layers of digital image processing software.*

resolution, freeze an image in the image store, or alter video timing parameters. Additionally, the device drivers must set up and monitor accelerator processors and special-purpose processors. These processors require a device driver to pass processing parameters to particular hardware registers. Such parameters include convolution mask values, LUT values, and geometric transformation parameters. It is the device driver's job to understand the hardware mechanisms of specific hardware devices intimately and to carry out the interface between them and the host computer.

Separating the device driver functions from the rest of the software allows us to place different hardware processor components into a system without rewriting all of the higher-level software. For instance, if we needed to replace a special-purpose processor board with a faster one from another vendor, we would only need to rewrite the device driver software. The high-level user interface, software application, and function library software can often remain unchanged.

Function Libraries/Toolkits

Digital image processing function libraries are collections of primitive software routines that carry out digital image processing operations. This is the software that actually processes an image. Function libraries are often referred to as *toolkits* because they are used by software developers as the foundation for a complete digital image processing application.

Function library software can reside and execute in either the host computer or an accelerator processor. Special-purpose processors generally do not require software because they embody the actual processing operation in hardware. If the function software executes in the host computer, then image data either is in the host computer's system memory or must be moved there from the image store. If an accelerator processor is available, then the function software is typically downloaded from the host computer to the accelerator. It is then directly executed on the accelerator processor. This speeds up the function software because the accelerator processor has direct, local access to the image store and is not involved in user interface or high-level tasks, like the host computer.

Spatial convolution is an example of a common function library routine. This function implements a pixel group process and is used in systems that do not contain a special-purpose group processor. The spatial convolution function requires the values of the convolution mask as input. The function's software then processes the entire input image's data and returns when complete. The resulting output image is left in the image store or system memory. By making successive calls to various function library routines, a software application can carry out a sequence of processing operations. For instance, a machine vision application may require calls to several functions, such as binary contrast enhancement, morphological operations, and finally, classification routines.

A digital image processing function library can contain hundreds of functions. Common functions include the following major groups of software routines:

1. Image Data Handling Functions
   - Digitization Control
   - Display Control
   - Image Store Control
   - Image Store Data Access
   - Open Image File
   - Save Image File
2. Image Data Processing Functions
   - Enhancement Operations
   - Restoration Operations
   - Analysis Operations
   - Compression Operations
   - Synthesis Operations

The functions of the image data-handling group control the operation of the digital image processing system hardware, discussed in Chapter 9. The image data-processing functions typically implement the operations discussed in Part II.

Software Applications and User Interfaces

The software application and user interface are the highest levels of software in a digital image processing system. This software is executed on the host computer. The user interface provides the controls and feedback to the user, allowing interaction with the system. The software application provides the functionality of the system to the user.

The software application is a software program that has a defined purpose of solving particular problems. The software application defines and provides the high-level commands and abilities of the system. For instance, in machine vision, a software application's purpose may be to provide the user with all of the software commands necessary to carry out machine vision tasks. The software application orchestrates the program's flow and calls upon the primitive functions of the function library to accomplish the user's directives.

The adjoining user interface complements the software application by providing the appropriate user–system interaction mechanisms. User interfaces are generally visual interfaces, called *graphical user interfaces* (*GUIs*). The user interacts with the visual elements of the GUI using a *pointing device*, like a mouse or trackball. The visual interface is composed of various *windows*, and the pointing device generally controls an arrow-shaped cursor that appears within the windows. Figure 10.12 illustrates a digital image processing software application's user interface, built around the Microsoft Windows environment.

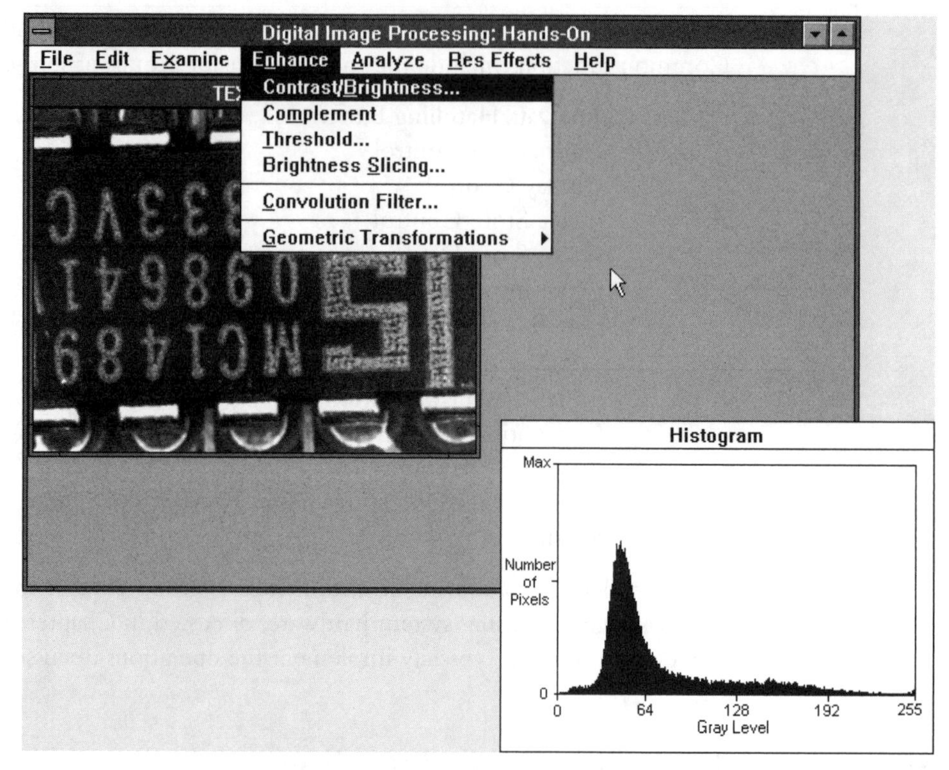

Figure 10.12 *The user interface of a digital image processing software application built on the Microsoft Windows environment.*

Some of the GUI windows will contain menu items to select the software application's digital image processing operations. Other windows can contain information resulting from applied operations, such as image analysis information like image histograms and profile plots. Often, the user interface windows are used to display input and output images, as well. This allows systems without a frame grabber to display images for evaluation.

We can summarize the overall software flow of a digital image processing system as follows:

1. The user visually interacts with the system using a graphical user interface.
2. The user interface communicates the user's directives to the software application and graphically returns the application's responses to the user.
3. The software application provides the operational paradigm of the system—it defines the operations that the system can do, and how they are implemented. In response to the user interface, the software application calls upon functions of the function library to accomplish the requested operations.

4. The function library executes the requested digital image processing operations upon the appropriate image data.
5. If an accelerator processor or special-purpose processing hardware is part of the system, the function library makes use of those facilities through the appropriate device driver. Device drivers interface arbitrary digital image processing hardware with the software portion of the system.

ISO/IEC-IPI Digital Image Processing Standard

Although there are numerous standards to define digital image compression techniques and image file formats, there has never been a standard to define specific digital image processing operations. In 1988, an imaging group formed under the *American National Standards Institute* (*ANSI*), to standardize imaging technology. The group and its resulting documents were dubbed *ANSI X3H3.8*. The ANSI X3H3.8 group was particularly interested in the standardization of computer-based handling and processing of digital images.

The ANSI group set out to define a programmer's imaging kernel (PIK) application programming interface. In late 1989, an international group convened under the *International Organization for Standardization* (*ISO*) and *International Electrotechnical Commission* (*IEC*) to set up an international imaging standard based, in part, on the ANSI X3H3.8 work. The ISO/IEC group has created the *Image Processing and Interchange* standard (*IPI*). The IPI standard is composed of three parts:

1. A Common Architecture for Imaging (CAI)—a description of the standard's common architecture.
2. The Programmer's Imaging Kernel System (IPI-PIKS) Application Programming Interface (API)—the definitions of digital image processing operations. This part is based on the ANSI X3H3.8 PIK work.
3. The Image Interchange Facility (IPI-IIF)—the definition for digital image interchange between software application programs.

The ISO/IEC IPI standard consists of over 1,000 pages of text. The IPI-PIKS part of the standard defines a wide range of digital image processing operations. They are arranged into the following functional groups:

- Analysis Operations
- Classification Operations
- Color Operations
- Complex Image Operations
- Detection Operations

- Edge Detection Operations
- Enhancement Operations
- Ensemble Operations
- Filtering Operations
- Geometric Operations
- Histogram Operations
- Morphological Operations
- Point Operations
- Presentation Operations
- Shape Operations
- Three-Dimensional Specific Operations
- Transform Operations

Each of these digital image processing operation groups consists of numerous specific operations, including virtually all of those discussed throughout the text of this book.

The ISO/IEC IPI international imaging standard is important to the digital image processing community, because it brings commonality and consistency to the use of digital image processing operations. As a result, improved algorithm uniformity will prevail across computing platforms, applications, and the imaging industry at large.

Image Operation Studies

PART IV
Processing in Action

We have encountered a wide variety of digital image processing operations throughout this book. Each operation was described in some depth, providing you with an understanding of its mechanics as well as its effects on various image types. Many of these operations were tagged with an IOS icon—like the one to the left—in the margin adjacent to the text describing them. These IOS (image operation study) icons refer to the supplemental installments located in this part of the book, "Processing in Action."

This compendium of image operation studies provides a comprehensive quick reference to digital processing operations commonly used to solve real-world problems. Operations from each of the five fundamental classes—enhancement, restoration, analysis, compression, and synthesis—are examined. Each study treats an independent image operation. First, the name and class of each operation are noted. A brief description of the operation, some of its common applications, and an implementation method are stated. Finally, "before" and "after" images demonstrate the visual effects of each operation.

Use this compendium as a quick reference to the effects of common digital image processing operations. Then, for more information, I encourage you to use Appendix A, "Further Reading/References," to search out follow-on texts and periodicals. The digital image processing field continues to change at a rapid pace; subscriptions to the relevant journals and trade magazines can be well worth the expense to stay current with the evolution of the technology.

Additionally, install and use the *Hands-On* Windows software application included with this book for further exposure to commonly used operations. The *Hands-On* application gives you the ability to experiment with a sampling of basic image enhancement and analysis operations using several included images from this book or your own images.

1 Complement Image

Class: Image Enhancement/Restoration

Description: Each image pixel is logically complemented. Black pixels map to white, white pixels map to black, and the intermediate grays are correspondingly mapped to their complemented values.

Applications: Complement operations are often useful in making the subtle brightness details in bright areas of an image more visible. Because the eye responds logarithmically to brightness changes, subtle brightness changes in the bright regions of an image may be undetectable. With the complement operation, these subtle brightness changes are transformed to dark regions, so that they become more clearly visible.

Implementation:

1. Define mapping function
2. Single-image pixel point process

The complement image operation uses the following mapping function to compute resulting pixel brightnesses:

$$O(x,y) = 255 - I(x,y)$$

Similar mapping functions can be used to complement only a portion of the gray scale.

Results:

Figure IOS.1a Original F-16 aircraft image.

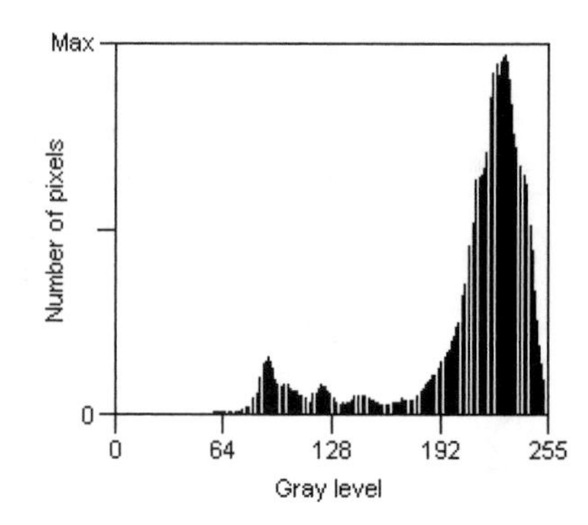

Figure IOS.1b Histogram of original image.

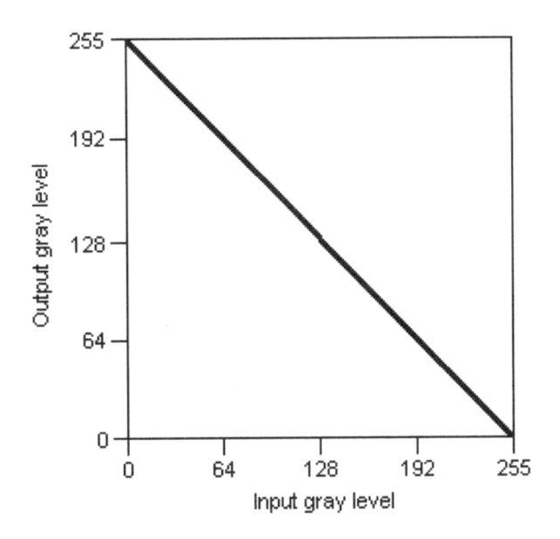

Figure IOS.1c *Complement image mapping function.*

Figure IOS.1d *Complemented image.*

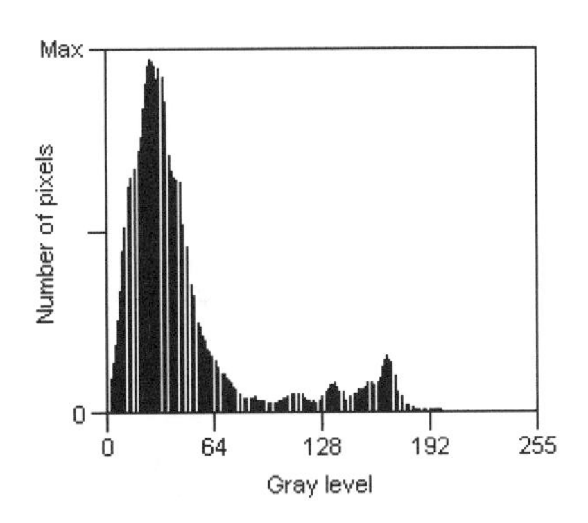

Figure IOS.1e *Histogram of complemented image illustrating complemented brightnesses.*

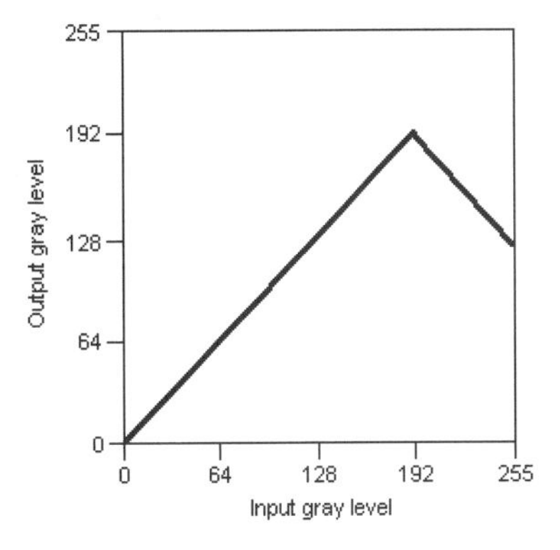

Figure IOS.1f *Partial complement image mapping function, where only the brightnesses above 192 are complemented.*

Figure IOS.1g *Partial-complemented image.*

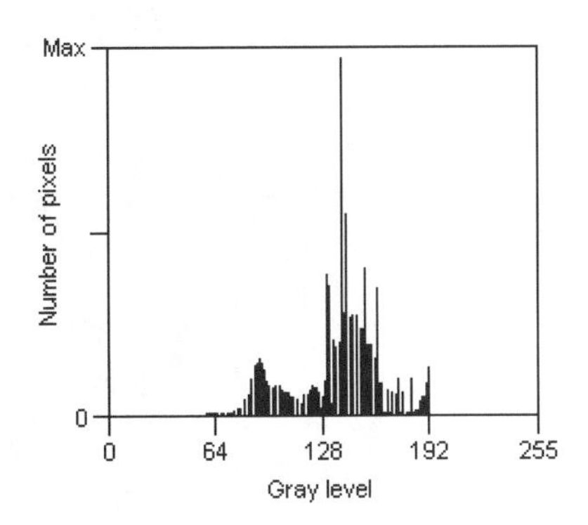

Figure IOS.1h *Histogram of partial-complemented image.*

IOS 2

2 Histogram Sliding and Stretching

Class: Image Enhancement/Restoration

Description: Each pixel is added to a constant slide value and multiplied by a constant stretch value. The resulting image histogram appears shifted left or right, and expanded or reduced in size.

Applications: Histogram sliding operations brighten or darken the resulting image, while stretching operations increase or decrease the contrast of the resulting image. Histogram sliding and stretching are used for general contrast enhancement of low-contrast images.

Implementation:

1. Define mapping function
2. Single-image pixel point process

The histogram slide and stretch operation uses the following mapping function to compute resulting pixel brightnesses:

$$O(x,y) = (I(x,y) + \text{slide value}) \times \text{stretch value}$$

Instead of sliding or stretching the histogram linearly using constant values, you can use a nonlinear mapping function instead. This allows a portion of the image brightness range to be mapped in one way while other portions are mapped in a different way or left unaffected.

Results:

Figure IOS.2a Original semiconductor chip image with low contrast and low dynamic range.

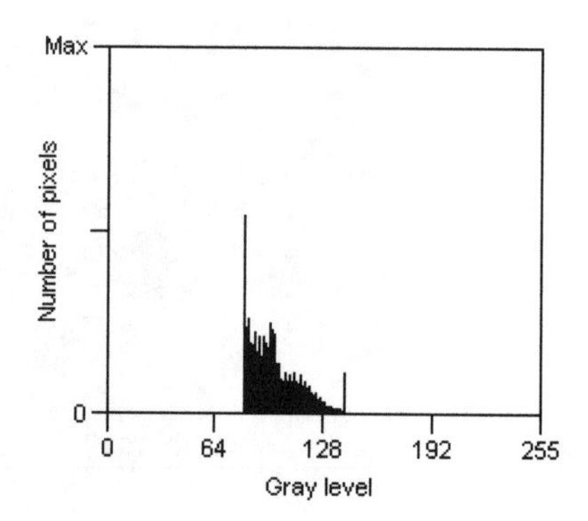

Figure IOS.2b Histogram of original image.

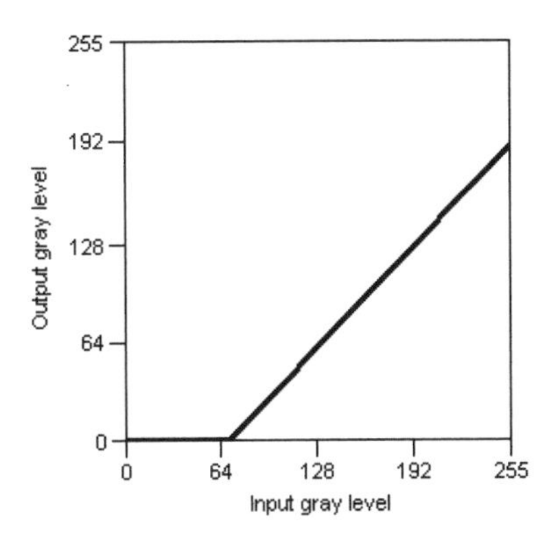

Figure IOS.2c Histogram slide-mapping function for a slide of −69.

Figure IOS.2d Histogram-slided image.

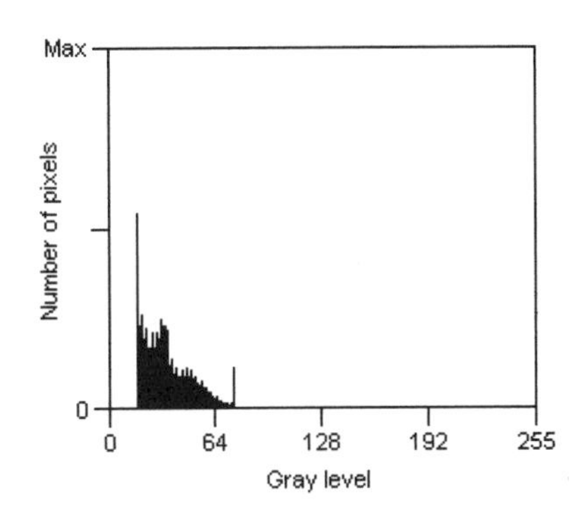

Figure IOS.2e Histogram of resulting image slided by −69.

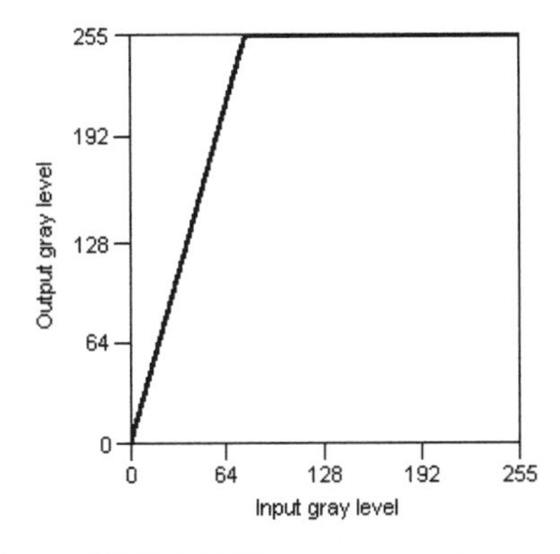

Figure IOS.2f Histogram stretch-mapping function for a stretch by a factor of 3.4.

Figure IOS.2g Histogram-stretched image.

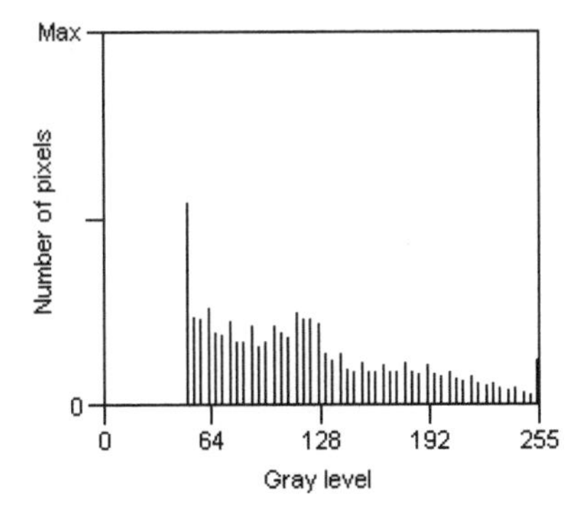

Figure IOS.2h Histogram of resulting image slided by −69 and then stretched by a factor of 3.4 showing well-balanced characteristics.

3 Binary Contrast Enhancement

Class: Image Enhancement/Restoration

Description: Each image pixel is determined to be above or below a predetermined brightness threshold value. If the pixel brightness is darker than the threshold, the resulting pixel brightness is set to 0. If the pixel brightness is brighter than the threshold, the resulting brightness is set to 255.

Applications: Binary contrast enhancement operations are used to create a very high-contrast image from a low-contrast image by highlighting an object of interest. This operation can separate an object from its background in a low-light-level image. The binary contrast enhancement operation works well as long as the object's brightnesses are greater than or less than the background brightnesses.

Implementation:

1. Define mapping function
2. Single-image pixel point process

The binary contrast enhancement operation uses the following mapping function to compute resulting pixel brightnesses:

$$O(x,y) = 0 \text{ if } I(x,y) < \text{threshold value}$$
$$= 255 \text{ if } I(x,y) \geq \text{threshold value}$$

A variable value can be used instead of a fixed threshold value. This technique is called adaptive thresholding. The adaptive threshold value is automatically determined throughout the image based on the image's local brightness characteristics. Typically, a spatial filtering operation such as a low-pass filter is used to determine the average brightness in a local region of the image. This average brightness value is then used to adapt the threshold value to distinguish foreground objects from their background.

Results:

Figure IOS.3a Original fingerprint image with low contrast.

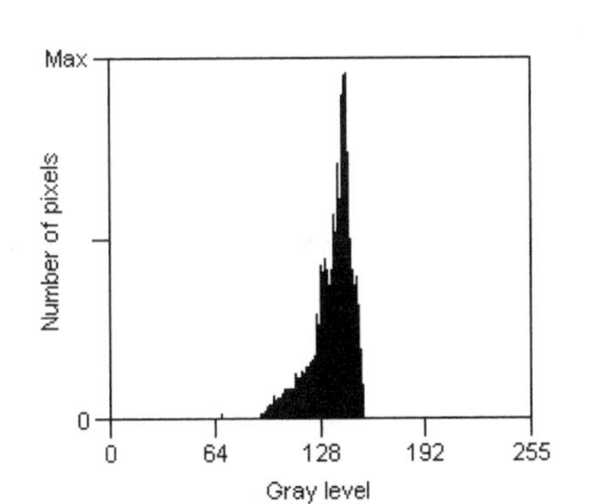

Figure IOS.3b Histogram of original image.

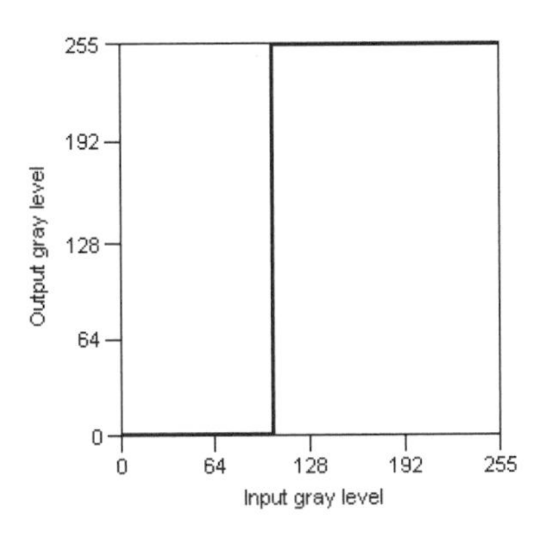

Figure IOS.3c Binary contrast enhancement mapping function with a brightness threshold of 106.

Figure IOS.3d Binary contrast enhancement image showing very high contrast.

Figure IOS.3e Original washer, nut, and bolt image with low contrast.

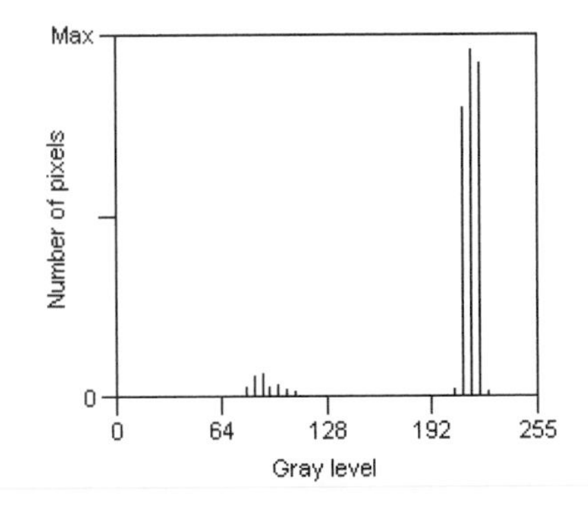

Figure IOS.3f Histogram of original bolt and nut image.

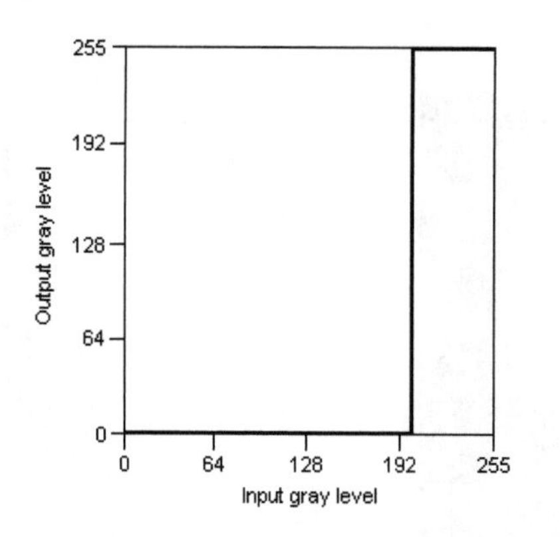

***Figure IOS.3g** Binary contrast enhancement mapping function with a brightness threshold of 210.*

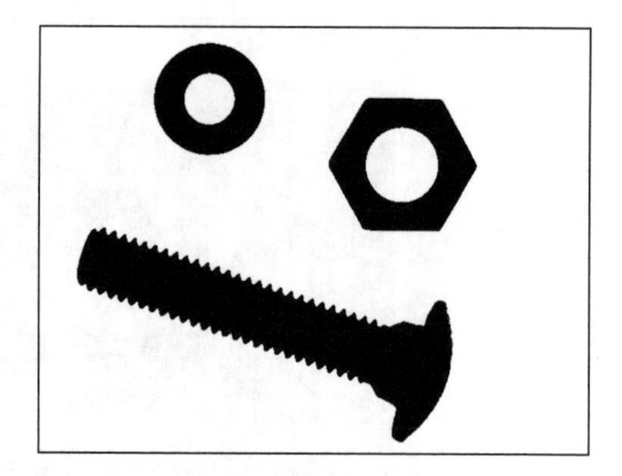

***Figure IOS.3h** Binary contrast enhancement image showing very high contrast.*

4 Brightness Slicing

Class: Image Enhancement/Restoration

Description: Each image pixel is determined to be between or outside a range of brightnesses defined by low and high brightness threshold values. If the pixel brightness is outside the thresholds, the resulting pixel brightness is set to 0. If the pixel brightness is between the thresholds, the resulting brightness is set to 255.

Applications: Brightness slicing operations are used to create a very high-contrast image that highlights an object of interest. This operation can separate objects from their background even when background brightnesses range above and below object brightnesses. The brightness slicing operation works well as long as the object's brightnesses are confined to a range that is distinct from the range of background brightnesses.

Implementation:

1. Define mapping function
2. Single-image pixel point process

The brightness slicing operation uses the following mapping function to compute resulting pixel brightnesses:

$O(x,y) = 255$ if $I(x,y) \geq$ low threshold value
and if $I(x,y) <$ high threshold value

$O(x,y) = 0$ otherwise

Results:

Figure IOS.4a *Original bank check image with low contrast and other objects of varying brightnesses.*

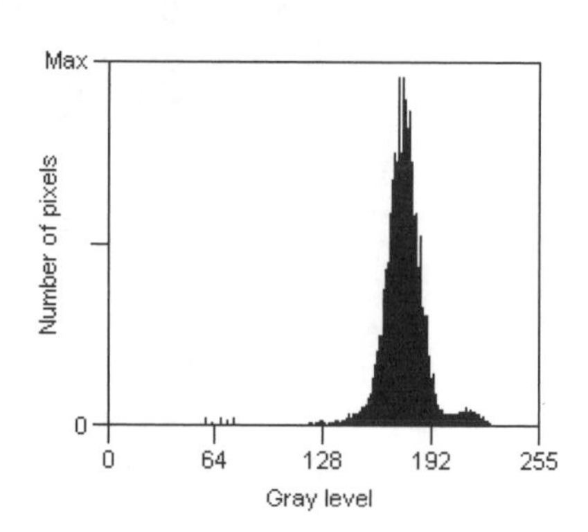

Figure IOS.4b *Histogram of original image.*

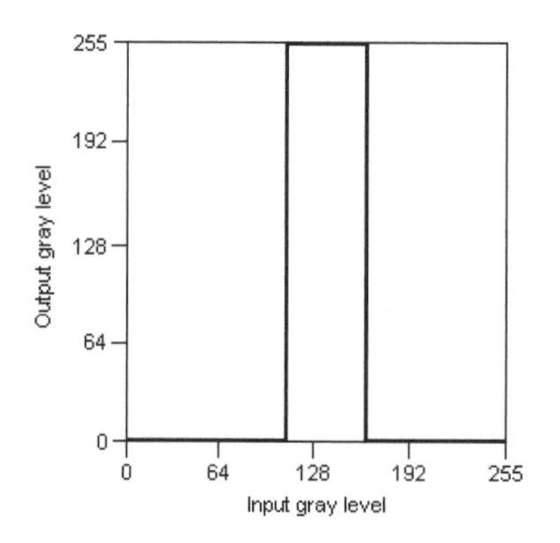

Figure IOS.4c *Brightness-slicing mapping function with a low threshold brightness of 113 and a high threshold brightness of 152.*

Figure IOS.4d *Brightness-sliced image showing the signature as white and everything else as black.*

5 Image Subtraction

Class: Image Enhancement/Restoration

Description: One image is subtracted from a second image, pixel by pixel. Where pixels have the same value in each image, the resulting pixel is 0. Where the two pixels are different, the resulting pixel is the difference between them.

Applications: Image subtraction can be used to remove common background image information from images of identical scenes. The resulting image shows only the foreground objects of interest; the static background elements are eliminated. In the case of *digital subtraction angiography*, a baseline image is subtracted from one where blood vessels are enhanced with an X-ray-opaque liquid. The resulting image shows only the enhanced blood vessels without the obscuring background imagery.

Illumination equalization is often used in biological sample imagery obtained from a microscope. A background image showing only the illumination nonuniformities is subtracted from a specimen image, yielding an image without the illumination inconsistencies.

Motion detection uses image subtraction to determine object motion between two images. Typically, two images of the same scene are subtracted, yielding an image that shows only the differences. Constant image information is removed and only objects that have moved between the two images will appear.

Implementation:

1. Dual-image pixel point process—subtraction

The image subtraction operation uses the following equation to combine two images:

$$O(x,y) = I_1(x,y) - I_2(x,y)$$

IOS 5

Results:

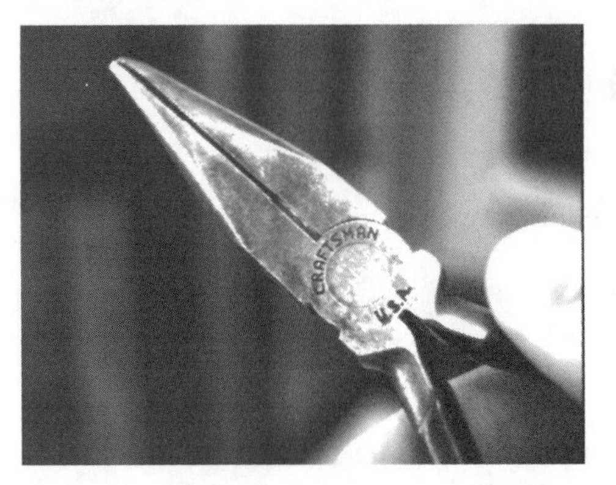

Figure IOS.5a *Original image 1 showing a pair of pliers in the foreground against a background.*

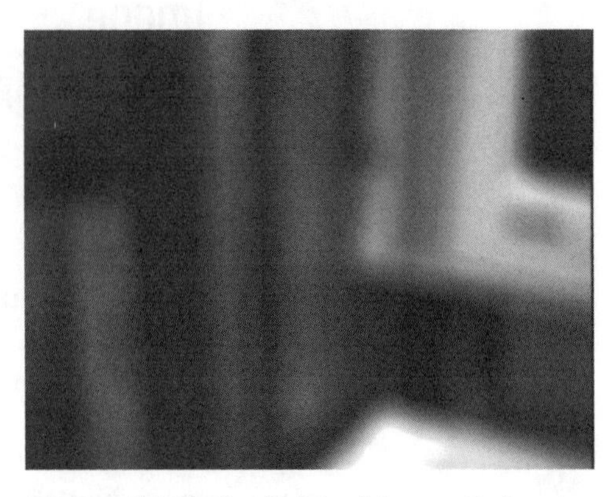

Figure IOS.5b *Original image 2 showing only the background.*

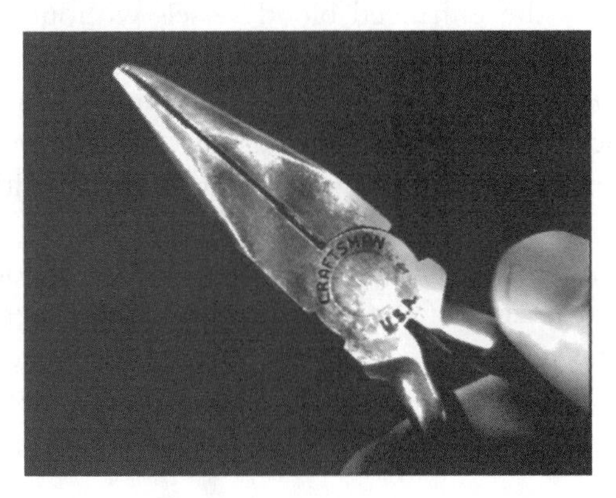

Figure IOS.5c *The difference image showing only the foreground pliers object with the background removed.*

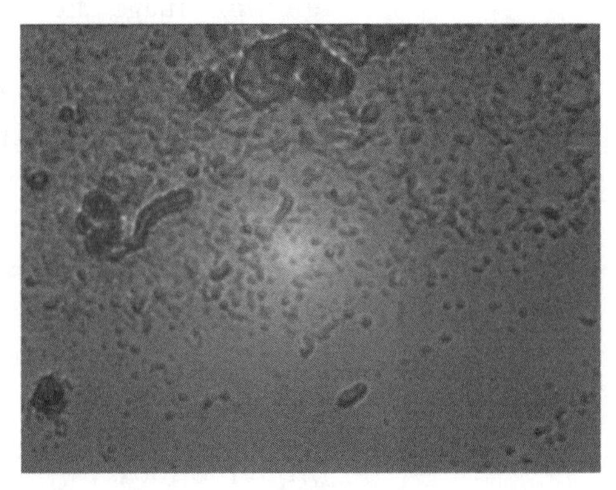

Figure IOS.5d *Original image 1 showing a microscopic specimen with uneven illumination.*

Figure IOS.5e *Original image 2 showing only the uneven illumination.*

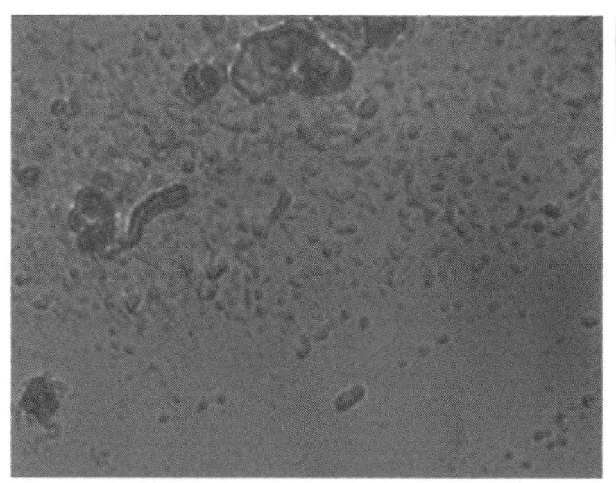

Figure IOS.5f *The difference image showing the specimen with the uneven illumination equalized.*

Figure IOS.5g *Original image 1 showing an image of an assembled printed circuit board.*

Figure IOS.5h *Original image 2 showing an image of the same printed circuit board assembly with a component left out.*

Figure IOS.5i *The difference image showing only the objects that appear differently—the missing component can be immediately identified.*

IOS 6

6 Image Division

Class: Image Enhancement/Restoration

Description: One image is divided by a second image, pixel by pixel. Where a pixel in the numerator image is greater than the corresponding pixel in the denominator image, the resulting pixel is large (bright). Where the opposite is true, the resulting pixel is small (dark).

Applications: Image division is applied to images of identical scenes. In the most common use, *spectral ratioing* highlights components of an image that appear differently when imaged through different spectral filters. For example, if it is necessary to detect red-colored objects in an image, a red-component image can be divided by a green-component image and a blue-component image. Where pixels have large red values and small green and blue values, a large result indicates that the pixel is more red than green or blue. When the red value is small and the green or blue value is large, a small value indicates the opposite.

Implementation:

1. Dual-image pixel point process—division

The image division operation uses the following equation to combine two images:

$$O(x,y) = I_1(x,y) \;/\; I_2(x,y)$$

Results:

Figure IOS.6a *Original image 1 of red and green apples through a red filter.*

Figure IOS.6b *Original image 2 of the same red and green apples through a green filter.*

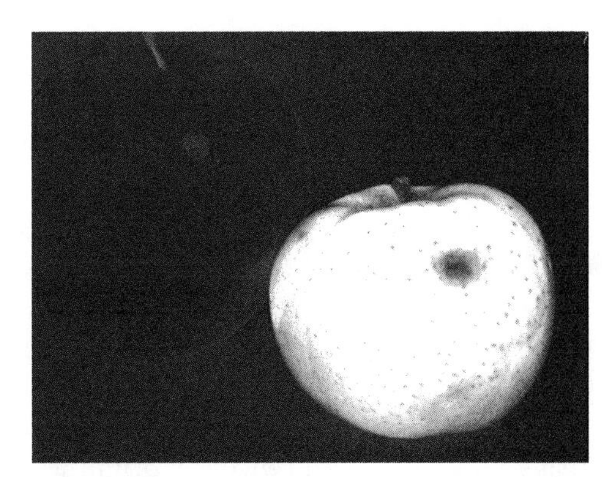

Figure IOS.6c *Ratio image; the result is brightness-scaled down not to exceed 255. This image shows the red apple significantly highlighted.*

7 Image Addition

Class: Image Enhancement/Restoration

Description: One image is added to a second image, pixel by pixel, combining the images into a single image. When two images of entirely different subjects are summed, a blend of the two will result. If two images of the same subject are summed, an identical image will result with twice the brightness.

Applications: Images of the same scene can be averaged to reduce random noise. Stationary objects between the two images will appear with doubled brightnesses in the summed image. Any randomly varying aspects contained in the images, such as noise, will statistically appear with less than doubled brightnesses, and are therefore reduced in brightness relative to the stationary subject matter. As long as the subject matter is stationary throughout the averaged images, only the randomly changing noise will be affected. If the subject matter changes between averaged images, motion blur will result as well.

Image addition and subtraction operations can be used together to perform image compositing operations. A foreground object is outlined in the first input image, and a mask image (or digital matte) is created: All pixels inside the outline are set to 0, and all pixels outside are set to 255. The mask image is then subtracted from the original image, yielding an image of only the object surrounded by black. The mask image is complemented and subtracted from a second input image (the background image), yielding an image of the background with a black hole exactly the size of the first image's object. The two resulting images are summed to produce a composite image of the object on the new background.

Implementation:

Noise reduction:

1. Dual-image pixel point process—addition
2. Repeat for all images in average
3. Define brightness scale map
4. Single-image pixel point process

Image compositing:

1. Outline object in "object image"
2. Create object mask image
3. Subtract object mask image from object image

4. Complement object mask image
5. Subtract complemented object mask image from "background image"
6. Add masked object and background images

The image addition operation uses the following equation to combine two images:

$$O(x,y) = I_1(x,y) + I_2(x,y)$$

Because the summation of two pixels can result in a brightness of up to 510, the resulting pixel brightness must be scaled not to exceed 255.

Results:

Figure IOS.7a *Original building image 1 with random noise.*

Figure IOS.7b *Image 2 with random noise.*

Figure IOS.7c *Image 3 with random noise.*

Figure IOS.7d *Image 4 with random noise.*

Figure IOS.7e *The average of images 1, 2, 3, and 4—the random noise is reduced.*

8 Low-Pass Filter

Class: Image Enhancement/Restoration

Description: Low-pass filtering smooths an image by attenuating high-spatial-frequency details. The convolution mask weighting coefficients are selected to vary the cutoff point where higher frequencies become attenuated. Further, the resulting low-pass filtered image can be brightness-scaled down and summed with the original image to create milder low-pass filter effects.

Applications: A low-pass filter can be used to attenuate image noise that is composed primarily of high-frequency components such as impulse spikes and wideband interference.

The characteristics of low-frequency objects of interest can be enhanced by removing high-frequency image information using low-pass filtering. In particular, edges and sharp lines and points are smoothed, while low-frequency attributes are left untouched. For example, an image of an object containing only low spatial frequencies with a distracting high-frequency grid pattern can be enhanced in this way. A case like this is a biological specimen imaged under a microscope with measurement reticles. The low-pass effect is to blur the sharp lines of the reticles while leaving the low-frequency object relatively unaffected.

Implementation:

1. Define spatial convolution mask
2. Pixel group process

The mask coefficients add to 1. All coefficients are positive, and must therefore be fractional. Three common low-pass filter masks are

$$\begin{matrix} \frac{1}{9} & \frac{1}{9} & \frac{1}{9} \\ \frac{1}{9} & \frac{1}{9} & \frac{1}{9} \\ \frac{1}{9} & \frac{1}{9} & \frac{1}{9} \end{matrix} \qquad \begin{matrix} \frac{1}{10} & \frac{1}{10} & \frac{1}{10} \\ \frac{1}{10} & \frac{1}{5} & \frac{1}{10} \\ \frac{1}{10} & \frac{1}{10} & \frac{1}{10} \end{matrix} \qquad \begin{matrix} \frac{1}{16} & \frac{1}{8} & \frac{1}{16} \\ \frac{1}{8} & \frac{1}{4} & \frac{1}{8} \\ \frac{1}{16} & \frac{1}{8} & \frac{1}{16} \end{matrix}$$

Mask 1 Mask 2 Mask 3

In practice, low-pass mask coeffiecients are generally normalized to integer values. Then a brightness divider is applied to the resulting output value to bring the result within a 0 to 255 brightness range.

The above masks can be converted to integer values with the appropriate brightness dividers, as shown:

$$\begin{matrix} 1 & 1 & 1 \\ 1 & 1 & 1 \\ 1 & 1 & 1 \end{matrix} \times \frac{1}{9} \qquad \begin{matrix} 1 & 1 & 1 \\ 1 & 2 & 1 \\ 1 & 1 & 1 \end{matrix} \times \frac{1}{10} \qquad \begin{matrix} 1 & 2 & 1 \\ 2 & 4 & 2 \\ 1 & 2 & 1 \end{matrix} \times \frac{1}{16}$$

Mask 1 Mask 2 Mask 3

Results:

Figure IOS.8a *Original building image.*

Figure IOS.8b *Low-passed image using mask 1.*

Figure IOS.8c *Low-passed image using mask 2.*

Figure IOS.8d *Low-passed image using mask 3.*

Figure IOS.8e *Low-passed image using mask 1 (all 1s), but with a center coefficient of 16 and a gain multiplier of 1/24. The effect is to add a brightness-scaled down, low-passed image to the original image. The resulting image shows a milder low-pass effect.*

9 Unsharp Masking Enhancement

Class: Image Enhancement/Restoration

Description: The unsharp masking enhancement operation sharpens an image by subtracting a brightness-scaled, low-pass-filtered image from its original. The resulting image appears with sharpened high-frequency details; low-frequency details are left untouched.

Applications: The unsharp masking enhancement produces a subtle and visually pleasing result. It is used primarily to enhance the subjective spatial appearance of an image's high-frequency details. Images suffering from poor spatial definition are best suited for the unsharp masking enhancement.

Implementation:

1. Define spatial convolution mask
2. Pixel group process
3. Define brightness scale map
4. Single-image pixel point process
5. Dual-image pixel point process—subtraction

The unsharp masking operation is implemented by first performing a low-pass filtering operation on the original image. The low-passed image is then brightness-scaled down to a desired level. The brightness-scaled image is subtracted from the original image. The resulting image contains sharpened edge detail, as shown in Figure IOS.9a.

Results:

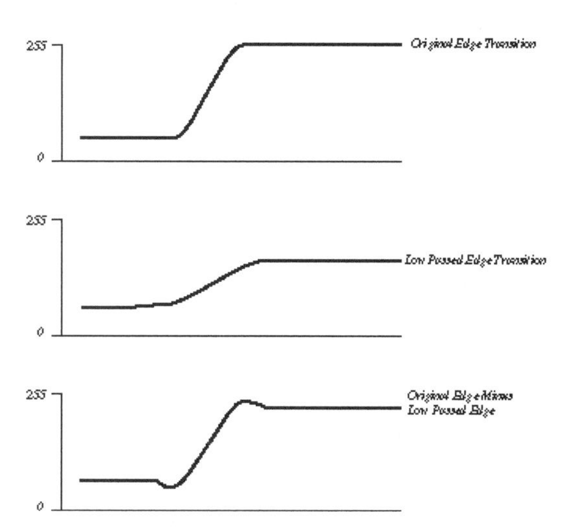

Figure IOS.9a *The unsharp masking edge enhancement operation illustrating the accentuation of edge details.*

Figure IOS.9b *Original building image.*

Figure IOS.9c *Low-passed image.*

Figure IOS.9d *Original image minus brightness-scaled low-passed image.*

10 High-Pass Filter

Class: Image Enhancement/Restoration

Description: High-pass filtering sharpens an image by accentuating high-spatial-frequency details. The convolution mask weighting coefficients are selected to vary the cutoff point where higher frequencies become accentuated. Further, the resulting high-pass filtered image can be brightness-scaled down and summed with the original image to create milder high-pass filter effects.

Applications: High-pass filters are used to highlight high-frequency details within an image, such as lines, points and edges. Images that appear spatially dull or poorly defined can be sharpened with high-pass filtering techniques.

Accentuating high-frequency image information with high-pass filtering can enhance the characteristics of high-frequency objects of interest. For example, an image of a spatially detailed object—say, a tree in front of a dirt-covered background—can be enhanced this way. The high-pass effect will sharpen the tree's features while leaving the dirt-covered background relatively unaffected.

Implementation:

1. Define spatial convolution mask
2. Pixel group process

The high-pass filter operation can produce results that are less than 0 or greater than 255. In these underflow and overflow cases, the resulting values are forced to 0 or 255, whichever is closest.

The mask coefficients add to 1. A large coefficient generally appears in the center of the mask, surrounded by smaller positive and negative coefficients. Three common high-pass filter masks are

$$\begin{matrix} -1 & -1 & -1 \\ -1 & 9 & -1 \\ -1 & -1 & -1 \end{matrix} \qquad \begin{matrix} 0 & -1 & 0 \\ -1 & 5 & -1 \\ 0 & -1 & 0 \end{matrix} \qquad \begin{matrix} 1 & -2 & 1 \\ -2 & 5 & -2 \\ 1 & -2 & 1 \end{matrix}$$

Mask 1 Mask 2 Mask 3

Results:

Figure IOS.10a *Original building image.*

Figure IOS.10b *High-passed image using mask 1.*

Figure IOS.10c *High-passed image using mask 2.*

Figure IOS.10d *High-passed image using mask 3.*

Figure IOS.10e *High-passed image using mask 1 (all −1s), but with a center coefficient of 20 and a gain multiplier of* $\frac{1}{12}$*. The effect is to add a brightness-scaled down, high-passed image to the original image. The resulting image shows a milder high-pass effect.*

11 Sobel Edge Enhancement

Class: Image Enhancement/Restoration

Description: The Sobel edge enhancement operation extracts all of the edges in an image, regardless of direction. It is implemented as the sum of two directional edge enhancement operations. The resulting image appears as an omnidirectional outline of the objects in the original image. Constant brightness regions become black, while changing brightness regions become highlighted.

Applications: The Sobel edge enhancement operation is used to produce the edges of the objects in an image. The edges are any sharp brightness transitions rising from black to white or falling from white to black. This operation is omnidirectional, which means that all black-to-white and white-to-black edge transitions are highlighted, regardless of their directions in the image. The Sobel operation is more immune to image noise than the Laplacian operation, and it provides stronger edge discrimination.

Implementation:

1. Define spatial convolution mask number 1
2. Pixel group process
3. Define spatial convolution mask number 2
4. Pixel group process
5. Dual-image pixel point process—addition

The Sobel edge enhancement operation can produce results that are less than 0 or greater than 255. In these underflow and overflow cases, the resulting values are forced to 0 or 255, whichever is closest.

The mask coefficients add to 0. Two masks are applied to a copy of the original image to highlight horizontal and vertical edges independently. The resulting images are added to create the final Sobel edge enhancement image. The Sobel edge enhancement masks are

$$\begin{matrix} -1 & 0 & 1 \\ -2 & 0 & 2 \\ -1 & 0 & 1 \end{matrix} \quad \text{and} \quad \begin{matrix} -1 & -2 & -1 \\ 0 & 0 & 0 \\ 1 & 2 & 1 \end{matrix}$$

Vertical mask   Horizontal mask

Results:

Figure IOS.11a *Original building image.*

Figure IOS.11b *Sobel edge enhancement.*

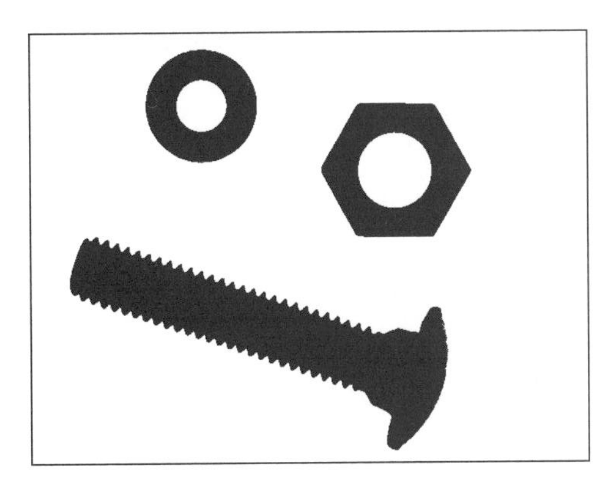

Figure IOS.11c *Original binary image of washer, nut, and bolt.*

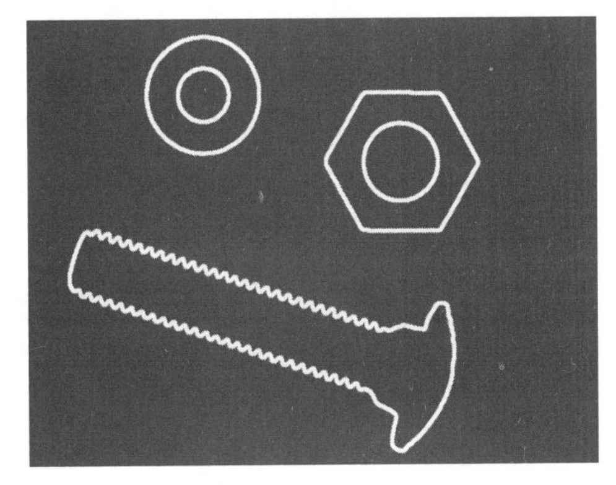

Figure IOS.11d *Sobel edge enhancement.*

12 Shift and Difference Edge Enhancement

Class: Image Enhancement/Restoration

Description: The shift and difference operation extracts the vertical, horizontal, or diagonal edges in an image. Each pixel is subtracted from its adjacent neighboring pixel in the chosen direction. The resulting image appears as a directional outline of the objects in the original image: Constant brightness regions become black, while changing brightness regions become highlighted.

Applications: The shift and difference operation is used to produce the edges of the objects in an image. The edges are any sharp brightness transitions rising from black to white. This operation is directional, which means that black-to-white edge transitions are highlighted only in a single direction of travel across the image. The shift and difference operation is the most computationally efficient of all edge enhancement operations, but it tends to accentuate noise in an image.

Implementation:

1. Shift image to left, up, or both
2. Dual-image pixel point process—subtraction

or

1. Define spatial convolution mask
2. Pixel group process

You can implement this operation by geometrically translating the original image by one pixel and subtracting the result from the original. Implementing the operation as a group process can often be a more efficient approach. The mask coefficients add to 0. The center coefficient is 1 and another coefficient is −1. Depending on where the −1 is, a horizontal, vertical, or diagonal edge enhancement occurs.

The shift and difference operation can produce results that are less than 0. This underflow condition occurs when a bright pixel is subtracted from a dark pixel. In these underflow cases, the resulting value is forced to 0. The net effect is that edges going from black to white are brightly highlighted. Edges going from white to black are not, because the bright value is subtracted from the dark value, so the underflow condition yields a black result.

The three common shift and difference masks are

Vertical			Horizontal			Diagonal		
0	0	0	0	−1	0	−1	0	0
−1	1	0	0	1	0	0	1	0
0	0	0	0	0	0	0	0	0

Results:

Figure IOS.12a *Original building image.*

Figure IOS.12b *Vertical shift and difference edge enhancement.*

Figure IOS.12c *Horizontal shift and difference.*

Figure IOS.12d *Diagonal shift and difference.*

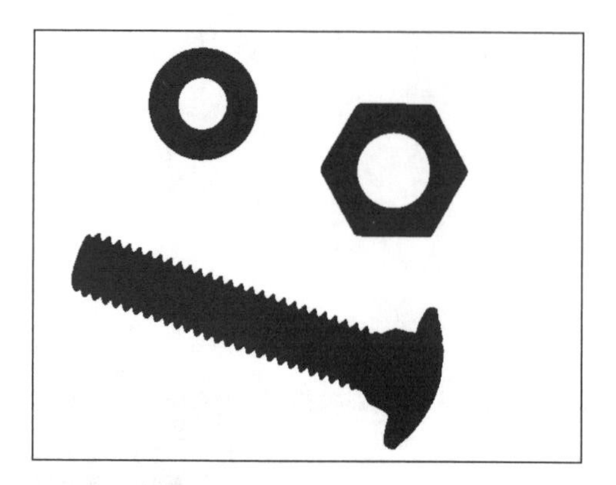

Figure IOS.12e *Original binary image of washer, nut, and bolt.*

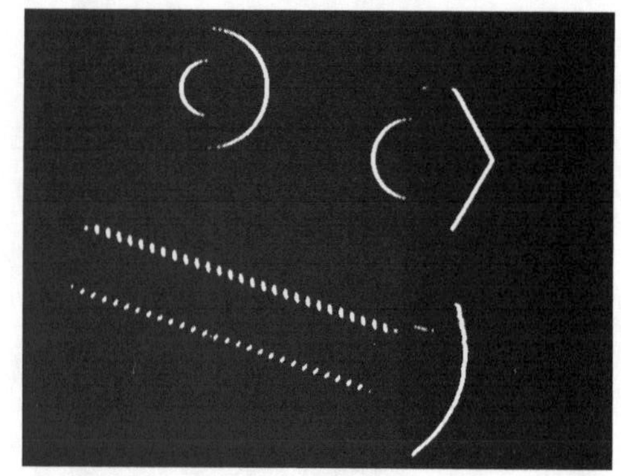

Figure IOS.12f *Vertical shift and difference edge enhancement.*

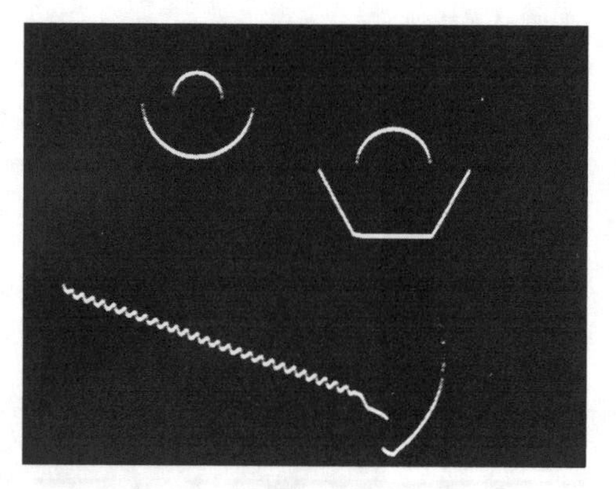

Figure IOS.12g *Horizontal shift and difference.*

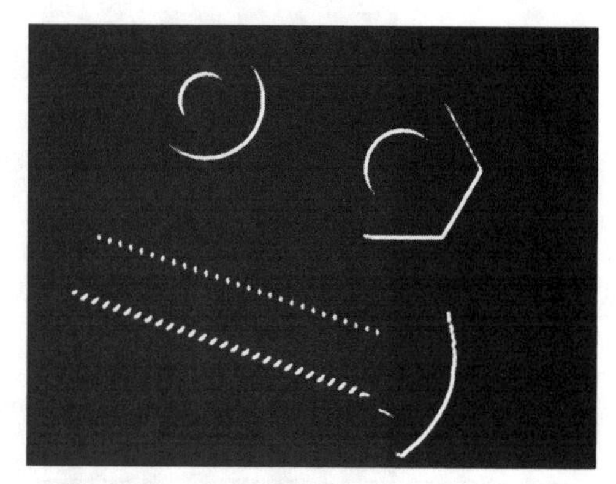

Figure IOS.12h *Diagonal shift and difference.*

13 Prewitt Gradient Edge Enhancement

Class: Image Enhancement/Restoration

Description: The Prewitt gradient edge enhancement operation extracts the north, northeast, east, southeast, south, southwest, west, or northwest edges in an image. The resulting image appears as a directional outline of the objects in the original image. Constant brightness regions become black and changing brightness regions become highlighted.

Applications: The Prewitt gradient edge enhancement operation is used to produce the edges of the objects in an image. The edges are any sharp brightness transitions rising from black to white. This operation is directional, which means that black-to-white edge transitions are highlighted only in a single direction of travel across the image. The Prewitt gradient operation is more immune to image noise than the shift and difference operation, and provides stronger directional edge discrimination.

Implementation:

1. Define spatial convolution mask
2. Pixel group process

The Prewitt gradient edge enhancement operation can produce results that are less than 0 or greater than 255. In these underflow and overflow cases, the resulting values are forced to 0 or 255, whichever is closest.

The mask coefficients add to 0. The eight directional masks are

$$\begin{matrix} 1 & 1 & 1 \\ 1 & -2 & 1 \\ -1 & -1 & -1 \end{matrix} \qquad \begin{matrix} 1 & 1 & 1 \\ -1 & -2 & 1 \\ -1 & -1 & 1 \end{matrix} \qquad \begin{matrix} -1 & 1 & 1 \\ -1 & -2 & 1 \\ -1 & 1 & 1 \end{matrix} \qquad \begin{matrix} -1 & -1 & 1 \\ -1 & -2 & 1 \\ 1 & 1 & 1 \end{matrix}$$

N mask — NE mask — E mask — SE mask

$$\begin{matrix} -1 & -1 & -1 \\ 1 & -2 & 1 \\ 1 & 1 & 1 \end{matrix} \qquad \begin{matrix} 1 & -1 & -1 \\ 1 & -2 & -1 \\ 1 & 1 & 1 \end{matrix} \qquad \begin{matrix} 1 & 1 & -1 \\ 1 & -2 & -1 \\ 1 & 1 & -1 \end{matrix} \qquad \begin{matrix} 1 & 1 & 1 \\ 1 & -2 & -1 \\ 1 & -1 & -1 \end{matrix}$$

S mask — SW mask — W mask — NW mask

Results:

Figure IOS.13a *Original building image.*

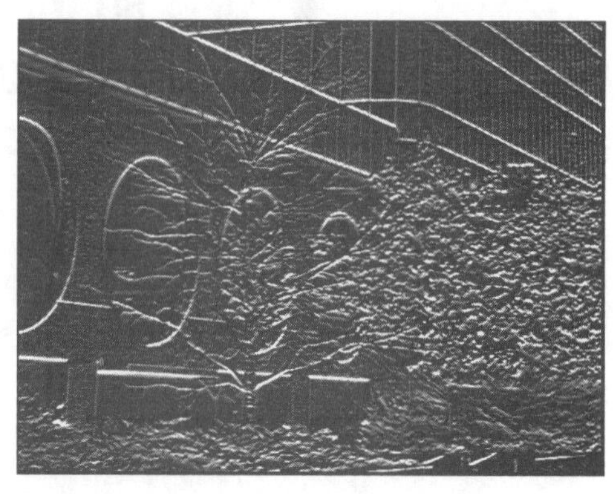

Figure IOS.13b *Prewitt northern direction gradient edge enhancement.*

Figure IOS.13c *Northeastern gradient.*

Figure IOS.13d *Eastern gradient.*

Figure IOS.13e *Southeastern gradient.*

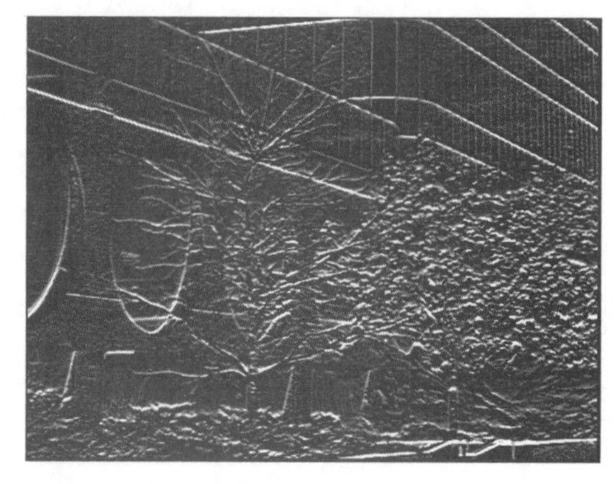

Figure IOS.13f *Southern gradient.*

Figure IOS.13g Southwestern gradient.

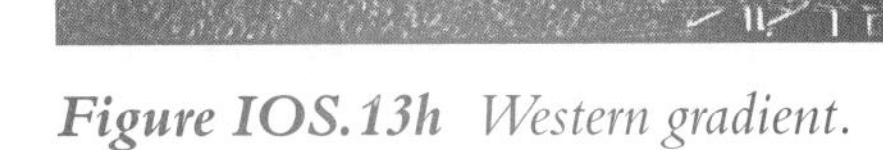

Figure IOS.13h Western gradient.

Figure IOS.13i Northwestern gradient.

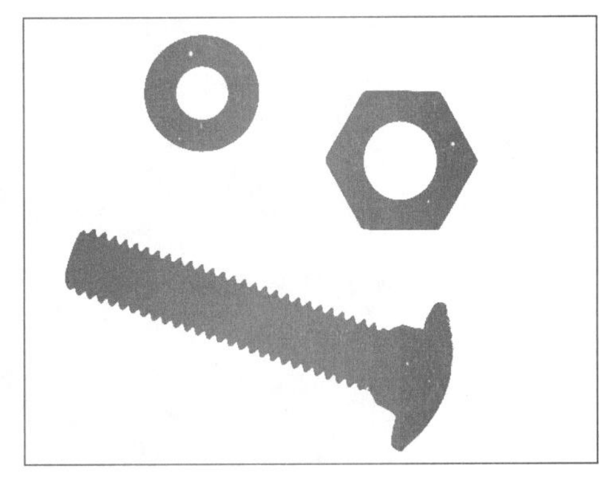

Figure IOS.13j Original binary image of washer, nut, and bolt.

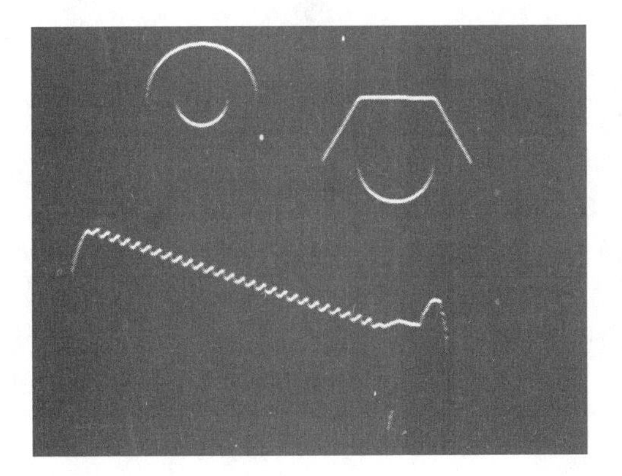

Figure IOS.13k Prewitt northern direction gradient edge enhancement.

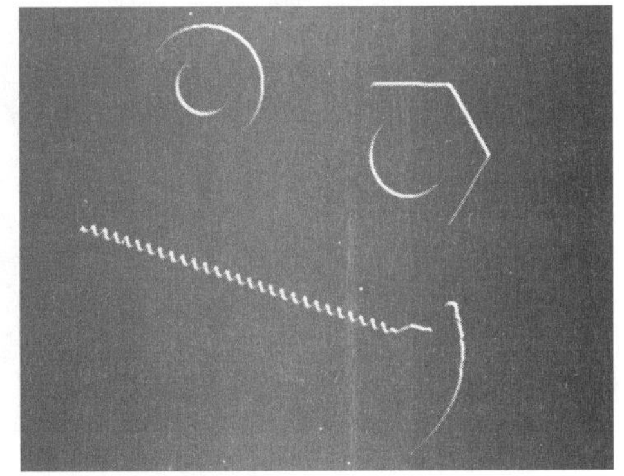

Figure IOS.13l Northeastern gradient.

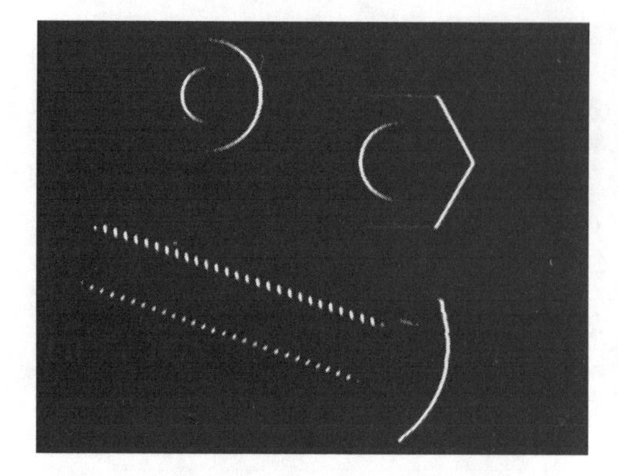

Figure IOS.13m *Eastern gradient.*

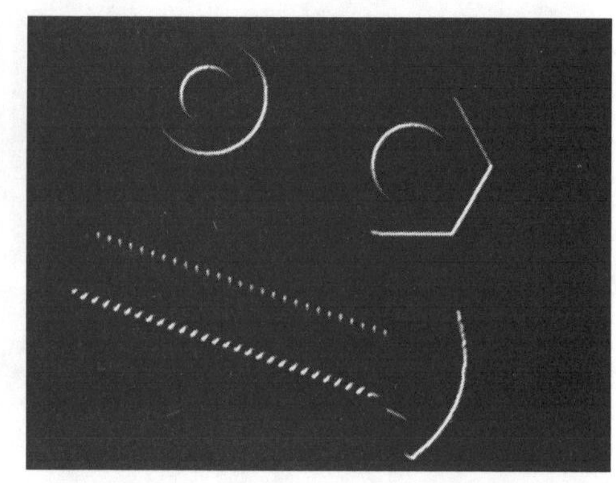

Figure IOS.13n *Southeastern gradient.*

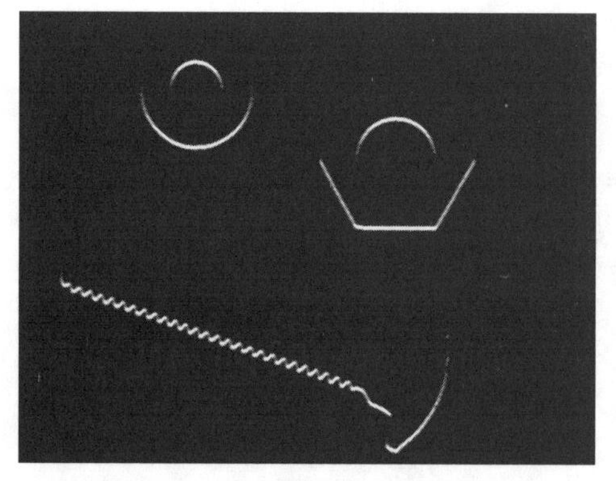

Figure IOS.13o *Southern gradient.*

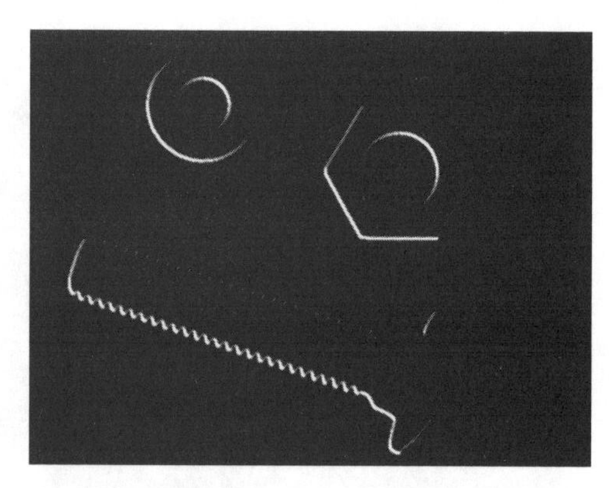

Figure IOS.13p *Southwestern gradient.*

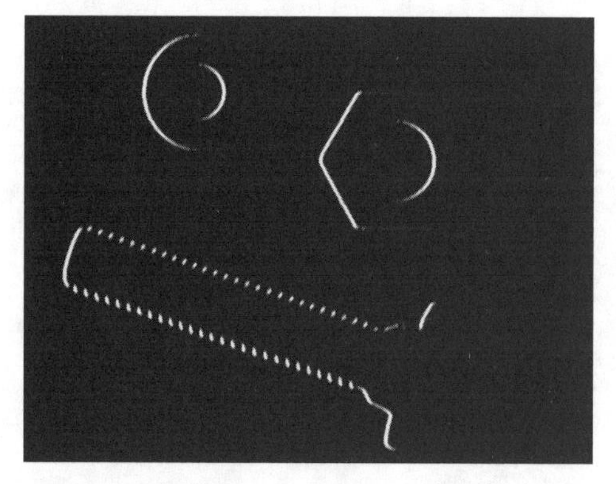

Figure IOS.13q *Western gradient.*

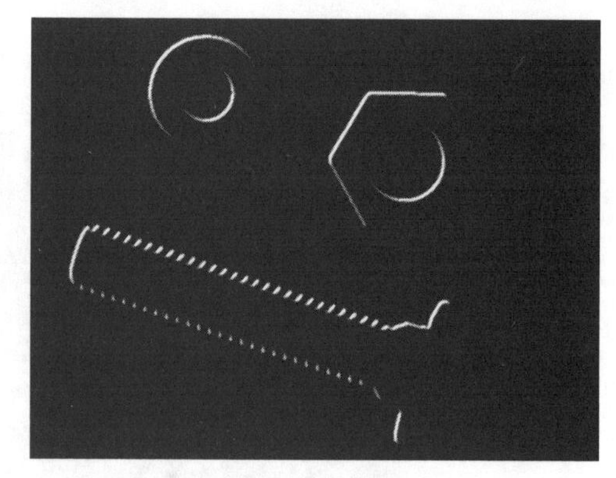

Figure IOS.13r *Northwestern gradient.*

14 Laplacian Edge Enhancement

Class: Image Enhancement/Restoration

Description: The Laplacian edge enhancement operation extracts all of the edges in an image, regardless of direction. The resulting image appears as an omnidirectional outline of the objects in the original image. Constant brightness regions become black, while changing brightness regions become highlighted.

Applications: The Laplacian edge enhancement operation is used to produce the edges of the objects in an image. The edges are any sharp brightness transitions rising from black to white or falling from white to black. This operation is omnidirectional, which means that all black-to-white and white-to-black edge transitions are highlighted, regardless of their directions in the image.

Implementation:

1. Define spatial convolution mask
2. Pixel group process

The Laplacian edge enhancement operation can produce results that are less than 0 or greater than 255. In these underflow and overflow cases, the resulting values are forced to 0 or 255, whichever is closest.

The mask coefficients add to 0. A large coefficient generally appears in the center of the mask, surrounded by smaller positive and negative coefficients. Three common Laplacian edge enhancement masks are

$$\begin{matrix} -1 & -1 & -1 \\ -1 & 8 & -1 \\ -1 & -1 & -1 \end{matrix} \qquad \begin{matrix} 0 & -1 & 0 \\ -1 & 4 & -1 \\ 0 & -1 & 0 \end{matrix} \qquad \begin{matrix} 1 & -2 & 1 \\ -2 & 4 & -2 \\ 1 & -2 & 1 \end{matrix}$$

Mask 1   Mask 2   Mask 3

Results:

Figure IOS.14a Original building image.

Figure IOS.14b Laplacian edge enhancement using mask 1.

Figure IOS.14c Laplacian using mask 2.

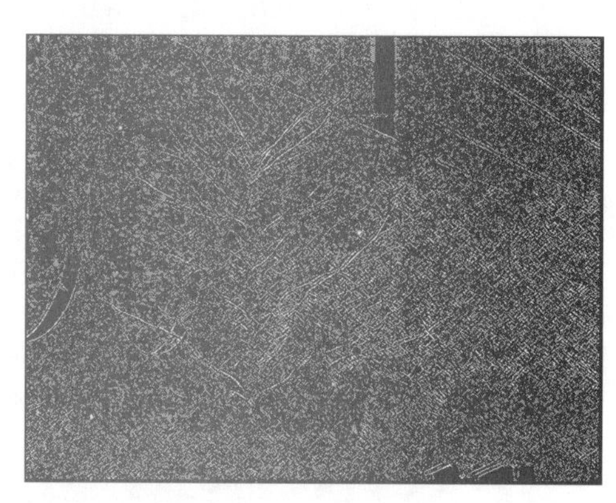

Figure IOS.14d Laplacian using mask 3.

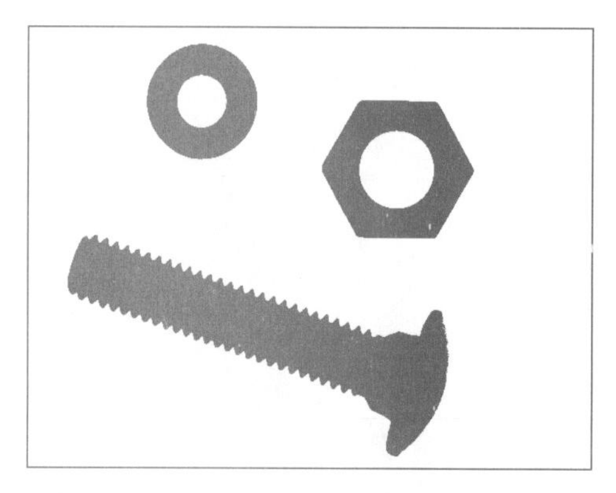

Figure IOS.14e Original binary image of washer, nut, and bolt.

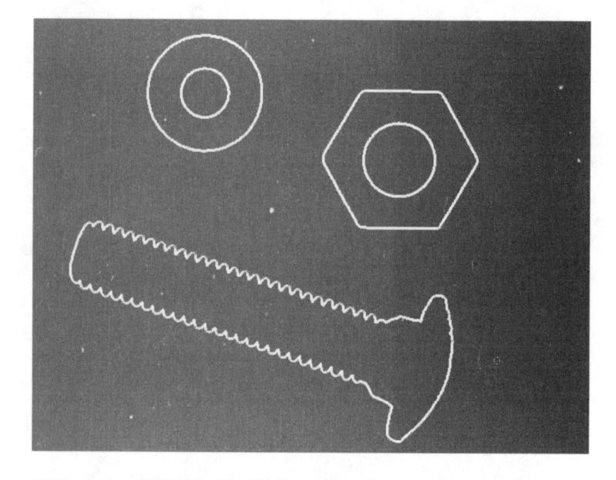

Figure IOS.14f Laplacian edge enhancement using mask 1.

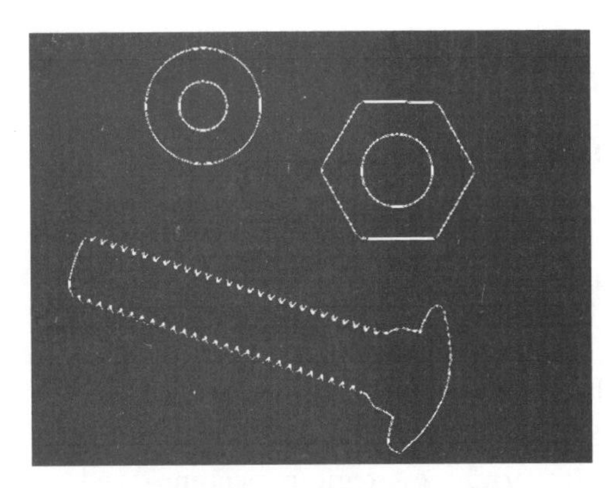

Figure IOS.14g *Laplacian using mask 2.*

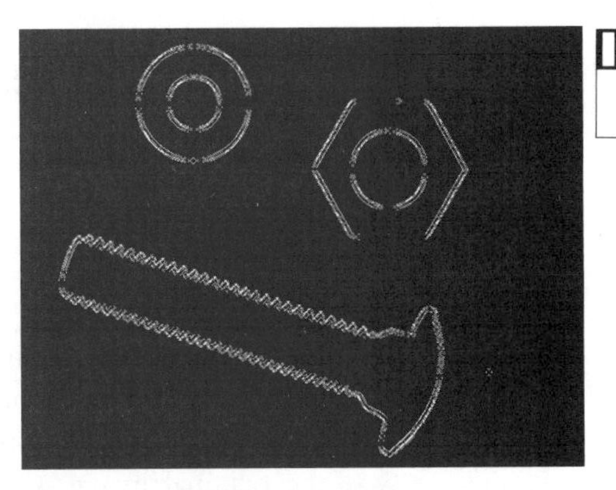

Figure IOS.14h *Laplacian using mask 3.*

15 Line Segment Enhancement

Class: Image Enhancement/Restoration

Description: The line segment enhancement operation enhances the vertical, horizontal, left-to-right diagonal, or right-to-left diagonal straight-line segments in an image. The resulting image appears as a directional outline of the objects in the original image. Constant brightness regions become black, while changing brightness regions become highlighted.

Applications: Line segment enhancement operations are used to produce the edges of the objects in an image. These operations produce edge-enhanced images where line segments are connected into more contiguous strings than other edge enhancement operations. Small gaps in edges tend to be filled by this operation. Multiple passes of the line segment enhancement operation further enhance the lines found in an image.

Implementation:

1. Define spatial convolution mask
2. Pixel group process

Effectively, the line segment enhancement operation is computed as the sum of two brightness slopes in an input pixel group. In vertical line segment enhancement, both the left-neighboring pixel and the right-neighboring pixel are subtracted from the center pixel. The two slopes are then summed. This operation is applied to the three rows of the input pixel group. The horizontal and diagonal line segment operations operate similarly.

The line segment enhancement operation can produce results that are less than 0 or greater than 255. In these underflow and overflow cases, the resulting values are forced to 0 or 255, whichever is closest.

The mask coefficients add to 0. The four line segment direction masks are

$$\begin{matrix} -1 & 2 & -1 \\ -1 & 2 & -1 \\ -1 & 2 & -1 \end{matrix} \qquad \begin{matrix} -1 & -1 & -1 \\ 2 & 2 & 2 \\ -1 & -1 & -1 \end{matrix} \qquad \begin{matrix} -1 & -1 & 2 \\ -1 & 2 & -1 \\ 2 & -1 & -1 \end{matrix} \qquad \begin{matrix} 2 & -1 & -1 \\ -1 & 2 & -1 \\ -1 & -1 & 2 \end{matrix}$$

Vertical mask   Horizontal mask   Left-to-right mask   Right-to-left mask

Results:

Figure IOS.15a *Original building image.*

Figure IOS.15b *Vertical line-segment enhancement.*

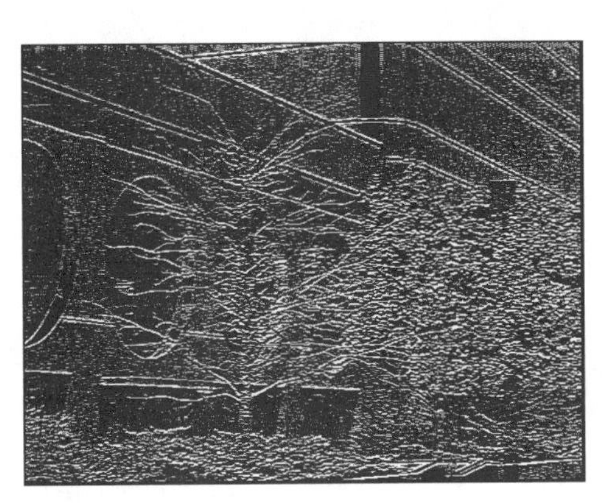

Figure IOS.15c *Horizontal line-segment enhancement.*

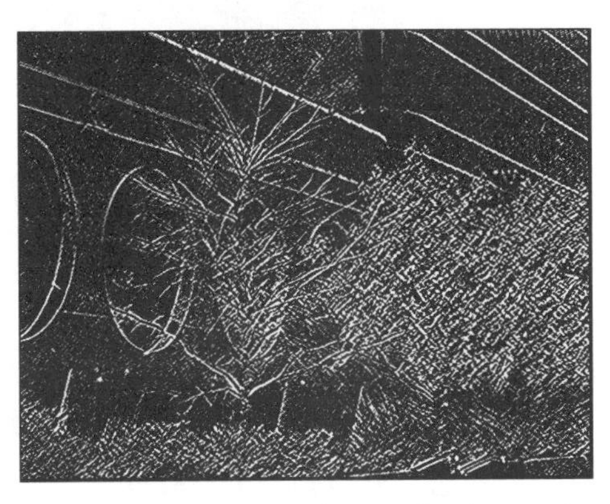

Figure IOS.15d *Left-to-right diagonal line-segment enhancement.*

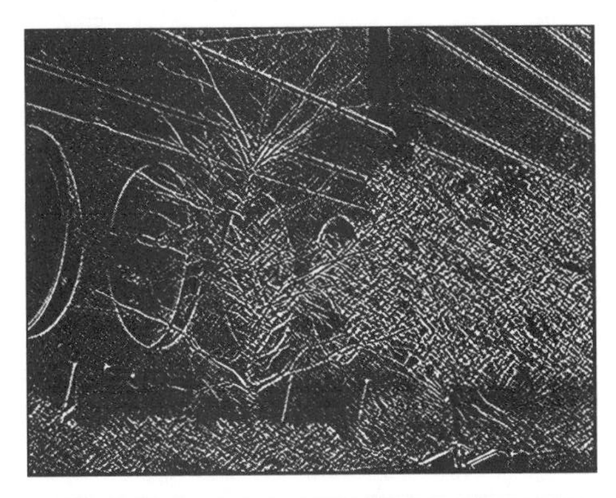

Figure IOS.15e *Right-to-left diagonal line-segment enhancement.*

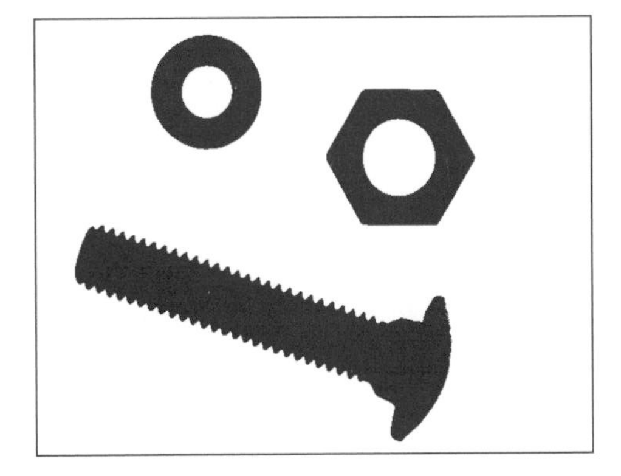

Figure IOS.15f *Original binary image of washer, nut, and bolt.*

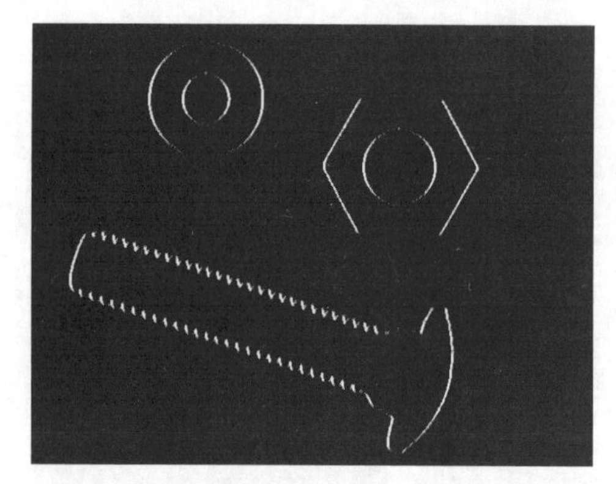

Figure IOS.15g *Vertical line-segment enhancement.*

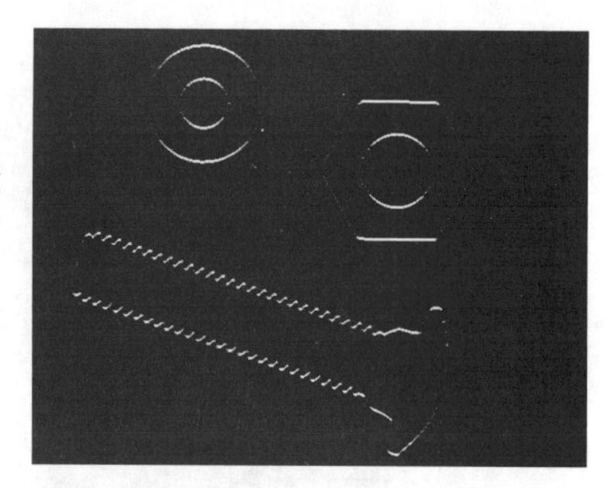

Figure IOS.15h *Horizontal line-segment enhancement.*

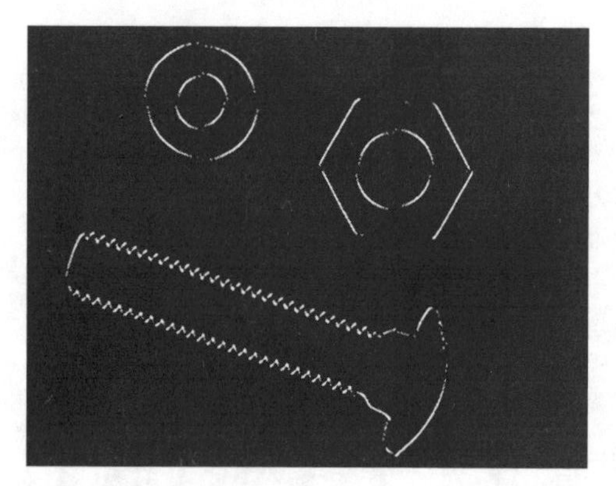

Figure IOS.15i *Left-to-right diagonal line-segment enhancement.*

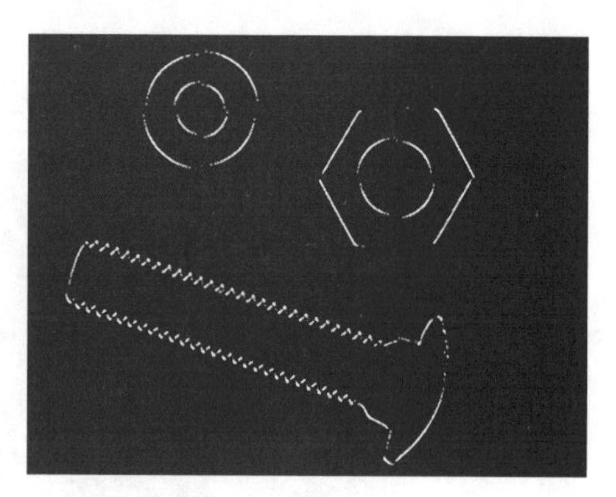

Figure IOS.15j *Right-to-left diagonal line-segment enhancement.*

16 Median and Ranking Filters

Class: Image Enhancement/Restoration

Description: The median filter is a ranking filter, where the fifth-ranked pixel brightness value is selected as the output pixel brightness from an input group of pixels. In a 3 × 3 group, the median-valued pixel brightness is the brightness where four pixels are brighter and four pixels are darker. The resulting image is free of pixel brightnesses that are at the extremes in each input group of pixels.

Other brightness ranks, such as maximum and minimum brightnesses, can also be used—see the discussions of gray-scale morphological operations.

Applications: The primary use of the median filter is in removing impulse noise spikes from an image. Impulse noise spikes appear as bright or dark pixels randomly distributed throughout the image. Because the spikes are significantly brighter or darker than their neighboring pixels, they generally end up at the top or bottom of the brightness ranking for a group of input pixels. When the median value is selected from the input pixel group, the bright or dark pixels are generally avoided, and hence eliminated in the output image.

Implementation:

1. Define brightness rank value
2. Nonlinear pixel group process

A nonlinear group process determines the brightness of the desired pixel-rank of the group of input pixels. In the case of the median filter, the fifth-ranked brightness is the output pixel value. The first-ranked brightness represents the maximum brightness of the group of input pixels; the ninth-ranked brightness represents the minimum value.

Brightness values

10	20	5
15	210	20
15	10	20

→

Brightness values in ascending order

5	10	10	15	15	20	20	20	210	
↑	↑	↑	↑	↑	↑	↑	↑	↑	
1st	2nd	3rd	4th	5th	6th	7th	8th	9th	← ranks
↑				↑				↑	
minimum				median				maximum	

Results:

Figure IOS.16a Original building image corrupted with impulse noise spikes.

Figure IOS.16b Median filtered image showing the elimination of all single-pixel impulses.

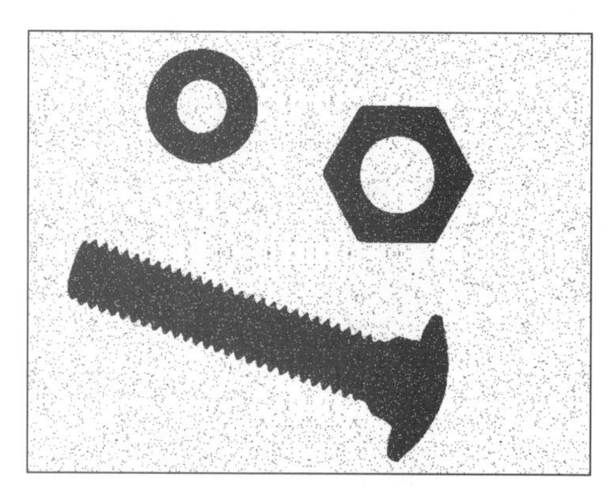

Figure IOS.16c Original binary image of washer, nut, and bolt corrupted with impulse noise spikes.

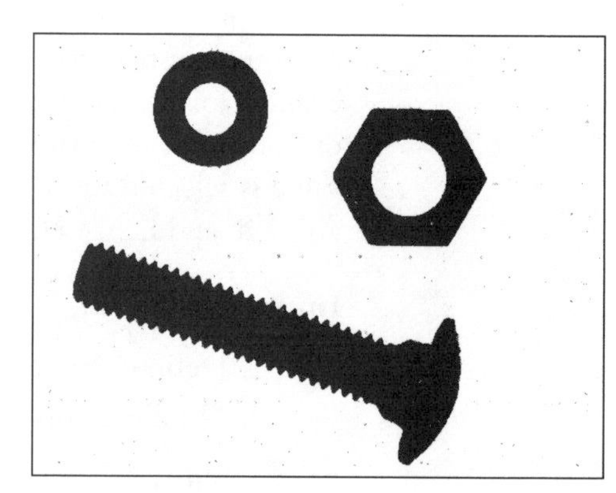

Figure IOS.16d Median filtered image showing the elimination of all single-pixel impulses.

17 Fourier Frequency Transform Filtering

Class: Image Enhancement/Restoration

Description: An image is transformed from its original spatial-domain brightness representation to a frequency-domain representation by a discrete Fourier transform operation. Desired frequency components are eliminated from the frequency image. The frequency image is then transformed back to the spatial domain. The resulting image is devoid of the removed image frequency components.

Applications: Frequency transform filtering operations are used for selective removal of periodic noise patterns from an image. Noise patterns of this type show as bright spots in the frequency image. Masking the bright spots eliminates the frequency from the frequency image, hence eliminating the noise in the spatial-domain image. Of course, actual image information at the masked frequencies is also removed.

Implementation:

1. Fast Fourier transform
2. Define frequency mask image using window
3. Dual-image pixel point process—multiply
4. Inverse fast Fourier transform

Because the Fourier transform models image spatial frequencies as though they were periodic, a windowing function should be used both before the transform and during the frequency masking process. The windowing function smoothly tapers the pixel brightnesses, removing any discontinuities that may lead to erroneous components in the transformed or inverse-transformed image. Appropriate windows include the triangular, Gaussian, Hamming, and Von Hann windows.

The discrete Fourier transform is typically implemented with the fast Fourier transform algorithm because of that algorithm's computational efficiency. Additionally, numerous other frequency transforms exist, such as the Hadamard, Haar, slant, Karhunen-Loeve, sine, and cosine transforms.

Results:

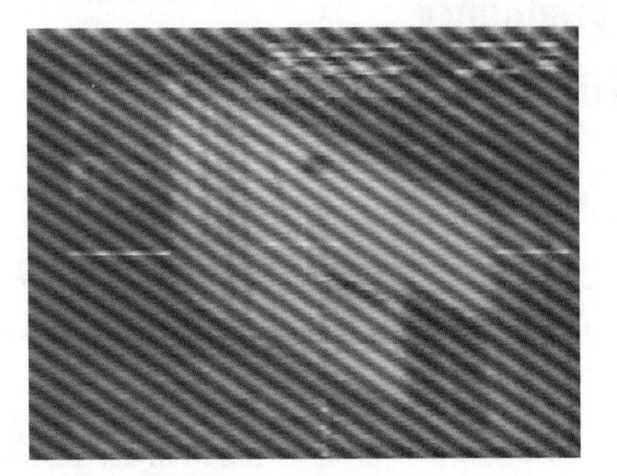

Figure IOS.17a Original image from the nose of an air-to-ground missile corrupted with periodic noise bands.

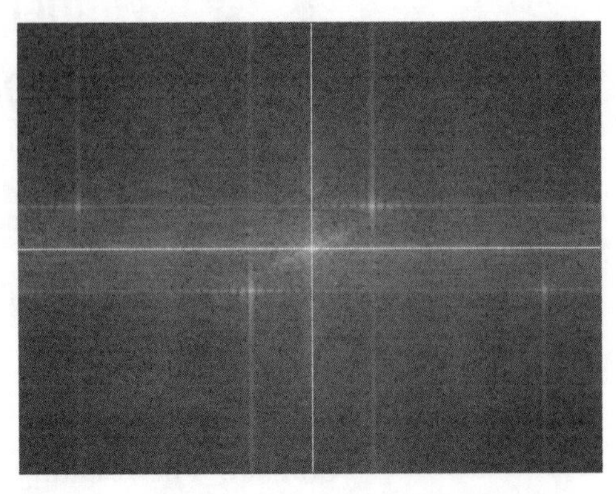

Figure IOS.17b Fourier transform of original image—the periodic noise shows as bright spots in the frequency image.

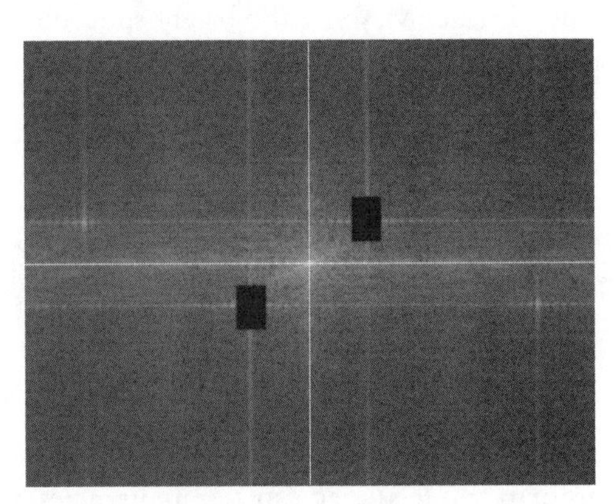

Figure IOS.17c Masked Fourier-transform frequency image with the noise spots eliminated.

Figure IOS.17d Inverse Fourier-transformed image showing the periodic noise removed—the image of a building with a window on the right and a vertical fire escape on the left becomes evident.

18 Translation Transformation

Class: Image Enhancement/Restoration

Description: The translation transformation shifts the spatial location of image pixels linearly from side to side and up and down. Pixels are translated by the distance T_x in the x direction and T_y in the y direction.

Applications: Translation transformations are used to register multiple images of the same or different scenes geometrically. Geometric adjustments from side to side and up and down can be carried out before a combination operation such as image addition, subtraction, division, or compositing.

The translation transformation is also used to correct geometric distortions created in the image acquisition process.

Implementation:

1. Define x and y translation values
2. Geometric transformation—translation

Translation transformations can be performed by a source-to-target mapping or target-to-source mapping. When noninteger translation values are used, interpolation techniques must be used to estimate intermediate pixel brightnesses. Appropriate interpolation is generally accomplished using the nearest neighbor or the bilinear method.

The translation transformation source-to-target mapping is

$$I(x,y) \rightarrow O(x',y')$$

where

$$x' = x + T_x$$

and

$$y' = y + T_y$$

Results:

Figure IOS.18a *Original scaled building image.*

Figure IOS.18b *Translation of (–50,–40).*

Figure IOS.18c *Translation of (50,40).*

19 Rotation Transformation

Class: Image Enhancement/Restoration

Description: The rotation transformation rotates the spatial location of image pixels linearly about the (0,0) point. Pixels are rotated clockwise by the angle θ.

Applications: Rotation transformations are used to geometrically register multiple images of the same or different scenes. Geometric adjustments of rotation about a point can be carried out prior to a combination operation such as image addition, subtraction, division, or compositing.

The rotation transformation is also used to correct geometric distortions created in the image acquisition process.

Implementation:

1. Define θ rotation angle
2. Geometric transformation—rotation

Rotation transformations can be performed using a source-to-target mapping or target-to-source mapping. When rotation angles are not a multiple of 90°, interpolation techniques must be used to estimate intermediate pixel brightnesses. Appropriate interpolation is generally accomplished using the nearest neighbor or the bilinear method.

The rotation transformation source-to-target mapping is

$$I(x,y) \rightarrow O(x',y')$$

where

$$x' = x\cos\theta + y\sin\theta$$

and

$$y' = -x\sin\theta + y\cos\theta$$

Results:

Figure IOS.19a *Original scaled building image.*

Figure IOS.19b *Rotation of 28° using nearest neighbor interpolation.*

Figure IOS.19c *Rotation of 28° using bilinear interpolation.*

Figure IOS.19d *Rotation of 200° using nearest neighbor interpolation.*

Figure IOS.19e *Rotation of 200° using bilinear interpolation.*

20 Scaling Transformation

Class: Image Enhancement/Restoration

Description: The scaling transformation linearly expands or reduces the spatial size of image pixels. Pixels are scaled by the factor S_x in the x direction and S_y in the y direction.

Applications: Scaling transformations are used to geometrically register multiple images of the same or different scenes. Geometric adjustments of size can be carried out prior to a combination operation such as image addition, subtraction, division, or compositing.

The scaling transformation is also used to correct geometric distortions created in the image acquisition process.

Implementation:

1. Define x and y scale values
2. Geometric transformation—scale

Scaling transformations can be performed using a source-to-target mapping or target-to-source mapping. When scale values greater than 1 are used, interpolation techniques must be used to estimate intermediate pixel brightnesses. Appropriate interpolation is generally accomplished using the nearest neighbor or the bilinear method. When scale values less than 1 are used, low-pass filter downsampling techniques must be used to avoid aliasing artifacts.

The scaling transformation source-to-target mapping is

$$I(x,y) \rightarrow O(x',y')$$

where

$$x' = xS_x$$

and

$$y' = yS_y$$

Results:

Figure IOS.20a *Original building image.*

Figure IOS.20b *Scaling by a factor of 0.2 without low-pass filter downsampling—notice the aliasing in the upper right of the image.*

Figure IOS.20c *Scaling by a factor of 0.2 with low-pass filter downsampling—aliasing artifacts are diminished.*

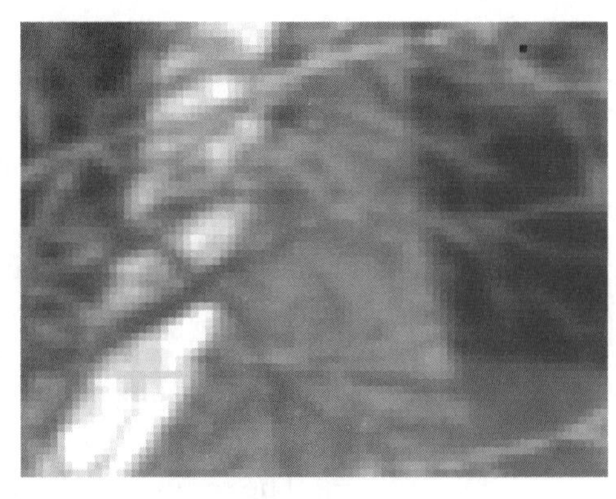

Figure IOS.20d *Scaling by a factor of eight using nearest neighbor interpolation.*

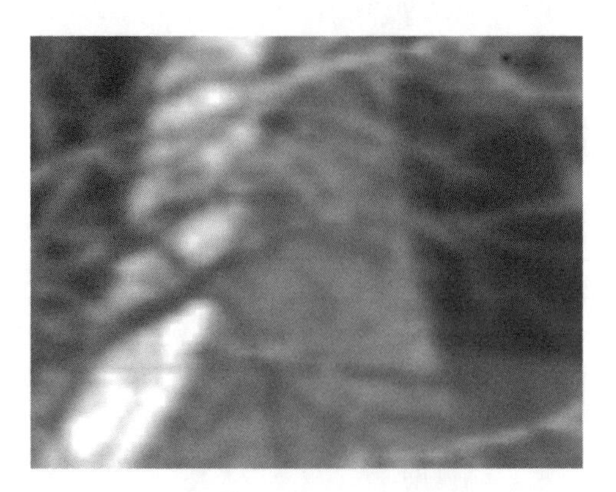

Figure IOS.20e *Scaling by a factor of eight using bilinear interpolation.*

21 Warping Transformation

Class: Image Enhancement/Restoration

Description: The warping transformation translates, rotates, and scales the spatial attributes of image pixels nonlinearly. Warping transformations are controlled by control points to stretch and pull the image to desired extents, as if the image were on a rubber sheet.

Applications: Warping transformations are used to geometrically register multiple images of the same or different scenes. Geometric adjustments of translation, rotation, scaling, and curvature can be carried out prior to a combination operation such as image addition, subtraction, division, or compositing.

The warping transformation is also used to correct geometric distortions created in the image acquisition process.

Implementation:

1. Define warping polynomial coefficients
2. Geometric transformation—warp

Warping transformations can be performed using a source-to-target mapping or target-to-source mapping. Generally, all warping transformations include some form of rotation and scaling, and therefore must use interpolation and low-pass filter downsampling techniques to estimate intermediate pixel brightnesses and avoid aliasing artifacts. Appropriate interpolation is generally accomplished using the nearest neighbor or the bilinear method.

The warping transformation source-to-target mapping for a third-order warp is

$$I(x,y) \rightarrow O(x',y')$$

where

$$x' = a_9x^3 + a_8y^3 + a_7x^2y + a_6y^2x + a_5x^2 + a_4y^2 + a_3x + a_2y + a_1xy + a_0$$

and

$$y' = b_9x^3 + b_8y^3 + b_7x^2y + b_6y^2x + b_5x^2 + b_4y^2 + b_3x + b_2y + b_1xy + b_0$$

Results:

Figure IOS.21a *Original building image.*

Figure IOS.21b *Warping of original image using bilinear interpolation and the corners as control points.*

Figure IOS.21c *Complex warping of original image, also using bilinear interpolation.*

22 Binary Erosion and Dilation

Class: Image Analysis

Description: The binary erosion operation uniformly reduces the size of white objects on a black background in an image. Using the omnidirectional structuring element, the reduction is by one pixel around the object's perimeter.

The binary dilation operation uniformly increases the size of white objects on a black background in an image. Using the omnidirectional structuring element, the object's size increases by one pixel around the object's perimeter.

Applications: The binary erosion operation is used to remove small anomalies such as single-pixel objects, called *speckle*, and single-pixel-wide spurs from an image. Multiple application of the erosion operation shrinks touching objects until they finally separate. This can be useful prior to object-counting operations.

Binary dilation operations are used to remove small anomalies, such as single-pixel holes in objects and single-pixel-wide gaps from an image. Multiple application of the dilation operation expands broken objects until they finally merge into one. This can be useful prior to object-counting operations.

Implementation:

1. Define morphological structuring element
2. Binary morphological group process

Structuring elements for omnidirectional, horizontal, vertical, and diagonal erosion using the hit or miss transformation are

1	1	1		X	X	X		X	1	X		X	X	1
1	1	1		1	1	1		X	1	X		X	1	X
1	1	1		X	X	X		X	1	X		1	X	X
Omnidirectional element				Horizontal element				Vertical element				Diagonal element		

Structuring elements for omnidirectional, horizontal, vertical, and diagonal dilation using the hit or miss transformation are

0	0	0		X	X	X		X	0	X		X	X	0
0	0	0		0	0	0		X	0	X		X	0	X
0	0	0		X	X	X		X	0	X		0	X	X
Omnidirectional element				Horizontal element				Vertical element				Diagonal element		

The effects of the binary erosion and dilation operations are exactly the reverse for black objects on a white background.

Results:

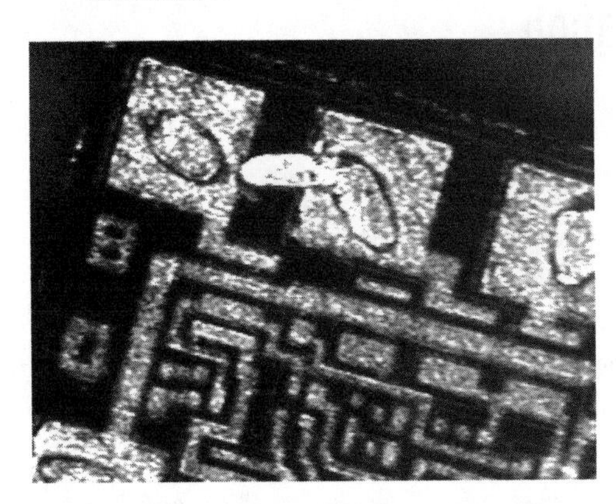

Figure IOS.22a *Original semiconductor chip image.*

Figure IOS.22b *Binary image after binary contrast enhancement operation.*

Figure IOS.22c *Eroded image.*

Figure IOS.22d *Erosion operation applied two times.*

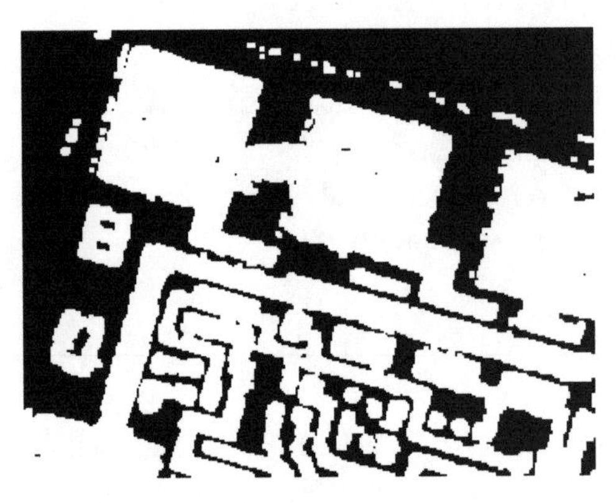

Figure IOS.22e *Dilated image.*

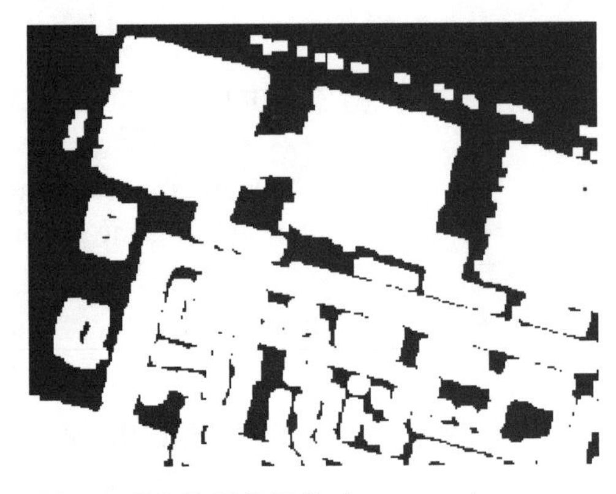

Figure IOS.22f *Dilation operation applied two times.*

23 Binary Opening and Closing

Class: Image Analysis

Description: The binary opening operation is a binary erosion operation followed by a binary dilation operation. Objects generally remain their original size.

The binary closing operation is a binary dilation operation followed by a binary erosion operation. Objects generally remain their original size.

Applications: In the binary opening operation, the erosion operation first eliminates small anomalies such as single-pixel objects and single-pixel-wide spurs from an image. The erosion operation also reduces object sizes. The following dilation operation then roughly expands objects back to their original size. The binary opening operation is useful to clean up images with noise and other small anomalies.

In the binary closing operation, the dilation operation first eliminates small anomalies such as single-pixel holes and single-pixel-wide gaps. The dilation operation also increases object sizes. The following erosion operation then roughly reduces objects back to their original size. The binary closing operation is useful to clean up images with object holes and other small anomalies.

Implementation:

1. Define morphological erosion (or dilation) structuring element
2. Binary morphological group process
3. Define morphological dilation (or erosion) structuring element
4. Binary morphological group process

Binary erosion and dilation structuring elements such as the ones for omnidirectional, horizontal, vertical, and diagonal erosion can be used, as well as others. Both the erosion and dilation operations used in a single opening or closing operation generally use the same structuring element.

To apply multiple opening operations, apply the multiple erosion operations first, followed by the multiple dilation operations. For multiple closing operations, the dilation operations are applied first, followed by the erosion operations.

The effects of the binary opening and closing operations are exactly the reverse for black objects on a white background.

Results:

Figure IOS.23a Original binary "chip" image.

Figure IOS.23b Opened image.

Figure IOS.23c Closed image.

24 Outlining

Class: Image Analysis

Description: The outlining operation operates on binary images by subtracting an eroded version of the original image from the original image. The resulting image shows a single-pixel-wide outline of the objects in the original image.

Applications: The outlining operation is useful as a precursor to object measurement operations. It produces the outline of the objects in an image. Using the omnidirectional erosion structuring element, this operation produces outlines of all objects, regardless of their orientation.

Implementation:

1. Define erosion morphological structuring element
2. Binary morphological group process
3. Dual-image pixel point process—subtraction

Alternately, other erosion structuring elements can be used, such as the ones for horizontal, vertical, and diagonal erosion. Use of these structuring elements produces directional object outlines.

Multiple erosion operations can be applied prior to subtracting the eroded image from the original. Additional erosions produce a resulting image with wider outline widths.

Results:

Figure IOS.24a *Original binary "chip" image.*

Figure IOS.24b *Eroded image.*

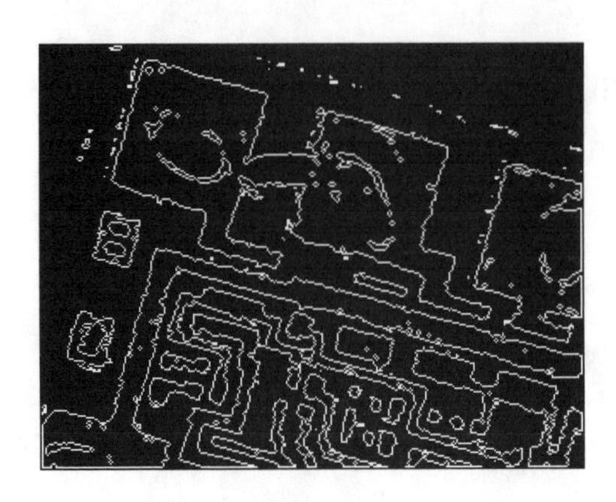

Figure IOS.24c *Original image minus eroded image yielding an outline image.*

25 Gray-Scale Erosion and Dilation

Class: Image Analysis

Description: The gray-scale erosion operation reduces the brightness (and, therefore, the size) of bright objects on a dark background in an image.

The gray-scale dilation operation increases the brightness (and, therefore, the size) of bright objects on a black background in an image.

Applications: The gray-scale erosion operation is used to eliminate small anomalies such as single-pixel bright spots from an image. Multiple application of the erosion operation shrinks touching objects by darkening their perimeters until they finally separate. This can be useful prior to object-counting operations.

Gray-scale dilation operations are used to eliminate small anomalies such as single-pixel dark spots from an image. Multiple application of the dilation operation expands broken objects by brightening their perimeters until they finally merge into one. This can be useful prior to object-counting operations.

Implementation:

1. Define morphological structuring element
2. Gray-scale morphological group process

Structuring elements for omnidirectional, horizontal, vertical, and diagonal gray-scale erosion and dilation are

Omnidirectional element	Horizontal element	Vertical element	Diagonal element
0 0 0	X X X	X 0 X	X X 0
0 0 0	0 0 0	X 0 X	X 0 X
0 0 0	X X X	X 0 X	0 X X

For the erosion operation, the output value is the minimum brightness value of the input pixel group summed with the corresponding structuring element values. For the dilation operation, the maximum brightness value is the output value.

Results:

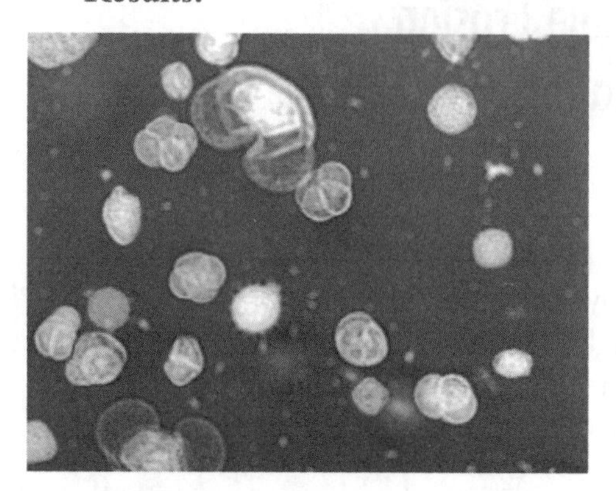

Figure IOS.25a *Original microscope specimen image of various pollens appearing as bright objects.*

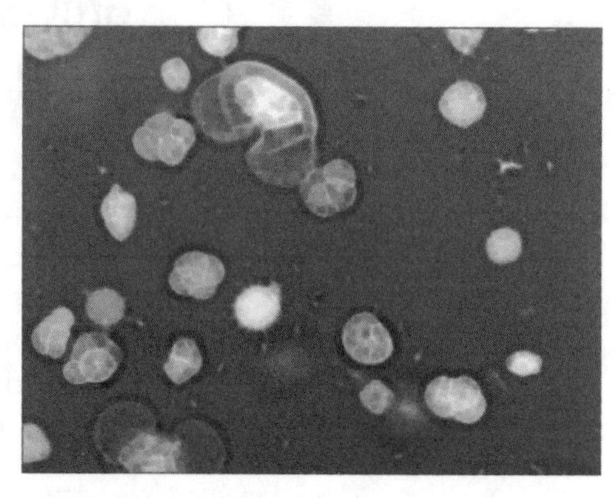

Figure IOS.25b *Eroded gray-scale image.*

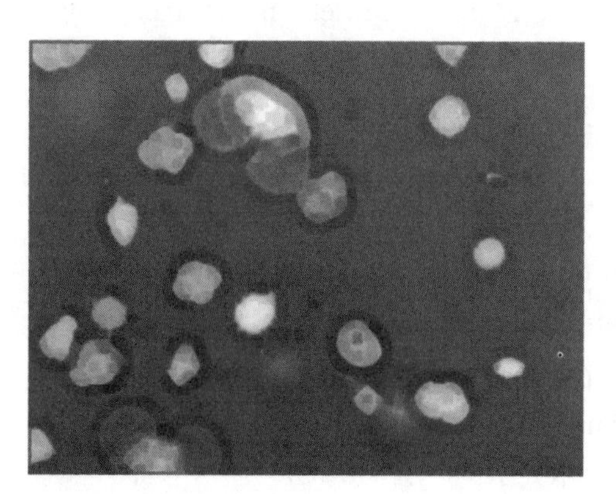

Figure IOS.25c *Erosion operation applied two times.*

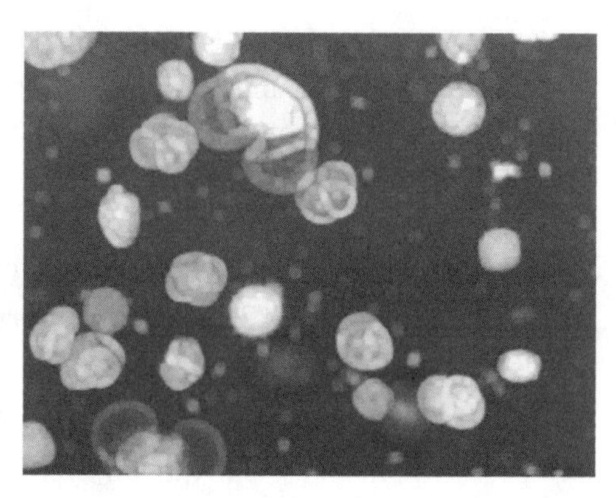

Figure IOS.25d *Dilated gray-scale image.*

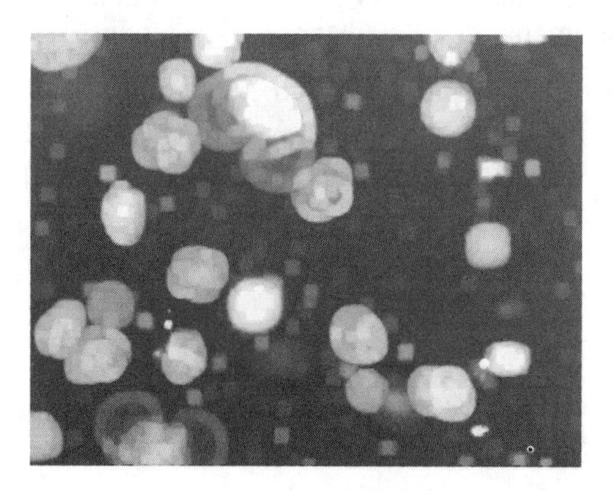

Figure IOS.25e *Dilation operation applied two times.*

26 Gray-Scale Opening and Closing

Class: Image Analysis

Description: The gray-scale opening operation is a gray-scale erosion operation followed by a gray-scale dilation operation. Objects tend to remain their original size.

The gray-scale closing operation is a gray-scale dilation operation followed by a gray-scale erosion operation. Objects tend to remain their original size.

Applications: In the gray-scale opening operation, the erosion operation first eliminates small anomalies such as single-pixel bright spots from an image. The erosion operation also reduces object brightnesses. The following dilation operation then roughly increases object brightnesses back to their original levels. The gray-scale opening operation is useful to clean up images with noise and other small anomalies.

In the gray-scale closing operation, the dilation operation first eliminates small anomalies such as single-pixel dark spots. The dilation operation also increases object brightnesses. The following erosion operation then roughly reduces object brightnesses back to their original levels. The gray-scale closing operation is useful to clean up images with object holes and other small anomalies.

Implementation:

1. Define morphological erosion (or dilation) structuring element
2. Gray-scale morphological group process
3. Define morphological dilation (or erosion) structuring element
4. Gray-scale morphological group process

Gray-scale erosion and dilation structuring elements such as the ones for omnidirectional, horizontal, vertical, and diagonal erosion can be used, as well as others. Both the erosion and dilation operations used in a single opening or closing operation generally use the same structuring element.

When applying multiple opening operations, apply the multiple erosion operations first, followed by the multiple dilation operations. For multiple closing operations, apply the dilation operations first, followed by the erosion operations.

Results:

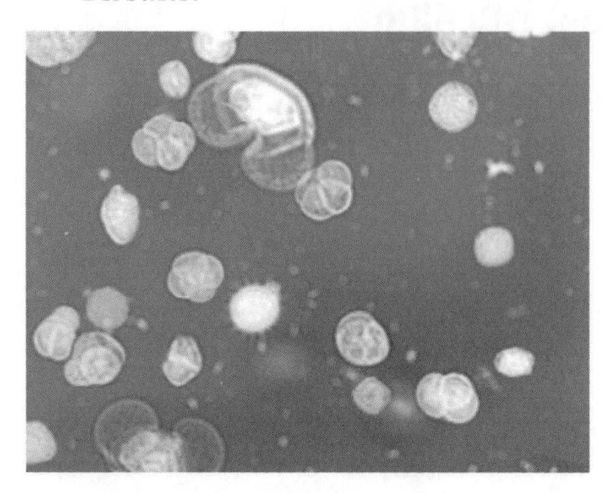

Figure IOS.26a *Original "pollen" image.*

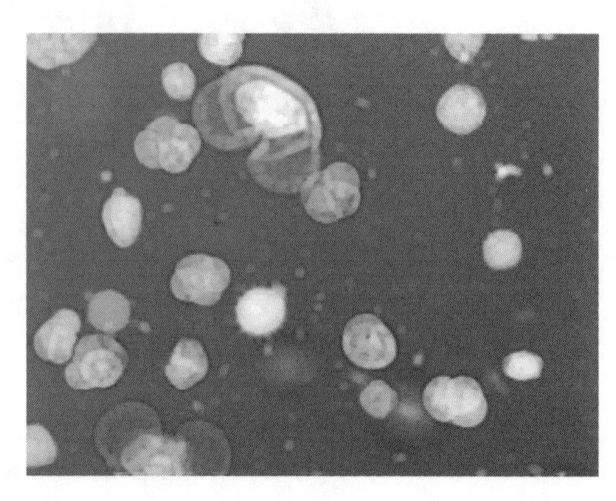

Figure IOS.26b *Opened gray-scale image.*

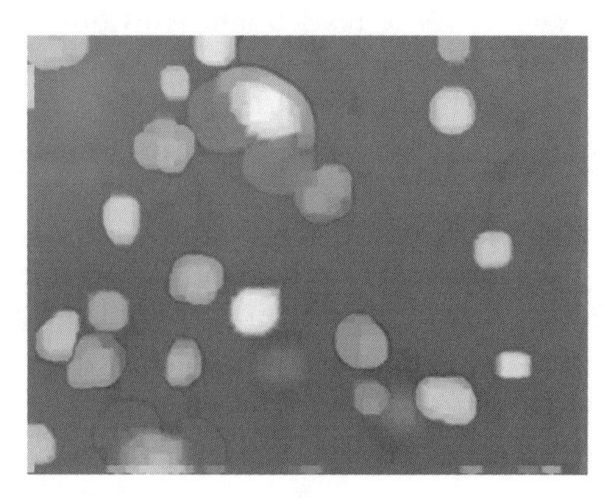

Figure IOS.26c *Opening operation applied four times.*

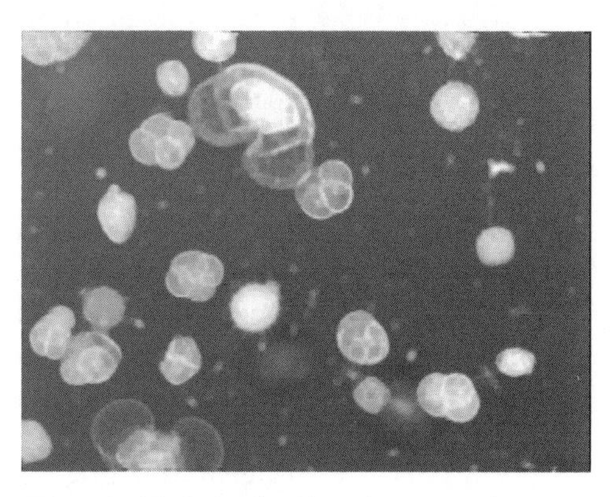

Figure IOS.26d *Closed gray-scale image.*

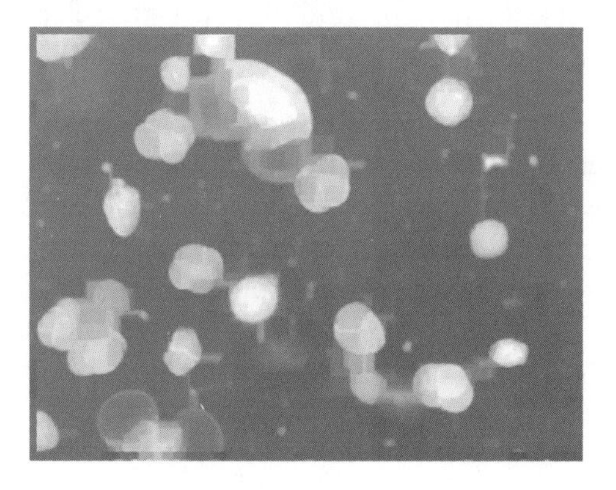

Figure IOS.26e *Closing operation applied four times.*

27 Morphological Gradient

Class: Image Analysis

Description: The morphological gradient operation operates on gray-scale images. Both eroded and dilated versions of the original image are created. Then the eroded version of the original image is subtracted from the dilated image. The resulting image shows the edges of the objects in the original image.

Applications: The morphological gradient operation is used to produces the edges of the objects in an image. Using the omnidirectional erosion and dilation structuring element, this operation produces edges of all objects, regardless of their orientation.

Implementation:

1. Define dilation morphological structuring element
2. Gray-scale morphological group process
3. Define erosion morphological structuring element
4. Gray-scale morphological group process
5. Dual-image pixel point process—subtraction

Alternately, other erosion and dilation structuring elements can be used, such as the ones for horizontal, vertical, and diagonal operations. Use of these structuring elements produces directional object edges.

Multiple erosion and dilation operations can be applied prior to subtracting the eroded image from the dilated image. Additional erosions and dilations produce a resulting image with wider edge widths.

Results:

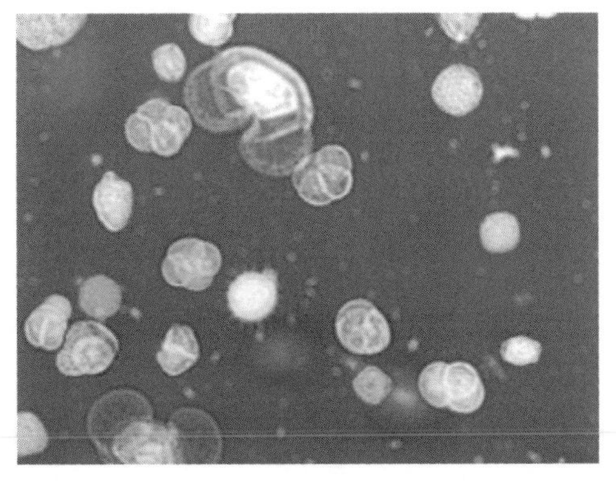

Figure IOS.27a *Original "pollen" image.*

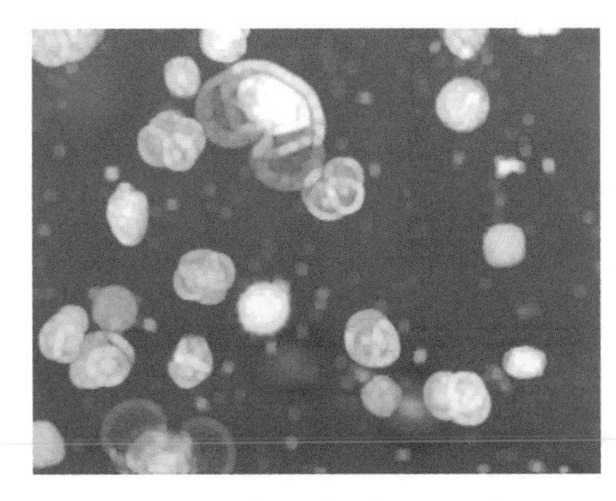

Figure IOS.27b *Dilated gray-scale image.*

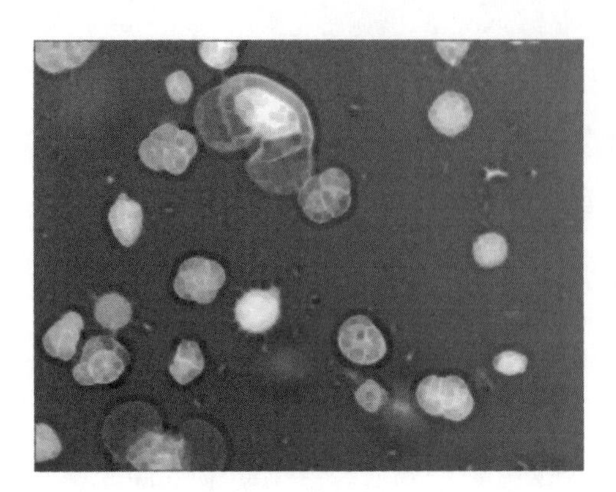

Figure IOS.27c *Eroded gray-scale image.*

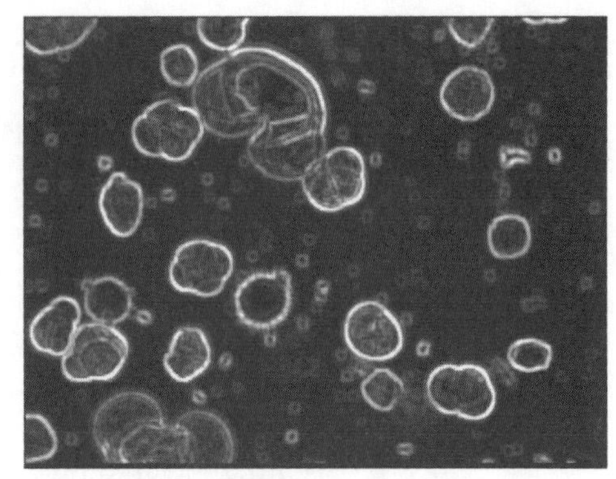

Figure IOS.27d *Dilated image minus eroded image, yielding a morphological gradient image.*

28 Object Shape Measures

Class: Image Analysis

Description: Imaged objects are measured to describe their characteristic attributes. Measures such as object perimeter, area, length, and width can often be sufficient for most classification tasks. The list of potential measures is endless and is generally dictated by the requirements of the application.

Applications: Once objects in an image are highlighted and separated with feature segmentation techniques, each object is be measured in some way prior to classification. The measures are chosen to enable the easiest discrimination between object characteristics of interest. For example, if a part can be classified as good or bad by its area alone, then a simple count of object pixels may be sufficient for classification purposes.

Object measures enable the automated understanding of images by machine. Careful selection of measurement parameters can provide expedient processing while maintaining required accuracy and consistency.

Implementation:

1. Preprocess image to highlight and separate imaged objects
2. Measure object parameters by counting appropriate object pixels or computing appropriate parameters

The performance of any shape measurements depends on the quality of the original image and how well objects are preprocessed. Object degradations such as small gaps, spurs, and noise can lead to poor measurement results, and ultimately to misclassifications.

IOS 28

Results:

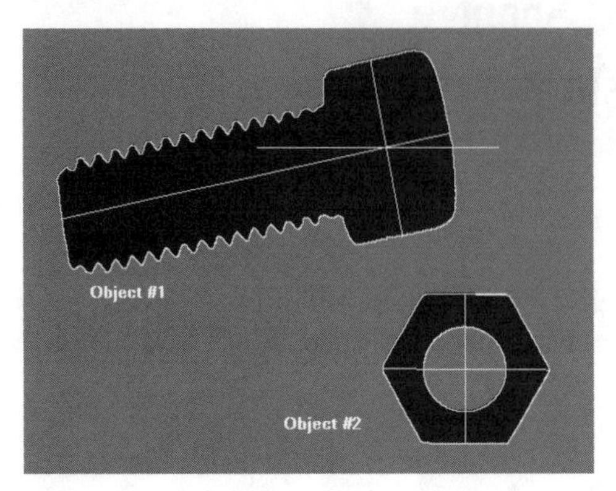

Figure IOS.28 *Binary image of bolt (object number 1) and nut (object number 2), overlaid with outlines and major and minor axes.*

Object #1 shape measures	
Perimeter:	1,397 pixels
Area:	50,970 pixels
Major axis length:	408 pixels
Major axis angle:	13°
Minor axis width:	177 pixels
Number of holes:	0
Total hole area:	N/A

Object #2 shape measures	
Perimeter:	504 pixels
Area:	8,135 pixels
Major axis length:	171 pixels
Major axis angle:	0°
Minor axis width:	147 pixels
Number of holes:	1
Total hole area:	5,430 pixels

29 Line Segment Boundary Description

Class: Image Analysis

Description: Imaged objects are measured to describe their characteristic attributes. When the classification task requires a specific understanding of an object's shape, object boundary descriptions are essential. Variable-length line segment boundary descriptions break the perimeter of an object into discrete line segments and record the length and angle of each segment relative to the previous line segment.

Applications: Object boundary descriptions are typically the ultimate means to describe arbitrarily-shaped object characteristics. The variable-length line segment boundary description enables the automated understanding of imaged objects by machine. With this boundary description technique, the object's perimeter can be concisely and accurately represented. The representation can be compared with stored representations to determine classification.

Implementation:

1. Preprocess image to highlight and separate imaged objects
2. Divide objects into fixed-length line segments
3. Combine line segments with similar angles
4. Record resulting line lengths and relative angles

The performance of any boundary description operation depends on the quality of the original image and how well objects are preprocessed. Object degradations such as small gaps, spurs, and noise can lead to poor boundary descriptions, and ultimately to misclassifications.

Results:

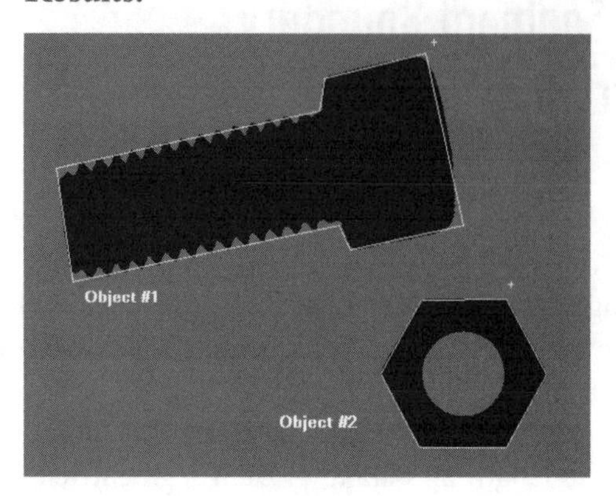

Figure IOS.29 *Binary image of bolt (object number 1) and nut (object number 2), overlaid with variable line segment outlines.*

Object #1 line segments	
Segment #1	108 pixels, 198°
Segment #2	33 pixels, 265°
Segment #3	280 pixels, 196°
Segment #4	112 pixels, 280°
Segment #5	278 pixels, 17°
Segment #6	40 pixels, 290°
Segment #7	115 pixels, 17°
Segment #8	175 pixels, 108°

Object #2 line segments	
Segment #1	86 pixels, 180°
Segment #2	83 pixels, 240°
Segment #3	81 pixels, 300°
Segment #4	83 pixels, 0°
Segment #5	86 pixels, 60°
Segment #6	85 pixels, 120°

30 Gray-Level Classification

Class: Image Analysis

Description: Each image pixel is determined to be inside or outside a number of brightness ranges, where each range is defined by low and high brightness threshold values. If the pixel brightness is between two thresholds, the resulting pixel brightness is set to a corresponding predetermined brightness value. If the pixel brightness is outside all threshold ranges, the resulting brightness is set to 0.

Applications: Gray-level classification operations can be used to separate objects with different gray levels into individual classes. This operation separates objects not only from the background, but also from one another. The gray-level classification operation works best when objects occupy distinct brightness ranges with clear delineations seen in the histogram. If this is not the case, pixels can be misclassified.

Implementation:

1. Define mapping function
2. Single-image pixel point process

The gray-level classification operation uses the following mapping function (for a four-class classification) to compute resulting pixel brightnesses:

$$O(x,y) = 64 \text{ if } I(x,y) \geq \textit{low threshold value}_1 \text{ and if } I(x,y) < \textit{high threshold value}_1$$

$$O(x,y) = 128 \text{ if } I(x,y) \geq \textit{low threshold value}_2 \text{ and if } I(x,y) < \textit{high threshold value}_2$$

$$O(x,y) = 192 \text{ if } I(x,y) \geq \textit{low threshold value}_3 \text{ and if } I(x,y) < \textit{high threshold value}_3$$

$$O(x,y) = 255 \text{ if } I(x,y) \geq \textit{low threshold value}_4 \text{ and if } I(x,y) < \textit{high threshold value}_4$$

$$O(x,y) = 0 \quad \text{otherwise}$$

Results:

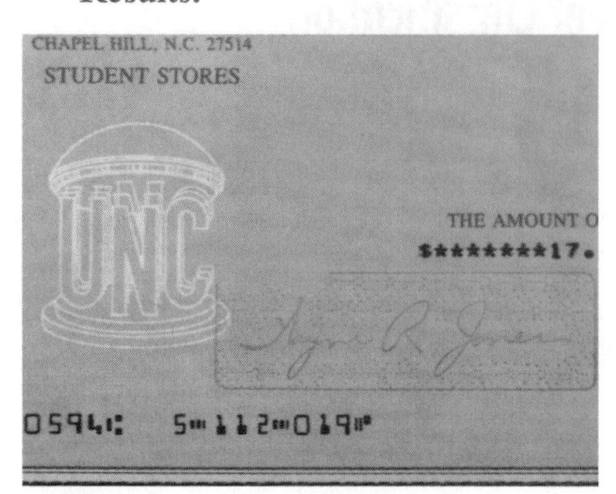

Figure IOS.30a Original bank check image showing four brightness classes composed of the text, signature, background, and logo pattern.

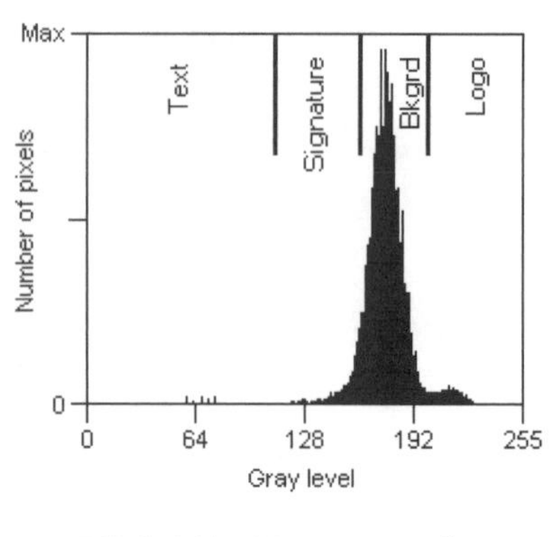

Figure IOS.30b Histogram of original image showing the four brightness class gray-scale regions.

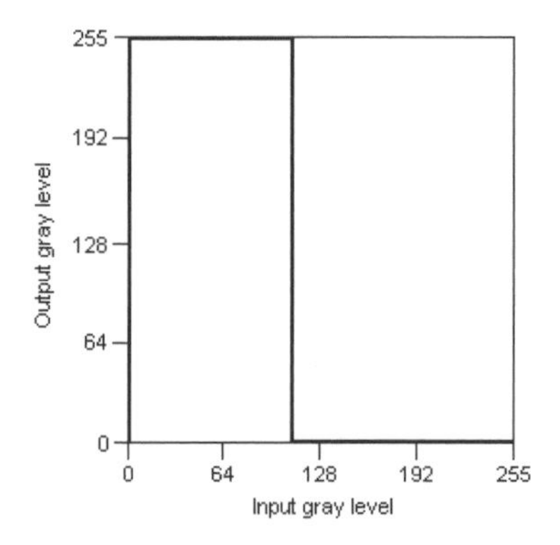

Figure IOS.30c Mapping function to isolate the text brightness class with a low threshold of 0 and a high threshold of 112.

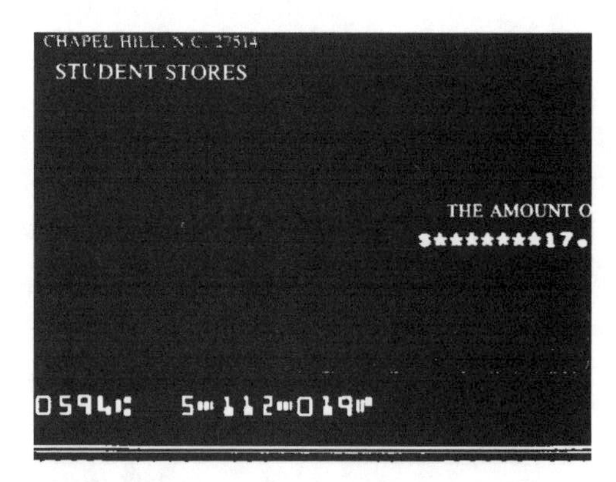

Figure IOS.30d The first class shown as white with everything else appearing as black.

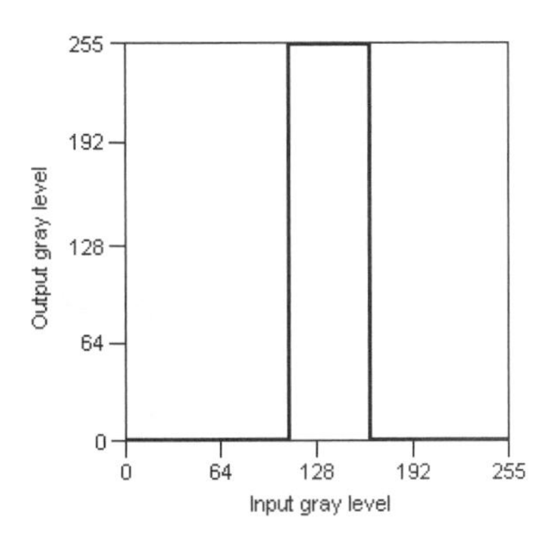

Figure IOS.30e *Mapping function to isolate the signature brightness class with a low threshold of 113 and a high threshold of 152.*

Figure IOS.30f *The second class shown as white with everything else appearing as black.*

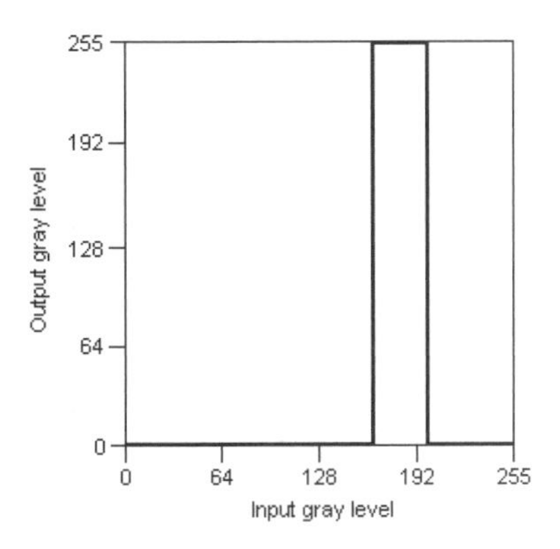

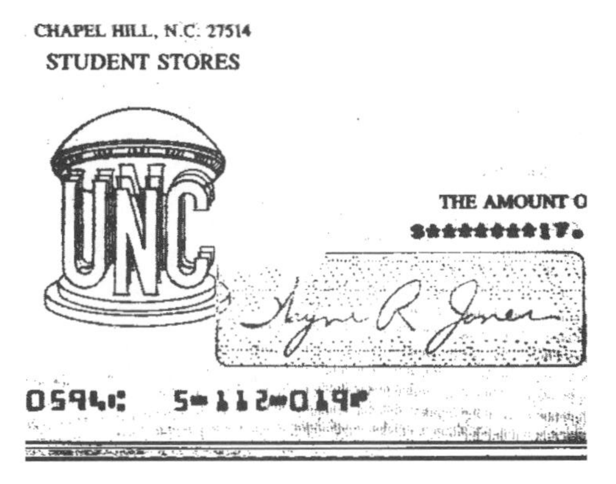

Figure IOS.30g *Mapping function to isolate the background brightness class with a low threshold of 153 and a high threshold of 199.*

Figure IOS.30h *The third class shown as white with everything else appearing as black.*

IOS 30

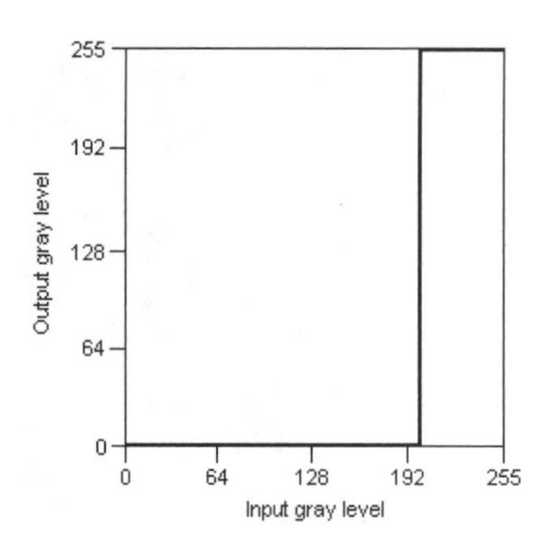

Figure IOS.30i *Mapping function to isolate the logo pattern brightness class with a low threshold of 200 and a high threshold of 255.*

Figure IOS.30j *The fourth class shown as white with everything else appearing as black.*

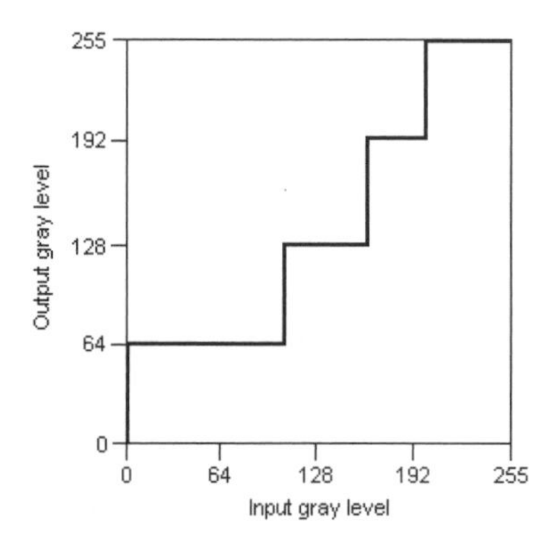

Figure IOS.30k *Gray-level classification mapping function to map each of the four brightness classes to distinct gray levels.*

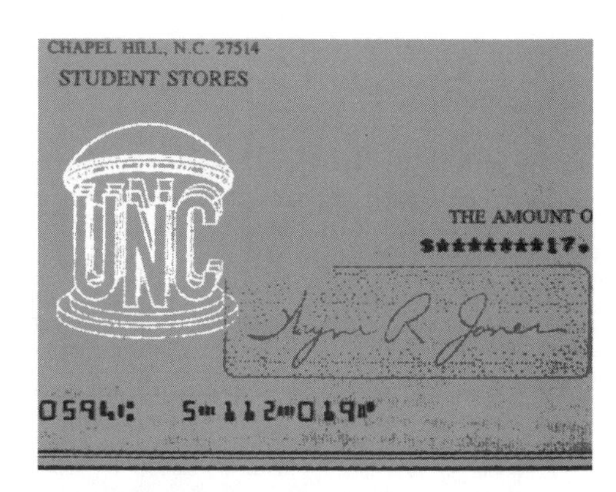

Figure IOS.30l *The resulting gray-level classified image showing the four classes as separate gray levels in a single image.*

31 Run-length Coding

Class: Image Compression

Description: The run-length image compression scheme is a lossless entropy coding technique. Across each line of an image, pixel brightnesses are sequentially compared and grouped together into runs of identical brightness. The resulting compressed image data is composed of a series of paired brightness and run-length values. Each brightness value represents the brightness of a sequential run of pixels, and each run-length value represents the number of pixels in the run. For example, a run of 25 pixels with the same brightness of 176 would be coded as 176|25. The 25 original image bytes are replaced with 2 run-length coded bytes.

Applications: Lossless image compression techniques are used to reduce the data size of image data files while preserving the exact image brightness data of the original image. Lossless schemes, like run-length compression, generally have smaller compression ratios than lossy compression schemes.

Implementation:

1. Scan across image lines for runs of identical-brightness pixels
2. Replace each run of constant-brightness pixels with a brightness|run-length code

Run-length coding is prone to data explosion when an image contains many brightness changes. Under the worst-case conditions, the compressed image file can become twice the size of the original image file. The conditions of data explosion can be detected and handled during compression to avoid negative consequences.

Run-length image compression techniques typically provide compression ratios from 4:1 to 10:1 or more on binary images, and about 1.5:1 on gray-scale images.

Results: Because run-length image compression is a lossless scheme, images before and after the operation are identical.

32 Huffman Coding

Class: Image Compression

Description: The Huffman image compression scheme is a lossless entropy coding technique. Pixel brightness values are replaced with variable-length codes based on their frequencies of occurrence in the image. An image histogram is computed, yielding the frequencies of occurrence for each of the brightnesses in the image. These frequencies of occurrence are used to assign smaller codes to the pixels with the most-frequent occurrence and longer codes to those with the least-frequent occurrence.

Applications: Lossless image compression techniques are used to reduce the data size of image data files while preserving the exact image brightness data of the original image. Lossless schemes, like Huffman compression, generally have smaller compression ratios than lossy compression schemes.

Implementation:

1. Compute image histogram
2. Assign variable-length codes based on brightness frequencies of occurrence
3. Replace pixel brightnesses with new codes

Huffman image compression techniques typically provide compression ratios between 1.5:1 and 2:1 on gray-scale images.

Results: Because Huffman image compression is a lossless scheme, images before and after the operation are identical.

33 Truncation Coding

Class: Image Compression

Description: The truncation image compression scheme is a direct lossy coding technique. It is carried out by reducing brightness resolution or spatial resolution. Brightness resolution reduction is accomplished by reducing the number of bits used to represent brightness values. Spatial resolution reduction is accomplished by reducing the number of pixels in an image.

Applications: Lossy image compression techniques are used to reduce the data size of image data files while sacrificing some of the original image quality. Lossy schemes like truncation compression generally have greater compression ratios than lossless compression schemes.

Implementation:

1. Reduce brightness resolution of all pixels from 8 bits per pixel to a smaller number

   or

1. Reduce spatial resolution of image by regularly discarding pixels

Truncation image compression techniques provide as much compression as desired. Of course, the image quality directly deteriorates as the compression ratio increases. Image quality degrades in the forms of brightness and spatial resolution loss, causing posterization and pixelation artifacts.

Dither noise can be added to decompressed brightness resolution-truncated images to improve subjective image quality. The random noise patterns break up the posterization effect. Similarly, interpolation techniques can improve decompressed spatial resolution-truncated images by softening the pixelation effect.

Truncation compression provides gray-scale image compression ratios between 2:1 and 8:1 with minimal image degradation. Image degradation is thoroughly subjective, based on the application's requirements.

Results:

Figure IOS.33a *Original image of birds.*

Figure IOS.33b *5-bit brightness-truncation coded image.*

Figure IOS.33c *Error image showing the difference between the 5-bit brightness-truncation coded image and original image.*

Figure IOS.33d *5-bit brightness-truncation coded image with a 3-bit dither noise added.*

Figure IOS.33e *Error image.*

Figure IOS.33f *4-bit brightness-truncation coded image.*

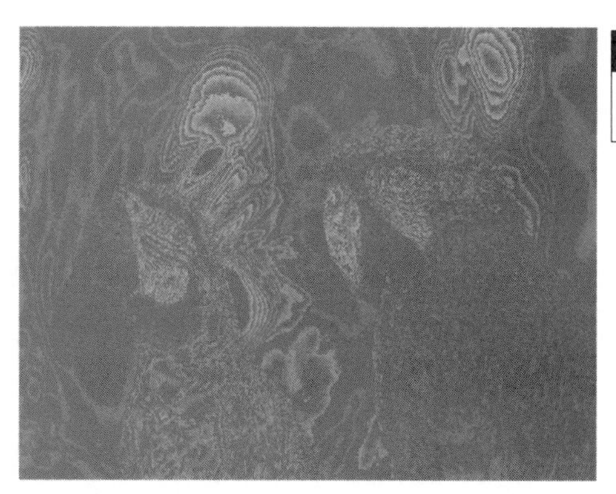

Figure IOS.33g *Error image.*

Figure IOS.33h *4-bit brightness-truncation coded image with a 4-bit dither noise added.*

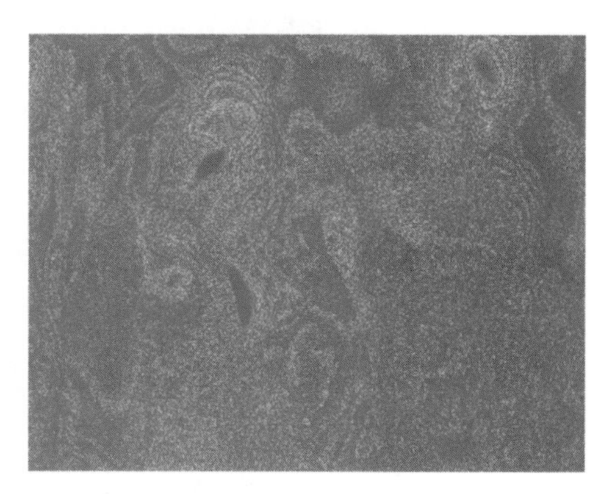

Figure IOS.33i *Error image.*

Figure IOS.33j *3-bit brightness-truncation coded image.*

Figure IOS.33k *Error image.*

IOS 33

Figure IOS.33l *3-bit brightness-truncation coded image with a 5-bit dither noise added.*

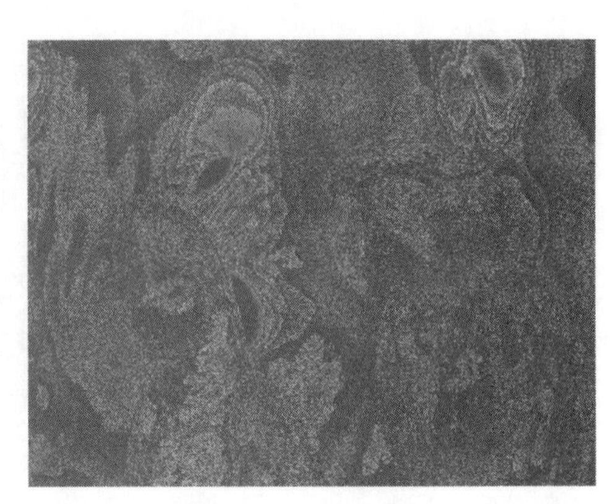

Figure IOS.33m *Error image.*

Figure IOS.33n *Spatial-truncation coded image with the resolution reduced by a factor of four (a factor of two in each axis) and nearest neighbor interpolation.*

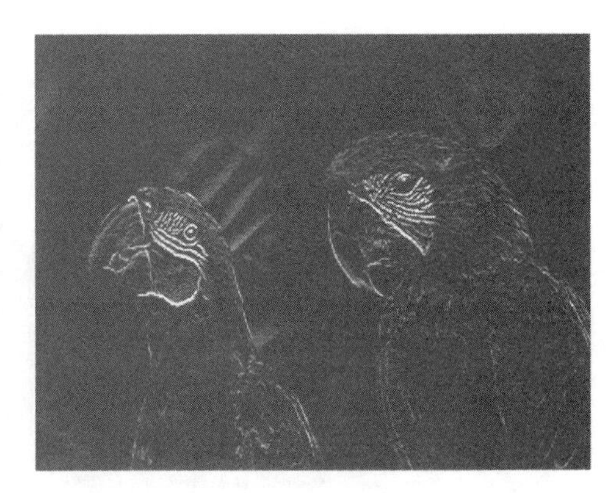

Figure IOS.33o *Error image.*

Figure IOS.33p *Spatial-truncation coded image with the resolution reduced by a factor of four and bilinear interpolation.*

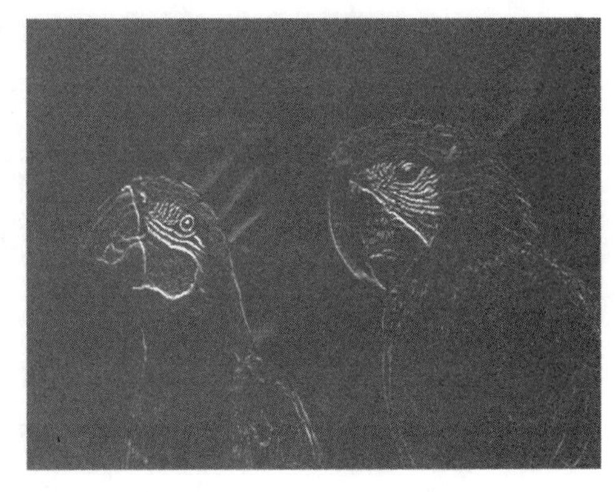

Figure IOS.33q *Error image.*

Figure IOS.33r *Spatial-truncation coded image with the resolution reduced by a factor of 16 (a factor of two in each axis) and nearest neighbor interpolation.*

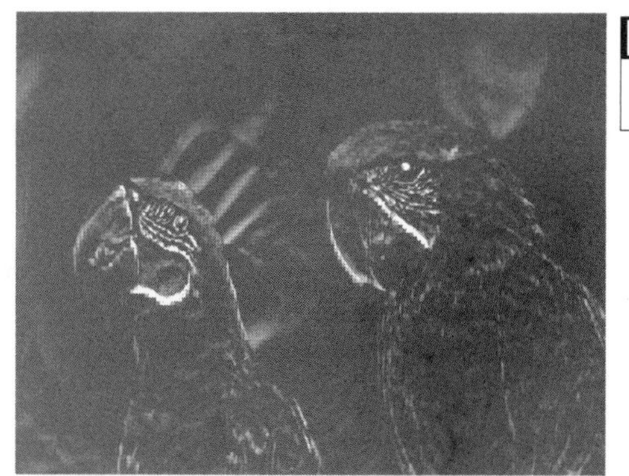

Figure IOS.33s *Error image.*

Figure IOS.33t *Spatial-truncation coded image with the resolution reduced by a factor of 16 and bilinear interpolation.*

Figure IOS.33u *Error image.*

34 Lossy Differential Pulse-Code Modulation Coding

Class: Image Compression

Description: Lossy differential pulse-code modulation (DPCM) image compression is a lossy coding technique. Each pixel is replaced by the difference value of it and its neighbor to the left. Because brightnesses statistically change slowly over most of an image, difference values can be represented by a lower-resolution value than the raw brightness value. When the full difference value cannot be represented with a lower difference resolution, degradation occurs in the compressed image.

Applications: Lossy image compression techniques are used to reduce the data size of image data files while sacrificing some of the original image quality. Lossy schemes like lossy DPCM compression generally have greater compression ratios than lossless compression schemes.

Implementation:

1. Compute pixel-by-pixel differences between current pixels and preceding neighbors
2. Truncate difference values to predetermined resolution

DPCM image compression causes image degradations that increase proportionally with increased compression ratios. Degradations in the form of image smearing of edge details can occur in extreme cases.

DPCM image compression techniques typically provide gray-scale image compression ratios of about 3:1 without serious image quality degradation.

Results:

Figure IOS.34a *Original "birds" image.*

Figure IOS.34b 6-bit lossy DPCM coded image.

Figure IOS.34c Error image showing the difference between the 6-bit lossy DPCM coded image and original image.

Figure IOS.34d 5-bit lossy DPCM coded image.

Figure IOS.34e Error image.

Figure IOS.34f 4-bit lossy DPCM coded image.

Figure IOS.34g Error image.

35 Discrete Cosine Transform Coding

Class: Image Compression

Description: Discrete cosine transform (DCT) image compression is a lossy coding technique. Pixel blocks—on the order of 8 × 8 pixels—are discrete-cosine transformed to a frequency-domain representation. Frequency components with minimal values are discarded, leaving only those components that contribute significantly to the image.

Applications: Lossy image compression techniques are used to reduce the data size of image data files while sacrificing some of the original image quality. Lossy schemes like discrete cosine transform compression generally have greater compression ratios than lossless compression schemes.

Implementation:

1. Break image into discrete 8 × 8 pixel blocks
2. Discrete cosine transform each block to the frequency domain
3. Discard minimally-valued frequency components

DCT image compression causes image degradations that increase proportionally with increased compression ratios. Degradations of the form of image blocking and high-frequency detail loss can occur in extreme cases.

DCT image compression techniques typically provide gray-scale image compression ratios of about 10:1 without serious image quality degradation.

Results:

Figure IOS.35a *Original "birds" image.*

Figure IOS.35b *5:1 compression-transformed coded image.*

Figure IOS.35c *Error image showing the difference between the 5:1 compression-transform coded image and original image.*

Figure IOS.35d *10:1 compression-transformed coded image.*

Figure IOS.35e *Error image.*

Figure IOS.35f *20:1 compression-transformed coded image.*

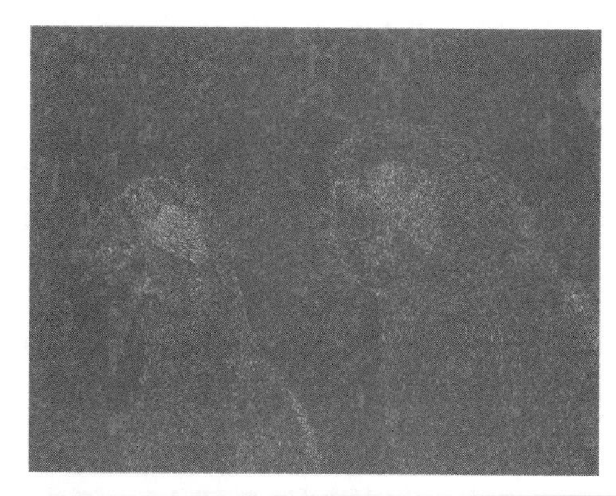

Figure IOS.35g *Error image.*

Figure IOS.35h *50:1 compression-transformed coded image.*

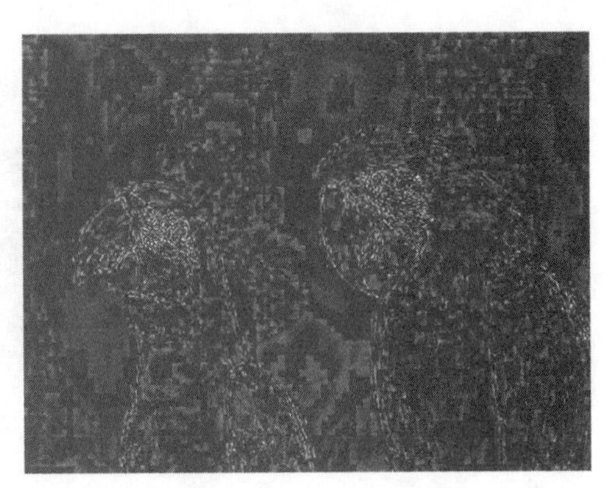

Figure IOS.35i *Error image.*

36 Tomographic Reconstruction

Class: Image Synthesis

Description: If multiple one-dimensional projections are imaged around an object, an internal cross-sectional image of the object can be computed. A three-dimensional volumetric image can then be created by stacking multiple cross-sectional image slices.

Applications: Computed tomography techniques are widely used in medical imaging applications such as Computed Tomography (CT), Magnetic Resonance Imaging (MRI), and Positron Emission Tomography (PET).

Implementation:

1. One-dimensional projections rotated and replicated across image frame
2. Dual-image pixel point process—addition
3. Repeat for all images in reconstruction
4. Define brightness scale map
5. Single-image pixel point process

One-dimensional image projections are accumulated around the object in the plane of the desired cross-sectional slice-image. Each projection is rotated to its original angle when acquired and expanded into its own image frame. Pixel interpolation techniques must be used to reduce rotation aliasing artifacts. All expanded projections are summed into one, creating a two-dimensional reconstruction of the original cross-section. The summed image is then brightness-scaled for display.

The resulting image quality is directly related to the number of projections used in the reconstruction operation. Generally, 180 or more projection images may be used to compute a high-quality cross-sectional slice-image.

Results:

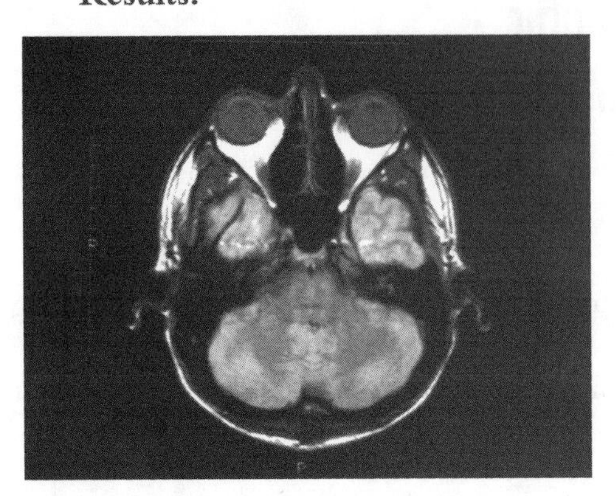

Figure IOS.36a *MRI cross-sectional slice-images. This transaxial view cuts through the subject horizontally relative to a standing subject (top view).*

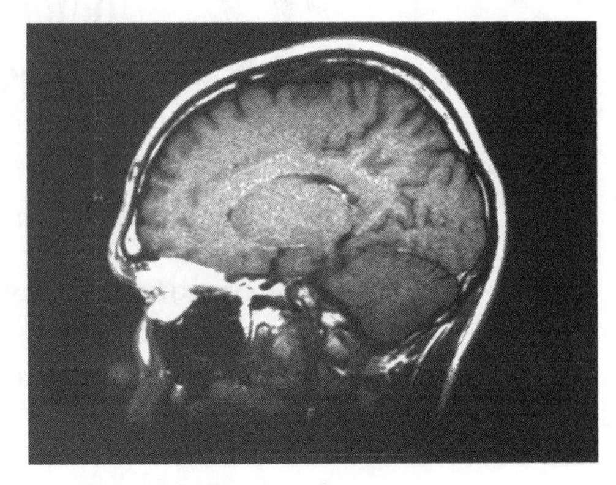

Figure IOS.36b *The sagittal view cuts through the subject vertically from front to back (side view).*

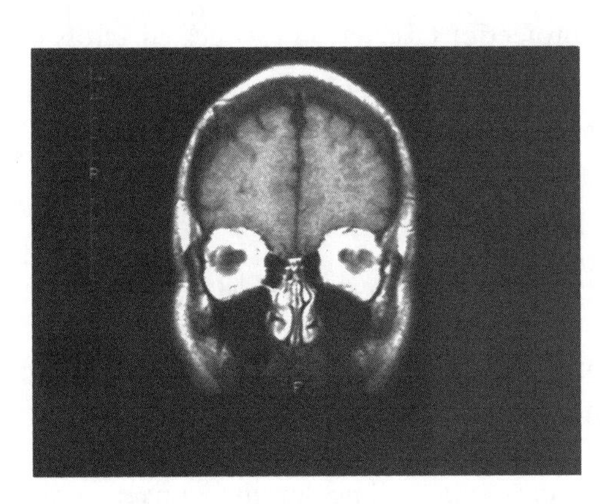

Figure IOS.36c *The coronal view cuts through the subject vertically from side to side (front view).*

37 Stereo Image Pairs

Class: Image Synthesis

Description: Stereo image pairs are two images of the same scene taken from a slightly different viewpoint. If the distance between the viewpoints is known, the relative depth of imaged objects can be computed (or perceived) based on the observed object parallax.

Applications: Stereo vision obviously plays an important role in human vision. It can also be used to compute the relative depth of distant objects in machine vision applications. Remote sensing applications that measure terrain elevation on Earth and on extraterrestrial bodies use stereo image pairs as an effective means for gathering their data.

Implementation:

1. Acquire images from separated viewpoints
2. Measure object parallax
3. Compute depth difference using known viewpoint distance and viewpoint separation

$$\text{Depth Difference} = \frac{\text{Viewpoint Distance}}{\text{Viewpoint Separation}} \times (\text{Right Separation} - \text{Left Separation})$$

where

Viewpoint distance = distance between the viewpoints and farthest object
Viewpoint separation = distance between the two viewpoints

Results: Two stereo image pairs are shown in reverse order. You can view these in stereo by gently crossing your eyes so that the left eye views the right image and the right eye views the left image. Stare blankly at the two images until the features of both images converge into one. Relax your focus, and you will perceive three-dimensionality. This procedure can be difficult for many people, and sometimes requires practice and patience.

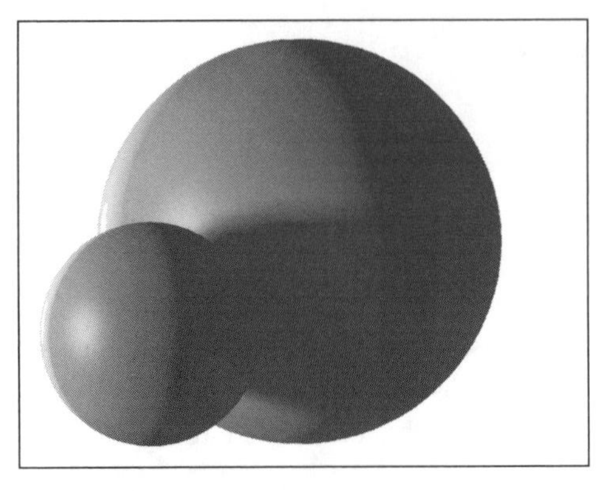

Figure IOS.37a *Stereo image pair of rendered spheres in reverse order for direct viewing—right image.*

Figure IOS.37b *Left image.*

Figure IOS.37c *Stereo image pair of items on a desk—right image.*

Figure IOS.37d *Left image.*

Figure IOS.37e *Stereo image pair of house—right image.*

Figure IOS.37f *Left image.*

38 Three-dimensional Model Shading

Class: Image Synthesis

Description: An important part of creating realistic synthetic images is the portrayal of lighting and shading. Shading operations show a rendered object as though it were illuminated by a natural light source. Three important shading techniques provide varying degrees of realism for shading three-dimensional objects.

Applications: When you shade a rendered three-dimensional object, significant depth information is added to the perceived image. In many applications, shading cues alone can provide enough information about an object's curvatures and form to convey its three-dimensional appearance accurately.

Implementation:

Flat shading:

1. Determine angle between light source(s) and surface
2. Compute cosine of angle of incident light
3. Shade surface with appropriate light value

Gouraud shading:

1. Determine angle between light source(s) and surface
2. Compute cosine of angle of incident light
3. Compute light at corners of neighboring surfaces
4. Smoothly shade surface pixels with interpolated light values based on surface corner light values

Phong shading:

1. Determine angle between light source(s) and surface
2. Compute cosine of angle of incident light
3. Compute light at corners of neighboring surfaces
4. Compute normal angles of surface corners
5. Smoothly shade surface pixels with light values based on surface corner light values and interpolated normal angles

Results: Several images of identical objects are shown, each rendered using a different shading method.

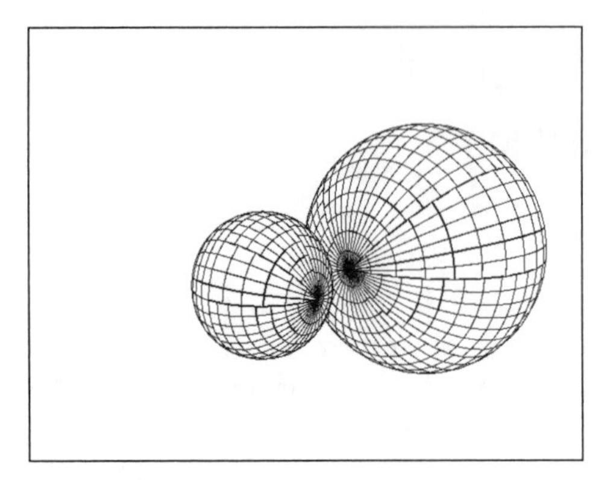

Figure IOS.38a *Three dimensional wireframe rendering of two spheres.*

Figure IOS.38b *Same rendering with flat shading.*

Figure IOS.38c *Same rendering with Gouraud shading.*

Figure IOS.38d *Same rendering with Phong shading.*

39 Three-dimensional Model Photo-realistic Rendering

Class: Image Synthesis

Description: Going a step past shading, photo-realistic rendering of a three-dimensional model includes the addition of effects like surface textures and finishes, multiple light sources, varied light source types, haze, object and light reflection, and natural optical distortions.

Applications: The addition of photo-realistic rendering techniques provides synthetic images of rendered objects with the natural appearance of a photograph. The goal of photo-realistic rendering is to portray a rendered object exactly as it will appear in reality.

Implementation: Numerous techniques of photo-realistic rendering have evolved, each a primary area of study unto itself. Each method generally attempts to model reality in a way convenient and efficient for computation with digital techniques.

Results: Several sophisticated photo-realistic image renderings are shown using surface texturing, multiple light sources, and various lighting effects.

Figure IOS.39a *Complex photo-realistic renderings—attic scene.*

Figure IOS.39b *Adobe house. (Images courtesy of Autodesk, Inc.)*

Figure IOS.39c *A-6 aircraft in flight.(Image courtesy of EDS/Unigraphics, Inc. and Lightwork Design, Ltd.)*

Figure IOS.39d *Building courtyard. (Image courtesy of Intergraph Corp.)*

Appendix A—Further Reading/References

The following books, journals, trade magazines, and professional societies are listed as references to aid in your follow-on study. Each provides excellent technical resources, providing specific coverage to a variety of digital image processing topics.

Book references are divided into three groups—general interest, professional, and academic—each offering a different degree of rigor in its treatment of the subject matter. Journal and trade magazine references cover periodical publications. Professional society references are groups that sponsor technical forums for the exchange and publication of technical advancements and provide introductory tutorial and overview materials.

Books

Low-level—General Interest

1. Morrison, M. 1993. *The Magic of Image Processing.* Carmel, Indiana: Sams Publishing.
2. Wegner, T.I. 1992. *Image Lab.* Corte Madera, California: Waite Group Press.

Mid-level—Professional

1. Baxes, G.A. 1984. *Digital Image Processing: A Practical Primer.* Englewood Cliffs, New Jersey: Prentice-Hall. Reprinted 1988, Denver: Cascade Press.
2. Dougherty, E.R. 1992. *An Introduction to Morphological Image Processing.* Bellingham, Washington: SPIE Optical Engineering Press.

3. Falk, D.S., Brill, D.R., and Stork, D.G. 1986. *Seeing the Light: Optics in Nature, Photography, Color, Vision, and Holography*. New York: John Wiley & Sons.

4. Galbiati, Jr., L.J. 1990. *Machine Vision and Digital Image Processing Fundamentals*. Englewood Cliffs, New Jersey: Prentice-Hall.

5. Green, W.B. 1989. *Digital Image Processing: A Systems Approach*. New York: Van Nostrand Reinhold.

6. Jensen, J.R. 1986. *Introductory Digital Image Processing: A Remote Sensing Perspective*. Englewood Cliffs, New Jersey: Prentice-Hall.

7. Lindley, C.A. 1991. *Practical Image Processing in C*. New York: John Wiley & Sons.

8. Rabbani, M. and Jones, P.W. 1991. *Digital Image Compression Techniques*. Bellingham, Washington: SPIE Optical Engineering Press.

9. Russ, J.C. 1992. *The Image Processing Handbook*. Boca Raton, Florida: CRC Press.

10. Sturge, J.M., Walworth, V.K. and Shepp, A. 1989. *Imaging Processes and Materials*. 8th ed. New York: Van Nostrand Reinhold.

High-level—Academic

1. Andrews, H.C. 1970. *Computer Techniques in Image Processing*. New York: Academic Press.

2. Andrews, H.C. and Hunt, B.R. 1977. *Digital Image Restoration*. Englewood Cliffs, New Jersey: Prentice-Hall.

3. Castleman, K.R. 1979. (2d ed. forthcoming.) *Digital Image Processing*. Englewood Cliffs, New Jersey: Prentice-Hall.

4. Chellappa, R. and Sawchuk, A.A. 1985. *Digital Image Processing and Analysis*. Vol. 1, *Digital Image Processing*. Vol. 2, *Digital Image Analysis* (collection of papers). Silver Spring, Maryland: IEEE Computer Society Press.

5. Foley, J. D. and Van Dam, A. 1990. *Fundamentals of Interactive Computer Graphics*. 2d ed. Reading, Massachusetts: Addison-Wesley.

6. Gonzalez, R.C. and Woods, R.E., 1992. *Digital Image Processing*. Reading, Massachusetts: Addison-Wesley.

7. Jain, A.K. 1989. *Fundamentals of Digital Image Processing*. Englewood Cliffs, New Jersey: Prentice-Hall.

8. Lee, H. and Wade, G. 1986. *Imaging Technology* (collection of papers). New York: IEEE Press.

9. Lim, J.S. 1990. *Two-Dimensional Signal and Image Processing*. Englewood Cliffs, New Jersey: Prentice-Hall.

10. Pratt, W.K. 1991. *Digital Image Processing*. 2d ed. New York: John Wiley & Sons.

11. Rosenfeld, A. and Kak, A.C. 1982. *Digital Picture Processing*. 2 vols. 2d ed. Orlando, Florida: Academic Press.

Journals and Trade Magazines

Advanced Imaging. Monthly. PTN Publishing Co., Melville, New York.

Computer Graphics World. Monthly. Pennwell Publishing Co., Westford, Massachusetts.

IEEE Computer Graphics and Applications. Bimonthly. IEEE Computer Society, Los Alamitos, California.

IEEE Transactions on Signal Processing. Monthly. IEEE Signal Processing Society, New York.

Journal of Electronic Imaging. Quarterly. The Society for Imaging Science and Technology and The International Society for Optical Engineering, Bellingham, Washington.

Optical Engineering. Monthly. The International Society for Optical Engineering, Bellingham, Washington.

SMPTE Journal. Monthly. Society of Motion Picture and Television Engineers, White Plains, New York.

Professional Societies

ACM SIGGRAPH, Association for Computing Machinery, Special Interest Group on Computer Graphics. New York.

AIIM, Association for Information and Image Management. Silver Spring, Maryland.

SPIE, The International Society for Optical Engineering. Bellingham, Washington.

IEEE Signal Processing Society, Institute of Electrical and Electronics Engineers Signal Processing Society. New York.

IEEE Computer Society, Institute of Electrical and Electronics Engineers Computer Society. Los Alamitos, California.

IS&T, The Society for Imaging Science and Technology. Springfield, Virginia.

SMPTE, Society of Motion Picture and Television Engineers. White Plains, New York.

Appendix B
Digital Image Processing: Hands-On User's Guide

What Is *Digital Image Processing: Hands-On?*

Digital Image Processing: Hands-On is provided as an educational supplement to this book. It is a self-installing, fully functional Windows application developed to be a compact demonstration of basic digital image processing operations. In addition to the application, several test images—taken from the text of this book—are provided for your use.

The *Hands-On* application was built as a Windows application using the Borland *ObjectWindows for C++* application framework. Data Translation's *Global Lab/Image Function Library* was used as the digital image processing engine for several of the operations.

Getting Started

System Requirements

Digital Image Processing: Hands-On requires the following computer system hardware and software:

1. Intel processor-based workstation with an 80386-class (or better) processor.
2. Microsoft Windows 3.1 (or later) operating system running in 386 Enhanced Mode.
3. Hard disk drive with at least 4 MB of free space.

4. At least 4 MB of available system RAM memory.
5. SVGA graphics adapter with the minimum capability of 640 × 480 × 256-color mode.
6. 3.5-inch floppy disk drive.

Making a Backup Copy

Before you install the *Digital Image Processing: Hands-On* software, it is strongly recommended that you make a backup copy of the original disk. Remember, however, the backup disk is for your personal use only. Any other use of the backup disk violates copyright law. Please take the time now to make the backup, using the following procedure:

1. Insert the *Hands-On* disk into drive A: of your computer (assuming that your floppy disk drive is drive A:).
2. At the C:\> prompt, type **DISKCOPY A: A:** and press <Enter>.

You will be prompted through the steps to complete the disk copy. Be sure to use a target disk of the same data format as the *Hands-On* distribution disk. When you are through, remove the new copy of the disk and label it immediately. Remove the original *Hands-On* disk and store it in a safe place.

Installing the Software

The *Digital Image Processing: Hands-On* disk contains all the necessary files in a compressed format. A Windows setup utility is also provided to automatically decompress and copy the files to your hard disk and install a Windows program group and icon. Start your Windows environment and follow these steps to complete the *Hands-On* installation:

1. Make the Windows Program Manager the active window by clicking on it.
2. From the Program Manager menu, select *File* and then *Run*—this will open the Run dialog box.
3. Insert the *Hands-On* disk into drive A: of your computer (assuming that your floppy disk drive is drive A:).
4. Type **A:\SETUP** in the Run command line box and press <Enter>.
5. Follow the setup instructions.

After the *Hands-On* installation is complete, remove the *Hands-On* disk and store it in a safe place.

In Case of Trouble—User Assistance and Information

John Wiley & Sons, Inc. is pleased to provide assistance to users of the *Digital Image Processing: Hands-On* software package. Should you have questions regarding the installation or use of this package, please call our technical support number at 212/850-6194 weekdays between 9 AM and 4 PM Eastern Time.

To place orders for additional copies of this book, including the software, or to request information about other Wiley products, please call 800/879-4539.

Using the Software

An Overview

Digital Image Processing: Hands-On is a compact demonstration of basic digital image processing operations. In addition to the application, several test images are provided for experimental use. The application can process and display uncompressed, gray-scale TIFF images up to a size of 320 pixels × 240 lines, like those provided. It displays a single image at a time.

Hands-On provides a sampling of the image enhancement and analysis operations discussed in this book. The user is referred to the body of the book for detailed information on the use of these operations.

When the *Hands-On* application is launched, a startup message is displayed; then the application window appears, as shown in Figure B.1. A smaller window, used for image display, appears within the application window. The main-menu bar at the top of the application window shows the available menu item selections. The *Hands-On* application can be enlarged to fill the entire display screen or collapsed to an icon by clicking on the maximize or minimize buttons at the upper right corner of the application window. Similarly, the application can be ended by double-clicking the control-menu box at the upper left corner of the application window.

The Main Menu

Digital Image Processing: Hands-On operations are carried out by clicking on main-menu items. When a menu item is clicked, a drop-down menu appears that provides additional items to choose from. The following are brief descriptions of these drop-down menus.

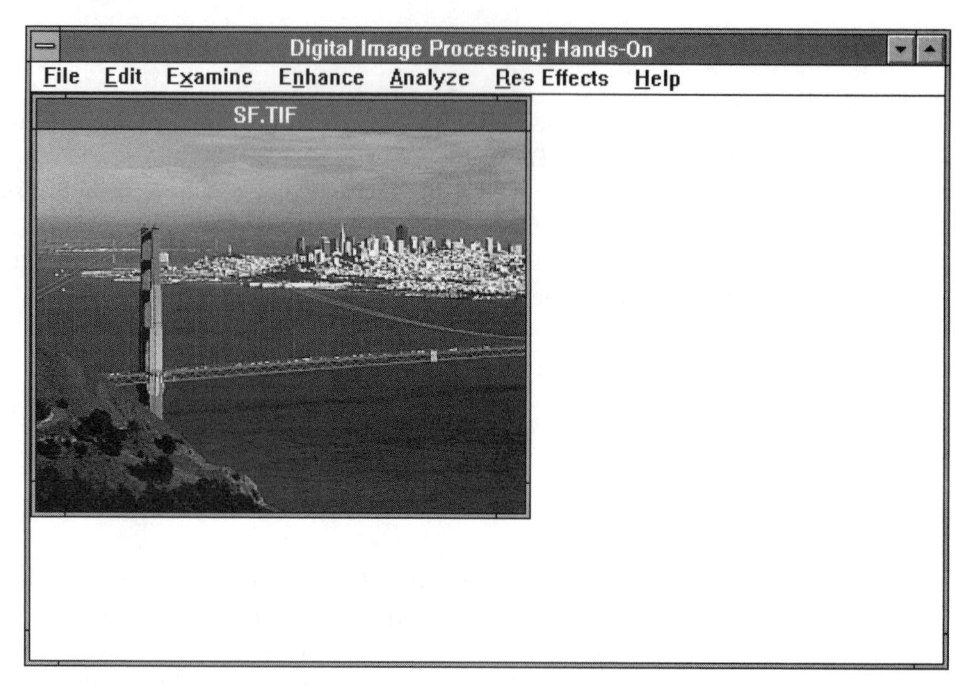

Figure B.1 *The* Digital Image Processing: Hands-On *application window.*

File

The *File* menu item provides items for accessing image files, printing images, and quitting the *Hands-On* application.

New Clears the image display window—used when the subsequently opened image is smaller than 320 pixels × 240 lines.

Open Image... Opens an uncompressed gray-scale TIFF image for processing.

Save Image... Saves the displayed image as an uncompressed gray-scale TIFF file.

Print Image Prints the displayed image.

Exit Ends the Hands-On application session.

Edit

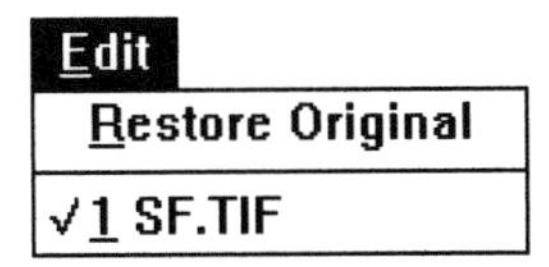

The *Edit* menu item provides items for restoring and displaying the filename of the image last opened.

Restore Original Restores the image as it appeared when first opened.

✔*1 SF.TIF* The last-opened image's filename.

Examine

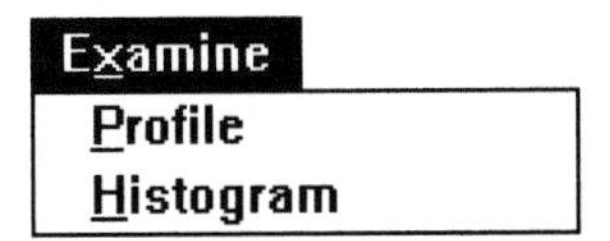

The *Examine* menu item provides items for displaying information about the characteristics of the displayed image.

Profile Displays a profile window that shows the pixel brightnesses along a line displayed in the image window. The profile line can be sized and moved as desired.

Histogram Displays a histogram window that shows the brightness histogram of the displayed image.

Enhance

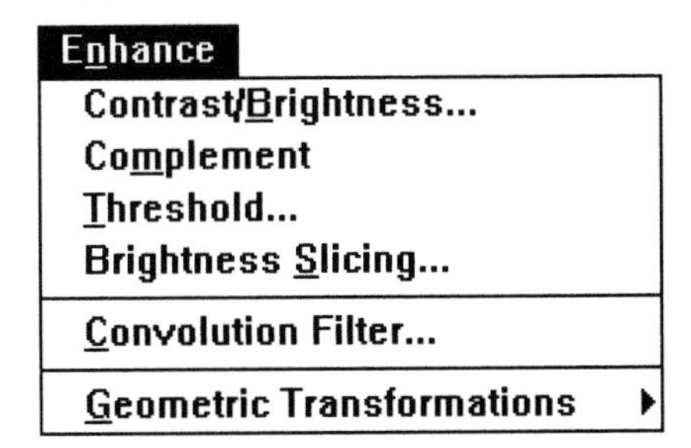

The *Enhance* menu item provides items for enhancing the visual quality of the displayed image.

Contrast/Brightness... Implements histogram stretch and slide contrast enhancement operations.

Complement Complements the displayed image to appear as a negative of the original.

Threshold... Implements a binary contrast enhancement that sets brightnesses below a selected threshold to black and brightnesses above the threshold to white.

Brightness Slicing... Implements a brightness slicing enhancement that sets brightnesses above and below selected thresholds to black and brightnesses between the two thresholds to white.

Convolution Filter... Implements a 3×3 spatial convolution filter using selected mask coefficients—low-pass, high-pass, edge enhancement, and other convolution operations can be carried out.

Geometric Transformations Implements selected *Scale...* and *Rotate...* operations about the center of the displayed image.

Analyze

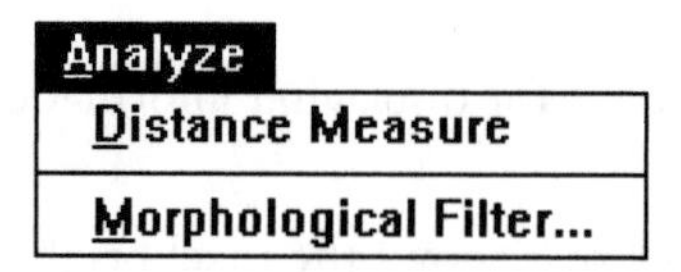

The *Analyze* menu item provides items for image feature measurement and morphological segmentation processing.

Distance Measure Displays a distance window that shows the (x,y) endpoint locations and length of a line displayed in the image window. The distance measure line can be sized and moved as desired.

Morphological Filter... Implements a 3×3 binary or gray-scale morphological operation using a selected structuring element—erosion, dilation, opening, and closing operations of varied orientations can be carried out.

Res Effects

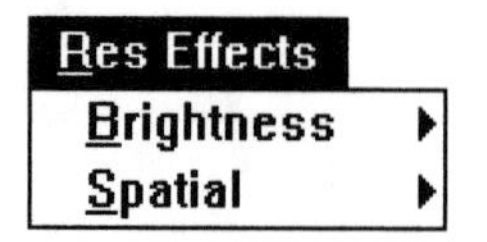

The *Res Effects* menu item provides items for reducing the resolution of the displayed image.

Brightness Reduces the displayed image's brightness resolution to any resolution between *8 Bits (256 levels)* and *1 Bit (2 levels)*.

Spatial Reduces the displayed image's spatial resolution to one of several resolutions between *320 Pixels* × *240 Lines* and *20 Pixels* × *15 Lines.*

Help

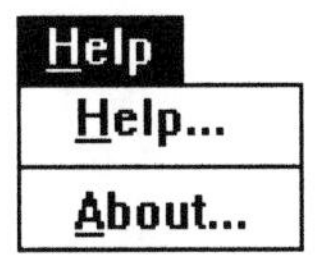

The *Help* menu item provides items for displaying brief *Help...* and *About...* dialogs regarding the *Hands-On* application.

Glossary

Throughout this book, new terms were highlighted when first encountered. Here, the most significant of these terms are presented with brief definitions. This section is intended as a quick reference to the nomenclature used in the book. When multiple definitions are possible, each term is defined in the digital image processing context.

Absolute chain code A pixel-by-pixel direction code used in a contour following operation that records an object boundary as a series of absolute directions.

Abstract data set A set of data, such as financial data, that has no spatial or physical significance.

Accelerator processor A hardware processor in a digital image processing system that is exclusively devoted to image processing tasks; typically based on a dedicated microprocessor.

Adaptive thresholding A method of binary contrast enhancement where the threshold value changes throughout an image based on local brightness characteristics.

Additive color The form of color creation based on the additive mixing of the red, green, and blue primary colors.

Affine transformation A linear geometric trasnformation such as translation, rotation, or scaling.

Aliasing See *spatial aliasing* or *temporal aliasing*.

Analog image processing The technique of processing images while they are in the form of a continuous electrical signal, typically a video signal.

Analog signal reconstruction The recreation of a continuous analog electrical signal (typically a video signal) from a series of discrete digital brightness values.

Analog-to-digital converter (A/D) A semiconductor device that converts an analog voltage level representing image brightness to a digital quantity.

Anti-aliasing filter An analog signal filter used prior to an A/D converter to remove frequency components from an image that are above one-half the sampling rate, thus removing the potential for spatial aliasing.

Aperture The light-gathering opening in a lens; also, the measure of the light-gathering opening in a lens.

Area photosensor A two-dimensional array of photodetectors used to acquire images; typically a semiconductor device.

Aspect ratio The ratio of an image's horizontal to vertical dimensions, generally stated as *x:y*.

Background color The selected color in an overlay video source used to control its overlaying upon another video source. Wherever the background color is *not* detected in the overlay source, the overlay video source is switched in.

Barrel distortion A common geometric lens distortion causing an acquired image to appear to pucker toward the center.

Basis function See *frequency component function*.

Bilinear interpolation A form of pixel interpolation that is based on a weighted average of the four pixels surrounding the pixel location of interest.

Binary contrast enhancement A point process that sets every pixel brightness in an image to either black (0) or white (255), depending upon whether the original pixel brightness is below or above a threshold value.

Binary image An image composed of only black (0) and white (255) brightnesses.

Binary morphological process The morphological process intended to operate on a binary image—neighborhood black and white pixel patterns are evaluated.

Bit-plane The view of a single bit of an image. A bit-plane represents the on or off level of a particular bit's contribution to each pixel's brightness.

Black-level reference The reference used by an A/D converter to establish the amplitude of a video signal to be converted to black (0).

Blanking level The amplitude of a video signal that is below black; intended to represent the blank video signal during sync intervals.

Block coding A form of lossless or lossy image compression that looks for repeating patterns of blocks of pixels and uses a codebook to store the patterns.

Boundary description A description of an object's perimeter; typically a series of chain codes or line segments having a length and direction.

Box filter A low-pass spatial filter composed of uniformly weighted convolution coefficients.

Brightness The quantity of light assigned to a pixel in a digital image. In comparison, *intensity* refers to the quantity of light actually reflected or transmitted from a physical scene. Also, one of the three color components of the HSB color space that controls how bright an HSB color appears.

Brightness contouring The effect of insufficient brightness resolution—gradual brightness changes appear to change abruptly, making subtle brightness changes in the image appear to have contours.

Brightness features Features of an object that relate to its brightness characteristics, such as bright and dark.

Brightness histogram A graphical representation of the number of pixels in an image at each gray level.

Brightness resolution The accuracy at which an image's pixel brightnesses are quantized, in number of bits or gray levels.

Brightness scaling The reduction of pixel brightness by a fixed divisor to eliminate pixel brightnesses above 255 resulting from an operation.

Brightness slicing A double binary contrast enhancement operation where pixel brightnesses below a lower threshold and above an upper threshold are set to black, while brightnesses between the two thresholds are set to white.

Brightness-weighted center of mass The (x,y) balance point of an object where there is equal brightness mass above, below, left, and right.

Boundary description A description of the perimeter shape characteristics of an object.

Cathode-ray tube (CRT) The image-display vacuum tube typically used in video display monitors.

CCIR Monochrome The European monochrome standard video format originally used by commercial television broadcasters and still used in monochrome video cameras.

CCIR Recommendation 601-1 A digital RGB component color standard video format for television production applications; both NTSC and PAL versions are defined.

CCITT Group 3/4 compression A standardized image compression scheme using run-length and Huffman coding techniques; used in facsimile image transmission.

CCITT Recommendation H.261 A standardized image compression scheme for motion image sequences; based on transform coding techniques.

Center of mass The (x,y) balance point of an object where there is equal mass above, below, left, and right.

Centroid See *center of mass.*

Chain code A pixel-by-pixel direction code used in a contour following operation that records an object boundary as a series of absolute or relative directions.

Charge-coupled device (CCD) A semiconductor photodetector device used in most video cameras.

Chrominance The color portion of a composite color video signal.

Chrominance filter An analog filter that removes the chrominance portion from a composite color video signal.

Classification See *object classification.*

Closing A binary or gray-scale morphological operation composed of a dilation operation followed by an erosion operation; small holes and gaps are filled, while objects tend to remain their original size.

CMY See *cyan, magenta, and yellow color space.*

CMYK Same as cyan, magenta, and yellow color space, but with the addition of black (K). Used in the printing industry for full-color printing.

Code overload The condition in a lossless differential pulse-code modulation compression scheme when the code size cannot handle the difference between two pixel brightnesses.

Codebook A list of patterns used in block coding to refer to the same pattern many times, hence effecting compression.

Color burst A color reference signal added to a composite color video signal that provides information for subsequent color decoding.

Color compression The compression of color images, typically first converting the image to a less redundant color space such as hue, saturation, and brightness.

Color decoder A circuit that breaks down a composite color video signal into a component form for digitization, typically from NTSC or PAL to RGB.

Color encoder A circuit that combines the parts of a component color video signal into a composite form for display, typically from RGB to NTSC or PAL.

Color features Features of an object that relate to its color characteristics, such as hue and saturation.

Color gamut The range, or spectrum, of colors that can be created by mixing a set of primary colors in differing proportions.

Color histogram Generally three histograms—one for each color component—representing the number of pixels in an image at each level.

Color key The selected color in a video source used to control the keying of it over another video source. Wherever the color-key color is detected in the video source, the second video source is switched in.

Color resolution Like brightness resolution but applied to each color component, the accuracy at which an image's pixels are quantized; measured in number of bits or gray levels.

Color space A representation of color controlled by a number of color components, such as RGB or HSB. Also, the range of colors that can be created by mixing a set of primary colors in differing proportions.

Color space conversion The conversion of a color using one color space, like RGB, to another, such as HSB.

Color vectors Two intermediate color signals derived from a composite color video signal and used to create a component form of the signal.

Combination function In a multiple-image pixel point process, the function that describes the way each pixel from one image is combined with the corresponding pixel of another image. Simple combination functions such as addition, subtraction, and division are common.

Complement image An image appearing as a negative of the original.

Component color A video signal composed of multiple physical signals, one for each color component—typically red, green, and blue.

Composite color A video signal composed of a single physical signal, into which is encoded both luminance and chrominance information.

Compression asymmetrical coding When the relative time is greater and/or efficiency is lower for an image compression operation than for its corresponding decompression operation.

Compression ratio The ratio of uncompressed image data size to compressed image data size—the higher the ratio, the greater the amount of compression.

Compression symmetry The relative time and efficiency of an image compression operation versus its corresponding decompression operation.

Computed tomography (CT) A tomographic image acquisition system using X-ray transmission for gathering cross-sectional slice-images from projection images; used primarily in medical imaging applications. Also, a generic reference to any computed tomography technique, including MRI and PET.

Computer-aided design (CAD) A computer-based tool for designing three-dimensional models of parts and assemblies.

Confocal laser scanning microscope A scanning microscope capable of creating slice-images of an object by focusing laser light to a single focal plane.

Continuous tone The smoothly and continuously varying brightness characteristics of a natural image.

Contrast enhancement The enhancement of the brightness attributes of an image.

Control points Multiple points placed on an image to control a warping transformation; the points represent the before and after locations of the transformation.

Convolution coefficient One of several weights used in the weighted average computation of the spatial convolution process.

Convolution mask The set of convolution coefficients used in the spatial convolution process.

Cosine transform A frequency transform that, similar to the Fourier transform, decomposes a spatial image into a set of sinusoidal frequency component functions.

Cosine^4 light falloff A phenomenon causing the reduction of image brightness in the outer field of a lens.

Cross-correlation A mathematical comparison function used in matched filtering to compare an image to a small reference image called an image mask.

Cross-sectional image A slice-image created by the reconstruction of a series of projection images taken around an object of interest.

Cross-sectional image stacking A method for creating a three-dimensional representation of an object by stacking several two-dimensional slice-images.

Cubic convolution A form of pixel interpolation that is based on a weighted average of 16 or more pixels surrounding the pixel location of interest.

Cyan, magenta, and yellow color space (CMY) A color representation used in the printing industry based on subtracting the primary colors (red, green, and blue) from white—which is the same as adding the secondary colors cyan, magenta, and yellow to black—to create the desired color.

Data bandwidth The quantity of image data per unit time that can be transferred to and from a memory—in particular, an image store memory—or processor.

Data channel error Data errors in a transport link (or archive mechanism) that, if not corrected, can cause severe image corruptions in some image decompression schemes.

Data explosion Results when an image compression scheme produces a compressed image with a greater data content than the original.

Decibel (dB) A unit of measure based on a logarithmic scale, where 1 bit of brightness quantization equals 6 dB.

Decompression The recreation of an original image from a compressed form of the image.

Decompression asymmetrical coding When the relative time is greater and/or efficiency is lower for an image decompression operation than for its corresponding compression operation.

Deconvolution A restoration operation where a previous optical or other process—such as image blurring due to misfocus—is reversed.

Degrees of freedom The mathematical flexibility and precision of the spatial convolution process; dependent on the number of elements in the kernel.

Delta modulation (DM) A form of lossy differential pulse-code modulation compression that compares pixel brightnesses with their preceding neighbors and codes the differences using only 1 bit.

Depth of field The range of distances between imaged object and lens where the object will appear in focus.

Depth of focus The range of distances between lens and photosensor where the imaged object will appear in focus.

Depth-slice image An image in a series of slice-images taken at incremental depths through the object.

Device driver The lowest level of software in a digital image processing system. A device driver interfaces the function library to physical devices such as frame grabbers, accelerator processors, and special-purpose processors.

Differencing An operation that subtracts one image from another, pixel by pixel. Typically, each image is of the same scene but acquired at different times or under different lighting conditions.

Difference image The image resulting from the pixel-by-pixel subtraction of one image from another.

Differential pulse code modulation (DPCM) A form of lossless or lossy predictive image compression that compares pixel brightnesses with their preceding neighbors and codes the differences.

Digital image An image composed of discrete pixels, each having an associated brightness value.

Digital image processing The technique of processing images while they are in the form of discrete digital brightness quantities.

Digital matte See *mask image.*

Digital signal processor (DSP) A semiconductor processor specifically intended for processing digital signals such as digital images.

Digital-to-analog converter (D/A) A semiconductor device that converts digital image brightness quantities to analog voltage levels.

Digitization Sampling and quantizing an analog video signal to create a digital image.

Dilation A binary or gray-scale morphological operation that increases the size of bright objects uniformly in relation to the background; small holes and gaps are filled.

Discrete cosine transform (DCT) A discrete version of the cosine transform intended for use on digital signals, such as digital images; commonly used in transform coding operations.

Discrete Fourier transform (DFT) A discrete version of the Fourier transform intended for use on digital signals, such as digital images; commonly used in frequency-domain filtering operations.

Dither noise A random noise source used in truncation coding schemes to reduce the visual appearance of pixel blocking and brightness contouring in the resulting decompressed image.

Dot pitch The size of the color phosphor dots in a color cathode-ray tube.

Downsampling A decrease in the sample rate of an image caused by a geometric transformation, such as scaling by a factor less than 1.

Dynamic range The brightness span of an image's gray scale.

Edge enhancement filter A spatial filter that increases the edge detail in an image.

Elevation data A data set much like a digital image except that pixel values represent terrain elevation.

Engine See *function library*.

Entropy coding A form of lossless image compression that exploits a measure of an image's actual versus theoretical information content.

Erosion A binary or gray-scale morphological operation that uniformly reduces the size of bright objects in relation to the background; small speckle and spurs are eliminated.

Even field The even-numbered lines in an interlaced scan video signal.

Explicit boundary description An object boundary description that explicitly describes the (x,y) location of each pixel in the perimeter of an object.

Fan beam projection A transmissive mode CT method where a single X-ray source is transmitted at an object and received by numerous detectors surrounding the opposite side of the object.

Fast Fourier transform (FFT) An algorithmically faster version of the discrete Fourier transform.

Feature extraction The second step of image analysis that seeks to measure the individual features of an object.

Feature measure A particular measure that describes some aspect of an object.

File interchange format A standardized image data format that provides portability of digital images between digital image processing systems.

Finite impulse response filter (FIR) Another term for *spatial convolution*.

First-order interpolation See *bilinear interpolation*.

First-order spatial moments The x and y sums of pixel brightnesses in an object, each multiplied by its respective x or y coordinate location in the image, resulting in measures that represent the object's mass.

Flash A/D converter A high-speed analog-to-digital converter typically used for real-time image digitization.

Flat shading A method of three-dimensional model shading, where surfaces are shaded with a constant shade value based on model lighting conditions.

Fourier descriptors A concise, yet abstract, object boundary representation based on the application of the Fourier transform to the sequence of boundary (x,y) coordinate locations.

Fourier transform A frequency transform that decomposes a spatial image into a set of sinusoidal frequency component functions.

Fractal compression A form of lossy image compression based on the scaling and rotation of fractal representations of pixel patterns.

Frame grabber A computer peripheral providing the acquisition, storage, and display of digital images.

Frame rate The rate at which video images are digitized, processed, or displayed from a digital image processing system.

Frame store See *frame grabber*.

Frequency component function A function of frequency, such as the sine and cosine functions, into which a frequency transform decomposes an image.

Frequency domain image The frequency form of an image; pixel brightnesses correspond to the spatial frequency content of the image.

Frequency domain filter A filter that acts on an image that has been converted from the spatial domain to the frequency domain.

Frequency domain processing Any operation that acts on an image that has been converted from the spatial domain to the frequency domain.

Frequency mask A mask image that, when multiplied with a frequency domain image, removes desired frequencies from the image.

Frequency transform An operation that decomposes an image from its spatial-domain form of brightnesses into a frequency-domain form of fundamental frequency components.

Function library A collection of core digital image processing software routines that run on either the host computer or an accelerator processor.

Fundamental classes of digital image processing The five primary areas of digital image processing operations—image enhancement, restoration, analysis, compression, and synthesis.

Fuzzy logic An advanced technique that adds fuzziness to the logical reasoning of object classification operations.

Gain The value by which every pixel in an image is multiplied in a histogram stretching operation.

Gamma response characteristic The nonlinear brightness display characteristic of a cathode-ray tube, where brightnesses displayed are not proportional to the video signals representing them.

Gaussian window A windowing function that gradually transitions an image's brightnesses to 0 in nonlinear steps.

Geometric distortion Spatial distortions caused by inaccuracies in an optical system, such as in a lens.

Geometric transformation An operation that transforms the geometric characteristics of an image to a new form, such as a rotation or scaling.

Gouraud shading A method of three-dimensional model shading where surfaces are shaded with an interpolated shade value based on model lighting conditions, providing smoother shading than the flat shading method.

Gray centroid See *brightness-weighted center of mass.*

Gray scale The number of gray levels available to represent the brightnesses in a digital image.

Gray-scale image A image composed of gray-level brightnesses.

Gray-scale morphological process The morphological process intended to operate on a gray-scale image; neighborhood minimum and maximum brightness values are evaluated.

Group process See *pixel group process.*

Haar transform A frequency transform that decomposes a spatial image into a set of nonsinusoidal frequency component functions.

Hadamard transform A frequency transform that decomposes a spatial image into a set of nonsinusoidal frequency component functions.

Hamming window A windowing function that gradually transitions an image's brightnesses to 0 in nonlinear steps.

Hanning window See *von Hann window.*

Hidden-line removal Similar to the wireframe synthetic image rendering method, except that surface edges that are not visible from the viewpoint are removed.

High-definition television (HDTV) A group of composite color standard video formats (varying somewhat by country) that provide significantly higher spatial resolution and wider aspect ratio than the current NTSC and PAL formats.

High-pass filter A spatial filter that accentuates the high-frequency detail or attenuates the low-frequency detail in an image.

Histogram See *brightness histogram* or *color histogram.*

Histogram sliding The addition of a constant brightness value to every pixel in an image, having the effect of sliding the image histogram to the left or right.

Histogram stretching The multiplication by a constant value of every pixel brightness in an image, having the effect of stretching or shrinking the image histogram.

Hit or miss transform The generalized implementation of the binary morphological process that evaluates a group of input pixels with a structuring element.

Horizontal sync The synchronizing signal in a video signal that conveys the beginning of a line.

Host computer The overseeing computer in a digital image processing system, generally responsible for user interface, system coordination, and often image processing activities.

Hue, saturation, and brightness color space (HSB) A color representation based on how humans perceive color; generally the best color space in which to carry out color image processing operations.

Hue One of the three color components of the HSB color space that controls the color spectrum from red through the yellows, greens, blues, and violets.

Hue, saturation, and lightness color space (HSL) Although distinct, an analogous color space to hue, saturation, and brightness.

Hue, saturation, and value color space (HSV) Although distinct, an analogous color space to hue, saturation, and brightness.

Hue, saturation, and intensity color space (HSI) Although distinct, an analogous color space to hue, saturation, and brightness.

Huffman coding A form of lossless entropy coding that reassigns brightness codes to variable-length codes, based on the frequency of occurrence of the brightnesses in an image.

Human visual system The components of the eye–brain system comprising human vision.

I(x,y) The nomenclature for referring to a pixel at location (x,y) in an input image.

Image acquisition The digitization and storage of a digital image.

Image analysis The processing of an image to extract quantitative object measurements and then classify the results.

Image archive The long-term storage of a digital image.

Image coding See *image compression.*

Image combination Any operation that combines two or more images—pixel by pixel, using a combination function—into a resulting image, such as differencing or spectral ratioing.

Image compositing An operation that cuts an object from one image and places it into another image, using a mask image to control the cut and placement processes.

Image compression The reduction of digital image data size by removing forms of data redundancy from the raw image data.

Image data processor The processor in a digital image processing system composed of one or more of the following: special-purpose processor, accelerator processor, or host computer.

Image data processor interface Hardware circuitry in a digital image processing system that provides the interface between working image storage and an image data processor.

Image enhancement The processing of an image to improve its visual qualities using known or unknown attributes of the image degradation.

Image mask A small reference image of an object of interest used in a matched filtering operation.

Image matching See *matched filtering.*

Image morphology See *morphological process.*

Image preprocessing The first step in image segmentation, where zero-information clutter is removed from an image.

Image restoration The processing of an image to improve its visual qualities, always using known attributes of the image degradation.

Image segmentation The first step of image analysis that seeks to simplify an image to its basic component elements, or objects. Image segmentation tasks are composed of image preprocessing, initial object discrimination, and object boundary cleanup.

Image store The hardware memory array that provides working image storage for a digital image to be acquired, processed, and displayed.

Image synthesis The creation of a unique new image from other images or non-image data.

Image transport The transfer of digital image data over a communications link.

Initial object discrimination The second step in image segmentation, where objects are grossly separated into groups of similar attributes.

Input image An image that is processed in a digital image processing operation.

Intensity The quantity of light actually reflected or transmitted from a physical scene. In comparison, *brightness* refers to the quantity of light assigned to a pixel in a digital image.

Interframe coding The technique used in motion compression of related image sequences, looking for similarities between an image frame and a previous reference frame.

Interlaced scan In a video signal where lines are interleaved into two fields, odd-numbered lines are first sequenced followed by even-number lines; common to all commercial television standard video formats such as RS-170 and NTSC.

Inverse filtering A restoration operation where a previous optical or other process, such as image blurring due to misfocus, is reversed.

Inverse frequency transform The reverse of a frequency transform: a frequency-domain image is transformed back to a spatial-domain image.

ISO/IEC IPI digital image processing standard An international standard defining a common architecture for image processing, programmer's imaging kernel, and image interchange facility.

Joint Bi-level Image Experts Group compression (JBIG) A standardized image compression scheme using run-length and Huffman coding techniques; intended to replace CCITT Group 3/4 compression.

Joint Photographic Experts Group compression (JPEG) A standardized image compression scheme using discrete cosine transform coding techniques.

Karhunen-Loeve transform A frequency transform that decomposes a spatial image into a set of nonsinusoidal frequency component functions.

Kernel The group of input image pixels used in the spatial convolution process.

Keying See *video keying.*

Kirsch edge enhancement A directional spatial edge enhancement filter that increases an image's edge detail in a particular direction.

Lambert shading See *flat shading.*

Laplacian edge enhancement An omnidirectional spatial edge enhancement filter that increases an image's edge detail.

Lateral inhibition Interactions between photoreceptors in the eye that provide contrast and edge enhancement.

Lempel-Ziv-Welch compression (LZW) A form of lossless block coding that looks for repeating patterns of blocks of pixels and codes the patterns in a codebook.

Lens The optical element of a camera. A lens focuses light, projecting it upon a photodetecting device.

Line The *y*-axis coordinate of a digital image; a line of pixels traverses an image horizontally.

Line segment boundary representation An object boundary description that divides the object perimeter into fixed-length line segments and describes the direction of each line.

Line segment enhancement A directional spatial line segment enhancement filter that emphasizes the line segments in an image in the horizontal, vertical, or diagonal direction.

Line-scan (or linear) photosensor A one-dimensional array of photodetectors used to acquire images in several passes to create a two-dimensional image; typically a semiconductor device.

Linear process The process of summing elements (such as pixel brightnesses) multiplied by constant weights (such as convolution coefficients).

Liquid crystal display (LCD) A solid-state image display device used in many video display monitors.

Look-up table (LUT) A hardware or software implementation of a pixel point process mapping function. For every possible input pixel brightness, a corresponding output brightness is stored in the look-up table.

Lossless compression A form of image compression where the data content of the original image is precisely preserved.

Lossy compression A form of image compression where the precise data content of the original image is not preserved; rather, the quality is maintained at some arbitrary level.

Low-pass filter A spatial filter that attenuates the high-frequency detail or accentuates the low-frequency detail in an image.

Luminance The brightness portion of a composite video signal.

Luminous brightness See *brightness.*

Mach band effect An effect in the eye that accentuates the edges of objects of differing intensity; caused by lateral inhibition interactions between photoreceptors.

Magnetic resonance imaging (MRI) A tomographic image acquisition system using magnetic excitation for gathering cross-sectional slice images from projection images; used primarily in medical imaging applications.

Major axis The endpoints of the longest line that can be drawn through an object.

Mapping function The function that converts input pixel brightnesses to output brightnesses in a pixel point process.

Mask image In compositing operations, the image used to cut an object from one image. A complemented mask image is then used to cut a hole in a destination image where the object is placed.

Matched filtering An operation that uses cross-correlation to compare an image with an image mask composed of a small reference image of the object of interest.

Measure invariance The ability of a feature measure to stay constant even when an identical object appears differently, such as rotated, scaled, or translated.

Median filter A nonlinear spatial filter based on the median brightness value of each input group of pixels; good for removing noise spikes and other single-pixel anomalies.

Memory data width The amount of data that can be transferred to or from an image store memory array each cycle time.

Minor axis The endpoints of the longest line that can be drawn through an object while maintaining perpendicularity with the major axis.

Model A description of a three-dimensional object used to render a synthetic image of the object.

Morphological gradient A gray-scale morphological operation implemented by subtracting an eroded version of an image from a dilated version of the same image; object boundaries are highlighted.

Morphological mask See *structuring element.*

Morphological process A group process that evaluates each pixel in a binary or gray-scale image along with its neighboring pixels. A resulting pixel brightness is determined by looking at the input pixel brightness patterns (binary image case) or minimum and maximum values (gray-scale image case).

Motion compression The compression of related image sequences, making use of similarities between an image frame and a previous reference frame.

Motion estimation The technique used in motion compression of related image sequences to code the movement of similar objects between an image frame and a reference frame.

Moving Picture Experts Group compression (MPEG) A standardized image compression scheme for motion image sequences, based on transform coding and motion estimation techniques.

Multispectral color space Any color space composed of several primary colors such as red, green, and blue, but typically reserved to describe non-RGB spaces, such as those using infrared and ultraviolet components.

Nearest neighbor interpolation The simplest form of pixel interpolation using the pixel closest to the pixel location of interest.

Neural network An advanced object classification technique that provides the ability for a classification algorithm to be trained based on actual image examples.

Non-image data A set of data that did not originate as an image, such as three-dimensional object models and elevation data.

Non-interlaced scan In a video signal, a scan where all lines are sequenced in order without any interleave; common to most computer display standard video formats such as SVGA.

Nonlinear spatial filter A spatial filter that is not based on the process of summing elements (pixel brightnesses) multiplied by constant weights (convolution coefficients), such as the median filter and the morphological process.

NTSC composite color National Television Systems Committee—The color standard video format found in the United States and other countries.

Nyquist rate One-half the sampling rate—a digitized image will contain no spatial frequency content above the Nyquist rate.

O(x,y) The nomenclature for referring to a pixel at location (x,y) in an output image.

Object Any element of interest in an image, such as a manufactured part on a conveyor or cell in a biological specimen; typically in image analysis operations.

Object boundary cleanup The third step in image segmentation, where object boundaries are reduced to single-pixel widths for subsequent feature extraction operations.

Object classification The third step of image analysis, which seeks to compare the feature measures of an object to known criteria and determine if the object belongs to a particular class of objects.

Object parallax The difference in the relative positions of two objects of different depth when viewed from two separated viewpoints.

Objective image enhancement The quantitative enhancement of an image for a degradation using known or measured information about the degradation.

Odd field The odd-numbered lines in an interlaced scan video signal.

Offset The brightness value added to every pixel in a histogram sliding operation.

Opening A binary or gray-scale morphological operation composed of an erosion operation followed by a dilation operation; small speckle and spurs are removed, while objects tend to remain their original size.

Operation A particular digital image processing task, such as edge enhancement, chosen from a fundamental class of operations.

Optical image processing The technique of processing images while they are in the optical form.

Optical resolution The overall capability of an imaging system to resolve spatial details in an imaged scene—not to be confused with *spatial density*.

Optical sectioning An optical technique for nondestructively creating depth-slice images of an object, typically using a confocal laser scanning microscope.

Optical transfer function (OTF) A measure of the optical resolving abilities of an optical system; typically a function of image contrast versus spatial frequency.

Outlining A binary morphological operation implemented by subtracting an eroded version of an image from the original image; object boundaries are highlighted.

Output image An image that results from a digital image processing operation.

Overlaying See *video overlaying*.

PAL composite color Phase alternation line—The color standard video format found primarily in European countries.

Parallel beam projection A transmissive mode CT method where multiple X-ray sources are transmitted at an object and received by similarly spaced detectors surrounding the opposite side of the object.

Peak detector See *top-hat transformation*.

Periodic noise An image noise that appears as bands across an image.

Permanent image storage The long-term storage of digital image data, generally for archive purposes.

Perspective distortion A common geometric image distortion where objects in an image appear trapezoidal rather than square, due to foreshortening.

Phase-locked loop (PLL) A circuit in a digital image acquisition system that generates a pixel clock that is precisely locked to the horizontal sync signal, providing a stable digitized image even when the video signal is noisy.

Phong shading A method of three-dimensional model shading where surfaces are shaded using interpolated normal angles based on model lighting conditions, providing smoother shading and better highlights than the Gouraud shading method.

Photo CD (PCD) A common digital image file format and compression standard providing a high-density CD-ROM-based image data archive mechanism; used particularly for photographic-type imagery.

Photo-realistic rendering Methods for producing highly natural-looking synthetic images using extensive graphical rendering techniques in addition to basic shading.

Photodiode A semiconductor photodetector device used in some video cameras.

Photometric distortion A form of brightness distortion caused by light response incongruencies of an image sensor, yielding pixel brightnesses that do not accurately represent the intensities of the original scene.

Phototransistor A semiconductor photodetector device used some in video cameras.

Picture element The smallest discrete spatial component of a digital image.

Pincushion lens distortion A common geometric lens distortion causing an acquired image to appear to bulge from the center.

Pipelined processor A special-purpose processor that is in line with the flow of pixel data, commonly between an image store and a video display monitor.

Pixel See *picture element.* Also, the x-axis coordinate of a digital image.

Pixel blocking effect The effect of insufficient spatial resolution—gradual spatial changes appear to change abruptly, making the image appear blocky.

Pixel brightness slope The difference in brightness between two adjacent pixels.

Pixel clock The heartbeat of the digital image acquisition system, providing the base timing for image digitization and display.

Pixel clock period The period of time that each pixel occupies in a video signal.

Pixel group process A process that evaluates each pixel in an image based on its brightness along with the brightnesses of its immediate neighbors. Group processes are used to implement spatial convolution and morphological processes.

Pixel interleaving The most common approach to time-slicing access to an image store memory between digitizer/display pixel flow and image data processor pixel flow.

Pixel interpolation The method of estimating the brightness of a pixel when it is geometrically transformed to a location between pixel grid locations.

Pixel point process—multiple image A process that evaluates each pixel in two or more images based solely upon their brightnesses. Each pixel is processed individually with its corresponding pixels from the other input images. The resulting brightness is determined by a mapping function that maps each of the possible input pixel brightness combinations to one of 255 output brightnesses.

Pixel point process—single image A process that evaluates each pixel in an image based solely upon its brightness. The resulting brightness is determined by a mapping function that maps each of the 255 possible input pixel brightnesses to one of 255 output brightnesses.

Pixel time See *pixel clock period.*

Pixelation See *pixel blocking effect.*

Point process See *pixel point process.*

Polygon See *surface.*

Polynomial An equation used to define a geometric transformation.

Positron emission tomography (PET) A tomographic image acquisition system using an introduced positron-emitting substance for gathering cross-sectional slice-images from projection images; used primarily in medical imaging applications.

Post-aliasing filter An analog signal filter that is used following a D/A converter to remove from a reconstructed image frequency components that are above one-half the sampling rate.

Posterization See *brightness contouring.*

Predictive coding A form of lossless or lossy image compression that predicts pixel brightness values based on earlier trends in the image data and then codes the error between the predictions and the actual brightnesses.

Preprocessing See *image preprocessing.*

Prewitt gradient edge enhancement A directional spatial edge enhancement filter that increases an image's edge detail in a particular direction.

Primary color One of three additive colors—red, green, and blue—used to create a color of the spectrum.

Process A mathematically specific technique used to implement a particular digital image processing operation.

Processing solution hierarchy The structure of a digital image processing solution: applications, fundamental classes, operations, and processes.

Progressive scan See *non-interlaced scan.*

Projection image A line image, acquired in the computed tomography process, where brightnesses proportionally represent the amount of material in the object across the line.

Pulse code modulation (PCM) The coding of pixel brightnesses as a binary value of 0 for black, 255 for white, and ascending values for the grays in between.

px64 See *CCITT Recommendation H.261.*

Quantization The process of converting discrete image samples to digital quantities following the sampling process.

Radiant intensity See *intensity.*

RAMDAC A D/A converter device that includes a static RAM for use as a look-up table, allowing the implementation of a pixel point pipelined processor.

Random access memory (RAM) A semiconductor memory device used in image store designs, special-purpose processors, accelerator processors, and host computers. RAM devices come in the forms of static RAM (SRAM), dynamic RAM (DRAM), video RAM (VRAM), and dynamic field store memory.

Raster The form of a video or scanned image where the image is in the form of a series of discrete horizontal lines.

Raster image file format An image file format that considers an image to be composed of multiple sequential lines, each containing a uniform number of pixels.

Real-time processing The ability to carry out a digital image processing operation, or series of operations, on entire images as quickly as new images are available—typically every 1/30th second in the case of a live video source.

Reconstruction See *analog signal reconstruction.*

Reconstruction from projections The creation of a two-dimensional cross-sectional slice-image from a series of projection images taken around an object of interest.

Red, green, and blue color space (RGB) A color representation used in color image sensors and displays, based on adding the primary colors (red, green, and blue) to black to create the desired color.

Reference frame The image frame used in motion compression of related image sequences to compare an image frame for similarities.

Relative chain code A pixel-by-pixel direction code used in a contour-following operation that records an object boundary as a series of directions relative to the previous boundary pixel's direction.

Rendering See *surface model rendering* and *volume rendering.*

Repetitive noise See *periodic noise.*

Resampling The alteration of a digital image's sample rate and orientation that occurs with most geometric transformations that involve scaling and rotation.

RGB See *red, green, and blue color space.*

RGB component color A component color standard video format conveyed as a red signal (R), a green signal (G), and a blue signal (B).

Robinson edge enhancement A directional spatial edge enhancement filter that increases an image's edge detail in a particular direction.

Rotation A geometric operation that rotates an image about a predetermined point through a desired angle.

RS-170 The North American monochrome standard video format originally used by commercial television broadcasters and still used in monochrome video cameras.

RS-170A An extension to the RS-170 standard video format to include color information in the video signal; synonymous with the *NTSC composite color* video format.

RS-330 An extension to the RS-170 standard video format to refine the electrical signal performance characteristics of the video signal.

RS-343 An extension to the RS-170 standard video format to provide for increased vertical resolution of the video signal.

Rubber sheet transformation See *warping transformation.*

Run-length coding A form of lossless image compression that looks for sequences of pixels with an identical brightness and codes them into a reduced description.

S-Video A component color standard video format conveyed as a luminance signal (Y) and a chrominance signal (C); identical to the Y/C format.

Sampling The process of dividing an analog video signal into discrete samples preceding the quantization process.

Sampling theorem A theorem that states that to fully represent the spatial details of an original continuous-tone image, the image must be sampled at a rate at least twice as fast as the highest spatial frequency contained in it.

Saturation One of the three color components of the HSB color space that controls the purity of a color from a pale to deep color.

Scaling A geometric operation that enlarges or shrinks the size of an image about a predetermined point.

Scanner A computer peripheral that optically scans an image and converts it to a digital form.

SECAM composite color Séquentiel à mémoire—The color standard video format found in France and most Middle Eastern and Eastern European countries.

Second-order interpolation See *cubic convolution.*

Secondary color One of three subtractive colors—cyan, magenta, and yellow—used to create a color of the spectrum.

Segmentation See *image segmentation.*

Sequential material removal A destructive technique for creating depth-slice images of an object; done by repetitively removing material from an object and imaging each step.

Shaded solid model Similar to the solid model synthetic image rendering method, except that surfaces are filled with brightnesses representing the appropriate shading for the lighting and viewing conditions.

Shape measure Any measure that describes some aspect of an object's shape characteristics, such as area, perimeter distance, and major axis length.

Shift and difference edge enhancement A directional spatial edge enhancement filter that increases an image's edge detail in the horizontal, vertical, or diagonal direction.

Simultaneous contrast An effect in the eye that makes objects of equal intensity appear different depending on the local intensities surrounding the object; caused by lateral inhibition interactions between photoreceptors.

Sine transform A frequency transform that, similar to the Fourier transform, decomposes a spatial image into a set of sinusoidal frequency component functions.

Skeletonization A binary morphological erosion-like operation that reduces objects within an image to a skeleton representation.

Slant transform A frequency transform that decomposes a spatial image into a set of nonsinusoidal frequency component functions.

Slice image An image representing a slice through a solid object, often acquired nondestructively through computed tomography techniques.

Sobel edge enhancement An omnidirectional spatial edge enhancement filter that increases an image's edge detail.

Software application A computer program used to interact with and implement digital image processing operations.

Solid model Similar to the hidden-line wireframe synthetic image rendering method, except that surfaces are filled with brightnesses to make them appear solid.

Source-to-target mapping The method of geometric transformation that transforms pixel locations from the input image to the output image; pixel locations in the output image can get missed using this method. See *target-to-source mapping.*

Spatial Relating to the two-dimensional nature of an image.

Spatial aliasing A form of aliasing that occurs when an image is sampled at a rate less than twice the rate of the highest spatial frequency in the image. The result is that the undersampled spatial frequency details are aliased to new, erroneous frequencies causing visual distortions.

Spatial convolution A group process used to implement a spatial filter. Each pixel in an image is evaluated using its neighboring pixels. The resulting pixel brightness is calculated as a weighted average of the group of pixels and corresponding convolution coefficients.

Spatial density The number of pixels in a digital image—not to be confused with *optical resolution.*

Spatial domain image The natural form of an image; pixel brightnesses correspond directly to spatial image brightnesses.

Spatial enhancement The enhancement of spatial detail within an image.

Spatial filter Any operation that accentuates or attenuates the appearance of spatial details in an image. A high-pass filter accentuates high-spatial-frequency details; a low-pass filter attenuates high-spatial-frequency details.

Spatial frequency The rate at which a spatial detail transitions from dark to light and back to dark. A fine detail has high spatial frequency content; a coarse detail has low spatial frequency content.

Spatial moments Statistical shape measures of an object that do not characterize an object specifically, but rather do so abstractly.

Spatial resolution The smaller of the *spatial density* and *optical resolution* measurements in an imaging system. Typically, spatial resolution can be assumed to be equal to the spatial density measurement.

Special-purpose processor A hardware processor, based on dedicated hardware circuits, that is exclusively devoted to a specific digital image processing task.

Spectral ratioing An operation that divides one image by another, pixel by pixel. Typically, each image is of the same scene but acquired through different spectral-band filters.

Square pixel The case where a pixel has equal dimensions in both the x and y directions.

Standard video format A standardized video signal that conforms to prescribed timing and electrical characteristics.

Stereo image pairing By using two images of the same scene, each taken from a different viewpoint, depth information can be perceived or computed.

Stereoscopy The ability to perceive depth through the evaluation of object parallax.

Structuring element The set of logical values (binary image case) or gray-level values (gray-scale image case) used in the binary or gray-scale morphological process.

Subjective image enhancement The nonquantitative enhancement of an image based on desired visual appeal.

Subtractive color The form of color creation based on the subtractive mixing of the cyan, magenta, and yellow secondary colors.

Super Video Graphics Array (SVGA) A group of video display standard video formats created and used by the personal computer industry.

Surface A component of a three-dimensional model used to create the description of an object.

Surface model rendering Any method used to render a three-dimensional model from a surface description.

Symmetrical coding When the relative time and efficiency of an image compression operation are identical to those of its corresponding decompression operation.

Sync extraction The separation of synchronizing information from a video signal to provide timing and control signals to a digital image acquisition system.

Sync insertion The addition of synchronizing information to a D/A converter output signal to create a standardized video signal for display on a video display monitor.

Tag Image File Format (TIFF) A common digital image file format providing convenient image data archive and transport between computing platforms.

Target-to-source mapping The method of geometric transformation that transforms pixel locations from the output image to the input image; unlike the source-to-target mapping, all pixels in the output image receive a transformed value using this method.

Temporal aliasing A form of aliasing that occurs with motion image sequences when the frame rate of the image acquisition system is not high enough. The apparent backward rotating wagon wheel phenomenon seen in old western films is a temporal aliasing artifact.

Temporal noise reduction A method of reducing random noise in an image using a multiple-image pixel point process to average, pixel by pixel, several identical images where only the noise portion has changed.

Temporal resolution The rate at which video images are digitized by a digital image processing system.

Texture features Features of an object that relate to its texture, such as smoothness and roughness.

Thresholding See *binary contrast enhancement.*

Time-slicing A method of providing two-port memory access to an image store memory. Digitizer/display pixel flow is interleaved in time with image data processor pixel flow.

Tolerance The allowed margin of error in the comparison of a feature measure with a known parameter.

Tomography A form of imaging that attenuates the effects of objects that lie in front of or behind a perpendicular plane of interest; now implemented using digital image processing methods. See *computed tomography.*

Toolkit See *function library.*

Top-hat transformation A gray-scale morphological operation implemented by subtracting an opened version of an image from the original image; bright object peaks are highlighted.

Transform Coding A form of lossy image compression that converts an image to the frequency domain using a frequency transform and discards minimally used spatial frequency components; typically, the discrete cosine transform is used.

Translation A geometric operation that shifts an image left, right, up, or down.

Transport link Any physical link for transferring image data between two or more digital image processing systems.

Triangular window A windowing function that linearly transitions an image's brightnesses to 0 in equal steps.

Trichromacy The ability to discern or create a wide spectrum of color by using three discrete color sensors or emitters.

Truncation coding A form of lossy image compression that discards a portion of an image's brightness or spatial data.

Two-port memory access An image store design methodology to provide two separate data paths in and out of image memory. One path provides incoming and outgoing pixel data flow from the digitizer and to the display circuitry; the other path provides pixel data flow to and from the image data processor.

Undersampling The situation when a digital image is sampled at a rate less than twice the highest frequency in the image—spatial aliasing can appear in an undersampled image.

Unit distance A known distance measure imaged along with an object to be classified that allows measure invariance in the presence of object-size scaling.

Unsharp masking enhancement A high-frequency accentuating filter created by subtracting a low-passed image from an original image.

Upsampling An increase in the sample rate of an image caused by a geometric transformation, such as scaling by a factor greater than 1.

User interface The highest level of software in a digital image processing system providing the look and feel of the system. The user interface allows the user to view images, control operations, and generally interact with the system.

Valley detector See *well transformation.*

Variable line segment boundary representation An object boundary description that divides the object perimeter into variable-length line segments and describes the direction and length of each line.

Vector image file format An image file format that considers an image to be composed of primitive graphical elements, such as lines, circles, and characters.

Vector quantization (VQ) A form of lossy block coding that looks for repeating patterns of blocks of pixels and codes the error between pixel blocks and patterns stored in a codebook.

Vertical sync The synchronizing signal in a video signal that conveys the beginning of a field (interlaced scan) or frame (non-interlaced scan).

Video camera An image acquisition device that converts an optical image to an electrical video signal (the analog image form).

Video display monitor An image display device that converts an electrical video signal (the analog image form) to an optical form.

Video keying A circuit for mixing two video signals. A color key is selected and is used to control the keying of a second video source over the first video source. Wherever the color-key color is detected in the first video source, the second video source is switched in, hence appearing to be cut into the first video signal.

Video overlaying A circuit for mixing two video signals. A background color is selected and is used to control the overlaying of a second video source over the first video source. Wherever the background color is *not* detected in the second video source, the second video source is switched in, hence appearing to overlay the first video signal.

Vidicon A vacuum tube photodetector device used in video cameras.

Viewing geometry The physical dimensions of a displayed image and its distance from the observer; the viewing geometry establishes the spatial resolution requirements of the display.

Viewplane The plane perpendicular to the line between the observer's viewpoint and the rendered three-dimensional model.

Viewpoint The point in space where a rendered three-dimensional model is to be viewed.

Visualization The process of "seeing" a three-dimensional model—the model is visualized by rendering it to a form that simulates its natural appearance.

Volume element A three-dimensional version of a pixel—a volume element is a cube with unit distances of width, height, and depth.

Volume rendering Any method used to render a three-dimensional model from a volume description.

Von Hann window A windowing function that gradually transitions an image's brightnesses to 0 in nonlinear steps.

Voxel See *volume element.*

Warping transformation A geometric operation that contorts an image, often using control points to determine how the resulting image is stretched and shifted.

Watershed edge detection A gray-scale morphological operation that finds and highlights the edges between objects.

Weber's law The relationship between actual intensity and perceived brightness, a characteristic that is close to logarithmic.

Weighted average The mechanism of the spatial convolution process. Pixel brightnesses in a group of input pixels (*kernel*) are multiplied by corresponding weights (*convolution coefficients*), and the results are summed.

Well transformation A gray-scale morphological operation implemented by subtracting an closed version of an image from the original image; dark object valleys are highlighted.

White balancing The calibration of red, green, and blue color component proportions so that mixing all three results in a pure white color without color casts.

White-level reference The reference used by an A/D converter to establish the amplitude of a video signal to be converted to white (255).

Window A component of a graphical user interface for viewing input and resulting images and data.

Windowing function A function applied to an image prior to applying a Fourier transform (or inverse Fourier transform) to reduce image edge discontinuities as seen by the transform's periodic interpretation of the image.

Wireframe The most primitive synthetic image rendering method that simply shows the edges of all the surfaces in a three-dimensional model.

Working image storage The short-term storage of digital image data, generally for intermediate storage during processing actions.

X3H3.8 The ANSI predecessor to the ISO/IEC IPI digital image processing standard's programmer's imaging kernel.

Y/C component color A component color standard video format conveyed as a luminance signal (Y) and a chrominance signal (C); identical to the *S-Video* format.

Zero-order interpolation See *nearest neighbor interpolation.*

Zero-order spatial moment The sum of pixel brightnesses in an object—in a binary image, the object's area; in a gray-scale image, the object's energy.

Index

D